D0076909

London Mathematical Society
Lecture Note Series 142

The Geometry of Jet Bundles

D. J. SAUNDERS

CAMBRIDGE UNIVERSITY PRESS

T t20 9K
KB

LONDON MATHEMATICAL SOCIETY LECTURE NOTE SERIES

Managing Editor: Professor J.W.S. Cassels, Department of Pure Mathematics and Mathematical Statistics, University of Cambridge, 16 Mill Lane, Cambridge CB2 1SB, England

The books in the series listed below are available from booksellers, or, in case of difficulty, from Cambridge University Press.

4 Algebraic topology, J.F. ADAMS
17 Differential germs and catastrophes, Th. BROCKER & L. LANDER
27 Skew field constructions, P.M. COHN
34 Representation theory of Lie groups, M.F. ATIYAH *et al*
36 Homological group theory, C.T.C. WALL (ed)
39 Affine sets and affine groups, D.G. NORTHCOTT
40 Introduction to H_p spaces, P.J. KOOSIS
42 Topics in the theory of group presentations, D.L. JOHNSON
43 Graphs, codes and designs, P.J. CAMERON & J.H. VAN LINT
45 Recursion theory: its generalisations and applications, F.R. DRAKE & S.S. WAINER (eds)
46 p-adic analysis: a short course on recent work, N. KOBLITZ
49 Finite geometries and designs, P. CAMERON, J.W.P. HIRSCHFELD & D.R. HUGHES (eds)
50 Commutator calculus and groups of homotopy classes, H.J. BAUES
51 Synthetic differential geometry, A. KOCK
54 Markov processes and related problems of analysis, E.B. DYNKIN
57 Techniques of geometric topology, R.A. FENN
58 Singularities of smooth functions and maps, J.A. MARTINET
59 Applicable differential geometry, M. CRAMPIN & F.A.E. PIRANI
60 Integrable systems, S.P. NOVIKOV *et al*
62 Economics for mathematicians, J.W.S. CASSELS
65 Several complex variables and complex manifolds I, M.J. FIELD
66 Several complex variables and complex manifolds II, M.J. FIELD
68 Complex algebraic surfaces, A. BEAUVILLE
69 Representation theory, I.M. GELFAND *et al*
74 Symmetric designs: an algebraic approach, E.S. LANDER
76 Spectral theory of linear differential operators and comparison algebras, H.O. CORDES
77 Isolated singular points on complete intersections, E.J.N. LOOIJENGA
78 A primer on Riemann surfaces, A.F. BEARDON
79 Probability, statistics and analysis, J.F.C. KINGMAN & G.E.H. REUTER (eds)
80 Introduction to the representation theory of compact and locally compact groups, A. ROBERT
81 Skew fields, P.K. DRAXL
82 Surveys in combinatorics, E.K. LLOYD (ed)
83 Homogeneous structures on Riemannian manifolds, F. TRICERRI & L. VANHECKE
84 Finite group algebras and their modules, P. LANDROCK
85 Solitons, P.G. DRAZIN
86 Topological topics, I.M. JAMES (ed)
87 Surveys in set theory, A.R.D. MATHIAS (ed)
88 FPF ring theory, C. FAITH & S. PAGE
89 An F-space sampler, N.J. KALTON, N.T. PECK & J.W. ROBERTS
90 Polytopes and symmetry, S.A. ROBERTSON
91 Classgroups of group rings, M.J. TAYLOR
92 Representation of rings over skew fields, A.H. SCHOFIELD
93 Aspects of topology, I.M. JAMES & E.H. KRONHEIMER (eds)
94 Representations of general linear groups, G.D. JAMES

95 Low-dimensional topology 1982, R.A. FENN (ed)
96 Diophantine equations over function fields, R.C. MASON
97 Varieties of constructive mathematics, D.S. BRIDGES & F. RICHMAN
98 Localization in Noetherian rings, A.V. JATEGAONKAR
99 Methods of differential geometry in algebraic topology, M. KAROUBI & C. LERUSTE
100 Stopping time techniques for analysts and probabilists, L. EGGHE
101 Groups and geometry, ROGER C. LYNDON
103 Surveys in combinatorics 1985, I. ANDERSON (ed)
104 Elliptic structures on 3-manifolds, C.B. THOMAS
105 A local spectral theory for closed operators, I. ERDELYI & WANG SHENGWANG
106 Syzygies, E.G. EVANS & P. GRIFFITH
107 Compactification of Siegel moduli schemes, C-L. CHAI
108 Some topics in graph theory, H.P. YAP
109 Diophantine Analysis, J. LOXTON & A. VAN DER POORTEN (eds)
110 An introduction to surreal numbers, H. GONSHOR
111 Analytical and geometric aspects of hyperbolic space, D.B.A.EPSTEIN (ed)
112 Low-dimensional topology and Kleinian groups, D.B.A. EPSTEIN (ed)
113 Lectures on the asymptotic theory of ideals, D. REES
114 Lectures on Bochner-Riesz means, K.M. DAVIS & Y-C. CHANG
115 An introduction to independence for analysts, H.G. DALES & W.H. WOODIN
116 Representations of algebras, P.J. WEBB (ed)
117 Homotopy theory, E. REES & J.D.S. JONES (eds)
118 Skew linear groups, M. SHIRVANI & B. WEHRFRITZ
119 Triangulated categories in the representation theory of finite-dimensional algebras, D. HAPPEL
120 Lectures on Fermat varieties, T. SHIODA
121 Proceedings of *Groups - St Andrews 1985*, E. ROBERTSON & C. CAMPBELL (eds)
122 Non-classical continuum mechanics, R.J. KNOPS & A.A. LACEY (eds)
123 Surveys in combinatorics 1987, C. WHITEHEAD (ed)
124 Lie groupoids and Lie algebroids in differential geometry, K. MACKENZIE
125 Commutator theory for congruence modular varieties, R. FREESE & R. MCKENZIE
127 New directions in dynamical systems, T.J. BEDFORD & J.W. SWIFT (eds)
128 Descriptive set theory and the structure of sets of uniqueness, A.S. KECHRIS & A. LOUVEAU
129 The subgroup structure of the finite classical groups, P.B. KLEIDMAN & M.W.LIEBECK
130 Model theory and modules, M. PREST
131 Algebraic, extremal & metric combinatorics, M-M. DEZA, P. FRANKL & I.G. ROSENBERG (eds)
132 Whitehead groups of finite groups, ROBERT OLIVER
133 Linear algebraic monoids, MOHAN S. PUTCHA
134 Number thoery and dynamical systems, M. DODSON & J. VICKERS (eds)
135 Operator algebras and applications, 1, D. EVANS & M. TAKESAKI (eds)
136 Operator algebras and applications, 2, D. EVANS & M. TAKESAKI (eds)
137 Analysis at Urbana, I, E. BERKSON, T. PECK, & J. UHL (eds)
138 Analysis at Urbana, II, E. BERKSON, T. PECK, & J. UHL (eds)
142 The geometry of jet bundles, D.J. SAUNDERS

London Mathematical Society Lecture Note Series. 142

The Geometry of Jet Bundles

D. J. Saunders

Honorary Research Fellow
Mathematics Faculty, The Open University

The right of the University of Cambridge to print and sell all manner of books was granted by Henry VIII in 1534. The University has printed and published continuously since 1584.

CAMBRIDGE UNIVERSITY PRESS
Cambridge
New York New Rochelle Melbourne Sydney

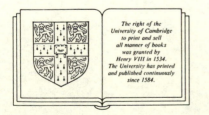

Published by the Press Syndicate of the University of Cambridge
The Pitt Building, Trumpington Street, Cambridge CB2 1RP
32 East 57th Street, New York, NY 10022, USA
10, Stamford Road, Oakleigh, Melbourne 3166, Australia

© Cambridge University Press 1989

First published 1989

Printed in Great Britain at the University Press, Cambridge

Library of Congress cataloging in publication data available

British Library cataloguing in publication data available

ISBN 0 521 36948 7

Sci
QA
614
S284
1989

Robert Manning Strozier Library

OCT 31 1989

Tallahassee, Florida

Contents

Introduction

This book is intended as an introduction to the language of jet bundles, for the reader who is interested in mathematical physics, and who has a knowledge of modern differential geometry.

Several ways of applying geometric techniques to physics are now well established in the literature: two major examples are the study of tangent and cotangent bundles in mechanics, and the use of connections on principal fibre bundles in field theories. More recently, the language of jets has appeared as a concise way of describing phenomena associated with the derivatives of maps, particularly those associated with the calculus of variations. In fact, a jet is no more than a generalisation of a tangent vector, and the geometrical theory of jet bundles includes the theories mentioned earlier as special cases. Generalisation, of course, sometimes introduces complexity: for instance, the coordinate representation used for jets bears some resemblance to the traditional coordinate representation used in the tensor calculus, but differs in that the transformation rules are no longer linear. In addition, many of the coordinate formulæ are symmetric in their indices, as a consequence of the commutativity of repeated partial differentiation, and this also introduces a certain complexity. On the other hand, the geometric nature of the theory introduces simplicity: there is, for instance, a clear geometric interpretation of the reason why the curvature of a connection is the obstruction to the integrability of the system of partial differential equations represented by the connection.

This book introduces those aspects of the theory of jet bundles which explain these local phenomena, although the theory itself is described in global terms. The first part of the book, comprising Chapters 1–3, sets out those elements of the theory of bundles and of linear structures which will be needed in subsequent chapters. Some of this material may be familiar to readers who are already acquainted with fibre bundles, although the perspective adopted here is one which ignores the existence of the structure group of the bundle.

The remainder of the book introduces the theory of jets. This is done in four distinct stages to make the task more manageable, although at a risk of some repetition. The basic definitions are given in Chapter 4, which describes first-order jets; the fundamental idea of prolongation also appears here, and is used in the specification of variational problems. Chapter 5, on second-order jets, introduces the idea of integrability, and also forms the setting for an intrinsic version of the Euler-Lagrange equations, constructed with the aid of a Cartan form. Higher-order jets are considered in Chapter 6, and a multi-index notation is adopted to deal with them; the global construction of a higher-order Cartan form also appears in this chapter. Finally, Chapter 7 uses the theory of calculus in infinite-dimensional Fréchet space to define infinite jets, and in this context proves the local exactness of the variational bi-complex; a consequence of this result is the Helmholtz condition in the inverse problem of the calculus of variations.

I should like to express my gratitude to colleagues with whom I have discussed this subject over the past few years. In particular, I should like to thank Mike Crampin, for his advice and encouragement, and Frans Cantrijn, who has read most of the manuscript and made many helpful suggestions. I am also indebted to the Research Advisory staff of the Open University's Academic Computing Service for their advice on the use of LaTeX.

D. J. Saunders
September 1988

Conventions

In this book, we suppose that all manifolds are real, and that manifolds and maps are smooth (that is to say, of class C^∞). We shall require the topology on each manifold to be Hausdorff, second-countable and connected. We shall assume, except in Chapter 7, that all our manifolds are finite-dimensional; it follows from these assumptions that our manifolds admit partitions of unity.

When using wedge products of cotangent vectors or of differential forms, it is always necessary to adopt a convention concerning the numerical factor to be employed: our convention will be such that, if α and β are cotangent vectors (or 1-forms), we may write

$$\alpha \wedge \beta = \alpha \otimes \beta - \beta \otimes \alpha$$

without any numerical factor.

Chapter 1

Bundles

In this chapter, we describe the basic structure upon which our study of jets will be based, namely that of bundles and sections. This structure is a generalisation of the more familiar structure of pairs of manifolds and maps, and allows more complicated topological arrangements. Although we shall be concerned primarily with local properties of jets, this more general description is still necessary for our discussion, because there are pairs of manifolds whose jet bundles do not themselves simplify to further pairs of manifolds.

1.1 Fibred Manifolds and Bundles

Many of the theories in modern mathematical physics can be described by considering smooth functions between differentiable manifolds. The domain of such a function might represent a region of space-time, and the codomain the possible states of the relevant physical system. Frequently, however, one considers not the function itself, but rather its *graph*: if the function is $f : M \longrightarrow F$ then its graph is the new function $\mathrm{gr}_f : M \longrightarrow M \times F$ defined by $\mathrm{gr}_f(p) = (p, f(p))$, and any function $\phi : M \longrightarrow M \times F$ which satisfies the condition $pr_1 \circ \phi = id_M$ is the graph of a uniquely-defined function f (namely, $f = pr_2 \circ \phi$). In this arrangement, the product manifold $M \times F$ is called the *total space*, because its local coordinate charts contain both dependent and independent variables for the function f. The domain M is also called the *base space*.

This way of looking at functions has two advantages. One is conceptual: the function may be thought of as a "field", in that for each point $p \in M$ there is a copy $\{p\} \times F$ of the codomain of f, and a single point in that copy gives the value of the field at p. This is a common way of picturing "vector fields", where the value of the field at a point may be represented by a vector attached to that point. The second advantage is more substantial, in that one may seek a generalisation of this arrangement where the total space as a whole is *not* diffeomorphic to the product of the base space and another manifold. For such a generalisation to be useful, however, there must nevertheless be a local product structure: each point of the total space must have a neighbourhood which "looks like" a product manifold. Such a structure is called a *fibred manifold.*

Definition 1.1.1 A *fibred manifold* is a triple (E, π, M) where E and M are manifolds and $\pi : E \longrightarrow M$ is a surjective submersion. E is called the *total space*, π the *projection*, and M the *base space*. For each point $p \in M$, the subset $\pi^{-1}(p)$ of E is called the *fibre over p* and is usually denoted E_p. ∎

As a shorthand, the same symbol E is sometimes used for the fibred manifold as for its total space. However this notation may be ambiguous, and in later chapters there will be many instances where the same manifold is the total space of two different fibred manifolds. We shall therefore denote the fibred manifold by the same symbol as we use for its projection, so that the shorthand for (E, π, M) will be π. Since the projection π of a fibred manifold (E, π, M) is a submersion, each connected component of the fibre E_p is a submanifold of E, and $\dim E_p = \dim E - \dim M$ is called the *fibre dimension* of π. We shall normally assume that both $\dim M$ and $\dim E_p$ are non-zero.

Example 1.1.2 If M and F are manifolds then $(M \times F, pr_1, M)$ is a fibred manifold. This is called a *trivial* fibred manifold; the word "trivial" has a technical meaning which is given in Definition 1.1.6. ∎

Example 1.1.3 Let $SL(2, \mathbf{R})$ be the three-dimensional manifold of real 2×2 matrices with determinant one, and let H be the subset $\operatorname{im} z > 0$ of the complex plane (regarded as a two-dimensional real manifold). Define a map $\pi : SL(2, \mathbf{R}) \longrightarrow H$ by

$$\pi \begin{pmatrix} a & b \\ c & d \end{pmatrix} = \frac{ai + b}{ci + d}.$$

A straightforward computation shows that the rank of π_* is 2 at each point of $SL(2, \mathbf{R})$. Since π is surjective, it follows that $(SL(2, \mathbf{R}), \pi, H)$ is a fibred manifold. ∎

Example 1.1.4 One of the simplest examples of a fibred manifold whose local product structure does not extend to a global product is the Möbius band. The total space (the Möbius band itself) may be constructed from the topological space $[0, 1] \times (0, 1)$ by identifying the points $(0, y)$ and

$(1, 1 - y)$ and giving the quotient space the structure of a 2-dimensional smooth manifold in a straightforward way. The image of the set of points $[0, 1] \times \{\frac{1}{2}\}$ under the quotient map is then diffeomorphic to the circle S^1, and the projection $[0, 1] \times (0, 1) \longrightarrow [0, 1] \times \{\frac{1}{2}\}$ passes to the quotient to give the Möbius band the structure of a fibred manifold over the circle. Each fibre is just a copy of the open interval $(0, 1)$, but the total space is not diffeomorphic to the Cartesian product $S^1 \times (0, 1)$ because of the "twist". ∎

The justification for describing a fibred manifold as having a local product structure comes from the properties of submersions. By using the implicit function theorem, we may see that for each point $a \in E$ there is a neighbourhood $U_a \subset E$, some other manifold V_a, and a diffeomorphism

$$t_a : U_a \longrightarrow \pi(U_a) \times V_a$$

which satisfies the condition that $pr_1(t_a(b)) = \pi(b)$ for all $b \in U_a$. The condition on t_a asserts that the fibres of π (when restricted to U_a) correspond to the fibres of the Cartesian product projection pr_1. A condition such as this, involving the composition of maps, is often expressed by using a "commutative diagram":

$$\begin{array}{ccc} U_a & \xrightarrow{\quad t_a \quad} & \pi(U_a) \times V_a \\ \Big\downarrow {\scriptstyle \pi|_{U_a}} & & \Big\downarrow {\scriptstyle pr_1} \\ \pi(U_a) & \xrightarrow[\quad id \quad]{} & \pi(U_a) \end{array}$$

where $\pi|_{U_a}$ denotes the restriction of π to U_a. Such a diagram is meant to assert that, when there is more than one route between two different nodes, then all such routes give the same result. In this case, the assertion is simply that the two maps $pr_1 \circ t_a$ and $id \circ \pi|_{U_a}$ are equal.

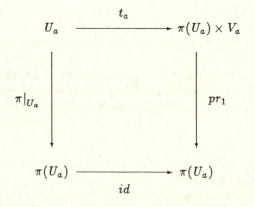

The existence of a local product structure on the total space of a fibred manifold allows us to use special local coordinate systems called *adapted coordinates*. These correspond to the product coordinates which may be constructed on a product manifold $M \times F$ from coordinates on the individual manifolds M and F.

Definition 1.1.5 Let (E, π, M) be a fibred manifold such that $\dim M = m$, $\dim E = m+n$, and let $y : U \longrightarrow \mathbf{R}^{m+n}$ be a coordinate system on the open set $U \subset E$. The coordinate system y is called an *adapted coordinate system* if, whenever $a, b \in U$ and $\pi(a) = \pi(b) = p$, then $pr_1(y(a)) = pr_1(y(b))$ (where $pr_1 : \mathbf{R}^{m+n} \longrightarrow \mathbf{R}^m$). ■

The meaning of this definition is that points in the same fibre $E_p \cap U$ have their first m coordinates equal, and are distinguished by their last n coordinates.

If $a \in E$ then adapted coordinates around a may be constructed from the local product structure in the following way. Starting with a coordinate system $x : W \longrightarrow \mathbf{R}^m$ around $\pi(a) = pr_1(t_a(a)) \in M$ (where W is chosen so that $W \subset \pi(U)$) and a coordinate system $u : V \longrightarrow \mathbf{R}^n$ around $pr_2(t_a(a)) \in V \subset V_a$, we define $y : t_a^{-1}(W \times V) \longrightarrow \mathbf{R}^{m+n}$ by

$$y = (x \circ pr_1 \circ t_a, u \circ pr_2 \circ t_a),$$

just as for product manifolds. Conversely, any adapted coordinate system $y : U \longrightarrow \mathbf{R}^{m+n}$ yields a coordinate system $x : \pi(U) \longrightarrow \mathbf{R}^m$ by setting $x(p) = pr_1(y(a))$, where $a \in E_p \cap U$; this is independent of the choice of a by Definition 1.1.5.

When dealing with the component functions of an adapted coordinate system, we shall usually adopt the following notation. If x^i $(1 \leq i \leq m)$ are the coordinate functions on M, then the coordinate functions on E will be labelled

$$(x^i, u^\alpha) \qquad 1 \leq i \leq m, \quad 1 \leq \alpha \leq n$$

so that the same symbol x^i will be used both for a function $\pi(U) \longrightarrow \mathbf{R}$ and for the composite function $U \longrightarrow \pi(U) \longrightarrow \mathbf{R}$. The latter function may also be written as the pullback $\pi^*(x^i)$, and this is the first of many occasions when the same symbol will be used to represent both an object and its pullback by a fibred manifold projection.

In many cases the idea of a fibred manifold without any additional restrictions, although useful, is slightly too general: for example, different fibres may have different topological structures. An example of this phenomenon may be constructed by taking the trivial bundle $(\mathbf{R} \times \mathbf{R}, pr_1, \mathbf{R})$

and deleting a single point. The result is a new fibred manifold where all the fibres except one are connected. If the fibred manifold is supposed to model a physical system then it may be unrealistic to allow the possible states of the system to depend on the choice of a particular point in space-time.

This problem may be resolved by insisting that the fibred manifold look rather more like a product than the definition of a submersion necessitates. The additional condition which such a fibred manifold must satisfy is expressed in terms of functions called *local trivialisations*, and the resulting object is called a *bundle*; after the present section, we shall be concerned almost entirely with bundles rather than more general fibred manifolds. We shall first describe what is meant by a *global* trivialisation.

Definition 1.1.6 If (E, π, M) is a fibred manifold then a (global) *trivialisation of* π is a pair (F, t) where F is a manifold (called a *typical fibre* of π) and $t : E \longrightarrow M \times F$ is a diffeomorphism satisfying the condition

$$pr_1 \circ t = \pi .$$

A fibred manifold which has at least one trivialisation is called *trivial.* ■

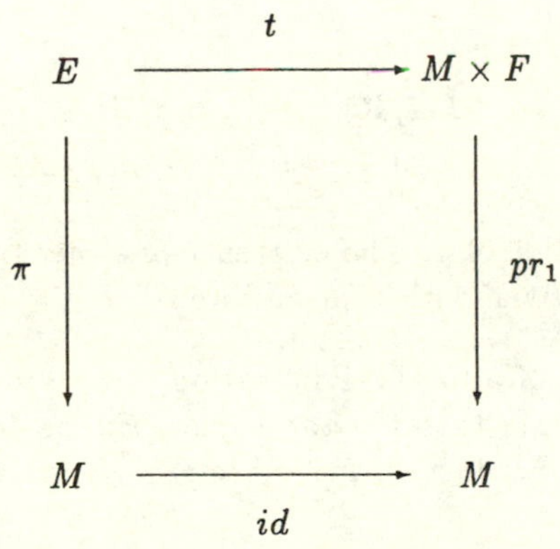

In particular, our original example $(M \times F, pr_1, M)$ is a trivial fibred manifold using the identity map as the trivialisation. However, suppose $g : M \times F \longrightarrow F$ satisfies the condition that, for each $p \in M$, the map $g_p : F \longrightarrow F$ defined by $g_p(q) = g(p, q)$ is a diffeomorphism. Then the map $t : M \times F \longrightarrow M \times F$ defined by $t(p, q) = (p, g_p(q))$ is another trivialisation, so it is important to be clear that requiring a fibred manifold to be trivial does not give its total space the structure of a Cartesian product in any particular way. Nevertheless, the typical fibres corresponding to two different trivialisations must clearly be diffeomorphic, so referring to a typical fibre of π rather than of the trivialisation is justified.

Example 1.1.7 If the circle S^1 is regarded as the unit circle in $\mathbf{R}^2$, then we may define the map $\rho_1 : SL(2,\mathbf{R}) \longrightarrow S^1 \subset \mathbf{R}^2$ by

$$\rho_1 \begin{pmatrix} a & b \\ c & d \end{pmatrix} = \left(\frac{a}{\sqrt{a^2+c^2}}, \frac{c}{\sqrt{a^2+c^2}} \right),$$

and then

$$\begin{aligned} t_1 : SL(2,\mathbf{R}) &\longrightarrow H \times S^1 \\ t_1(A) &= (\pi(A), \rho_1(A)) \end{aligned}$$

is a diffeomorphism. Consequently t_1 is a trivialisation of the fibred manifold $(SL(2,\mathbf{R}), \pi, H)$. However, we may also define the map $\rho_2 : SL(2,\mathbf{R}) \longrightarrow S^1$ by

$$\rho_2 \begin{pmatrix} a & b \\ c & d \end{pmatrix} = \left(\frac{b}{\sqrt{b^2+d^2}}, \frac{d}{\sqrt{b^2+d^2}} \right),$$

and then

$$\begin{aligned} t_2 : SL(2,\mathbf{R}) &\longrightarrow H \times S^1 \\ t_2(A) &= (\pi(A), \rho_2(A)) \end{aligned}$$

is another trivialisation of π. The existence of either trivialisation allows us to assert that π is trivial with typical fibre S^1. ∎

In the definition of a local trivialisation, the word "local" refers to the base manifold M rather than the total space E: the definition is concerned with expressing, in product form, subsets of E which are the unions of complete fibres of π.

Definition 1.1.8 If (E, π, M) is a fibred manifold and $p \in M$ then a *local trivialisation of* π *around* p is a triple (W_p, F_p, t_p) where W_p is a neighbourhood of p, F_p is a manifold and $t_p : \pi^{-1}(W_p) \longrightarrow W_p \times F_p$ is a diffeomorphism satisfying the condition

$$pr_1 \circ t_p = \pi|_{\pi^{-1}(W_p)} .$$

A fibred manifold which has at least one local trivialisation around each point of its base space is called *locally trivial* and is known as a *bundle*. ∎

$$\begin{array}{ccc} \pi^{-1}(W_p) & \xrightarrow{\;t_p\;} & W_p \times F_p \\ {\scriptstyle \pi|_{\pi^{-1}(W_p)}}\big\downarrow & & \big\downarrow {\scriptstyle pr_1} \\ W_p & \xrightarrow[id]{} & W_p \end{array}$$

It is worth noting that the existence of these local trivialisations around each point of M automatically implies that the map π is a submersion.

The concept of a typical fibre is also appropriate for bundles, although this is not quite immediate from the definition.

Lemma 1.1.9 *If (E, π, M) is a bundle then there is a manifold F such that, for each local trivialisation (W_p, F_p, t_p) of π, the manifolds F and F_p are diffeomorphic.*

Proof Notice first that if (W_p, F_p, t_p) and (W'_p, F'_p, t'_p) are both local trivialisations around the same point p then the manifolds F_p and F'_p must be diffeomorphic. So choose a fixed point $p \in M$ and a fixed local trivialisation (W_p, F_p, t_p), and put $F = F_p$. Let W be the set of points $q \in M$ such that there exists a local trivialisation (W_q, F, t_q) around q. Then W is non-empty, and is open because each W_q is open. On the other hand, $M - W$ must be open since it is the union of the open sets of points $r \in M$ where the local trivialisations (W_r, F_r, t_r) involve manifolds F_r which are *not* diffeomorphic to F. Therefore $M - W$ must be empty, because M is connected. ∎

On the total space of a bundle, adapted coordinate systems may be constructed from local trivialisations using coordinate systems on the base space and the typical fibre: this apparently unnecessary remark is useful when considering bundles with additional structure (such as vector bundles).

A trivial fibred manifold is obviously a bundle (and will be called a *trivial bundle*). The Möbius band is an example of a bundle which is not trivial. Further examples of bundles may be constructed from the manifolds of tangent and cotangent vectors associated with a given base manifold.

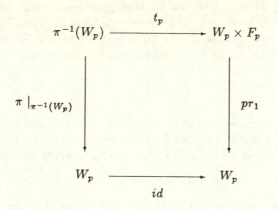

Example 1.1.10 Let TM denote the tangent manifold to the m-dimensional manifold M, and let $\tau_M : TM \longrightarrow M$ denote the map which associates to each tangent vector the point of M at which it is located. Then (TM, τ_M, M) is a bundle with typical fibre $\mathbf{R}^m$. To demonstrate this, it is convenient to use local coordinates. So let $\xi \in TM$ have the representation

$$\xi = \xi^i \left.\frac{\partial}{\partial x^i}\right|_p$$

where $p = \tau_M(\xi)$, the functions x^i are coordinate functions around p, and the summation convention is employed for the repeated index i. If $\gamma : \mathbf{R} \longrightarrow M$ is a curve whose tangent at zero is ξ then the real numbers ξ^i satisfy

$$\xi^i = (x^i \circ \gamma)'(0).$$

We may then define a coordinate system $(x^i, \dot{x}^i)$ on TM by writing (as usual) x^i instead of $x^i \circ \tau_M$, and setting $\dot{x}^i(\xi) = \xi^i$. To show that the fibred manifold constructed in this manner is locally trivial, let W_p be the coordinate neighbourhood of p on which the functions x^i are defined, and let $t : \tau_M^{-1}(W_p) \longrightarrow W_p \times \mathbf{R}^m$ be given by $t(\eta) = (\tau_M(\eta), \dot{x}(\eta))$. The map t is a diffeomorphism because it is the composition of the coordinate diffeomorphism $(x \circ \tau_M, \dot{x})$ on $\tau_M^{-1}(W_p)$ with the map $(x^{-1}, id_{\mathbf{R}^m})$. (The fact that TM has the topological properties which we require of a manifold, and that τ_M is therefore a bundle, is a consequence of a more general result which we shall give in Proposition 1.1.14.) ■

Example 1.1.11 If $M = \mathbf{R}^m$ then $TM \cong \mathbf{R}^m \times \mathbf{R}^m$ and the tangent bundle τ_M is trivial. Indeed, if $x : M \longrightarrow \mathbf{R}^m$ is a global coordinate system on a manifold M then $(\tau_M, \dot{x}) : TM \longrightarrow M \times \mathbf{R}^m$ is a global trivialisation. ■

Example 1.1.12 The tangent bundle (TS^1, τ_{S^1}, S^1) is trivial, even though the circle S^1 does not have a global coordinate system. To see this, let $\theta_1 : W_1 \longrightarrow \mathbf{R}$, $\theta_2 : W_2 \longrightarrow \mathbf{R}$ be two angle coordinate systems on S^1 whose domains W_1, W_2 together cover S^1, and such that if $p \in W_1 \cap W_2$ then $\theta_1(p) = \theta_2(p) \pm \pi$. Given a tangent vector $\xi \in TS^1$, suppose that ξ is determined by the curve γ, and put

$$\dot{\theta}(\xi) = (\theta_1 \circ \gamma)'(0)$$

if $\tau_{S^1}(\xi) \in W_1$,

$$\dot{\theta}(\xi) = (\theta_2 \circ \gamma)'(0)$$

if $\tau_{S^1}(\xi) \in W_2$. If it happens that $\tau_{S^1}(\xi) \in W_1 \cap W_2$ then $(\theta_1 \circ \gamma)'(0) = (\theta_2 \circ \gamma)'(0)$, because θ_1 and θ_2 differ by a constant, and so this procedure gives a well-defined mapping $\dot{\theta} : TS^1 \longrightarrow \mathbf{R}$. The map $(\tau_{S^1}, \dot{\theta}) : TS^1 \longrightarrow S^1 \times \mathbf{R}$ is then a global trivialisation. ■

Example 1.1.13 The tangent bundle (TS^2, τ_{S^2}, S^2), where S^2 denotes the 2-sphere, is *not* trivial. To see this, suppose that there were a global trivialisation $t : TS^2 \longrightarrow S^2 \times \mathbf{R}^2$. Choose a non-zero element $v \in \mathbf{R}^2$, and define $X : S^2 \longrightarrow TS^2$ by

$$X(p) = t^{-1}(p, v).$$

Then $X(p)$ is a non-zero tangent vector in T_pS^2 which depends smoothly on p, and so X is a non-vanishing smooth vector field on S^2: but this contradicts the famous Hairy Ball Theorem. ∎

An important property of any bundle is that the manifold structure on its total space E is completely determined by the manifold structures on its base space M and typical fibre F. For a trivial bundle $(M \times F, pr_1, M)$ this is a familiar result, but it applies equally to the case where the bundle is not trivial. The reason for this is that, if (W, F, t) is a local trivialisation, then t transports the manifold structure from $W \times F$ to the "strip" $\pi^{-1}(W)$ of E, and where the strips overlap the manifold structures are the same. In fact this technique can be used to construct a manifold structure on E when it is not given *a priori*.

Proposition 1.1.14 *Let M and F be manifolds, E a set, and $\pi : E \longrightarrow M$ a function such that, for each $p \in M$, $\pi^{-1}(p)$ has the structure of an n-dimensional manifold. Suppose also that, for each $p \in M$, there is a neighbourhood W_p of p and a bijection $t_p : \pi^{-1}(W_p) \longrightarrow W_p \times F$ satisfying:*

1. $pr_1 \circ t_p = \pi|_{\pi^{-1}(W_p)}$;
2. *for each $q \in W_p$, $pr_2 \circ t_p|_{\pi^{-1}(q)} : \pi^{-1}(q) \longrightarrow F$ is a diffeomorphism.*

Then E may be given a unique structure as a manifold such that π becomes a bundle and the maps t_p become local trivialisations.

Proof Let $a \in \pi^{-1}(p)$ and let $x : W \longrightarrow \mathbf{R}^m$ be a coordinate system around p and $u : V \longrightarrow \mathbf{R}^n$ be a coordinate system around $pr_2(t_p(a)) \in F$. Then, with our usual understanding about domains of functions being sufficiently small, the map $y_p = (x, u) \circ t_p$ is a "coordinate system" around a. We shall show that, whenever the domains of y_p and y_q have non-empty intersection then $y_q \circ y_p^{-1}$ is smooth (and hence a diffeomorphism). Since each map (x, u) is a diffeomorphism, it will be sufficient to show that

$$t_q \circ t_p^{-1} : (W_p \cap W_q) \times F \longrightarrow (W_p \cap W_q) \times F$$

is smooth. To do this, we note first that for each $r \in W_p \cap W_q$, the map $t_q \circ t_p^{-1}\big|_{\{r\}\times F}$ induces a diffeomorphism of F with itself. A consequence of this is that the map

$$\begin{aligned} (W_p \cap W_q) \times F &\longrightarrow F \\ (r, c) &\longmapsto pr_2\left(t_q \circ t_p^{-1}\Big|_{\{r\}\times F}(c)\right) \end{aligned}$$

is also smooth. But this latter map is just the second component of $t_q \circ t_p^{-1}$, and the first component of $t_q \circ t_p^{-1}$ is simply $pr_1 : (W_p \cap W_q) \times F \longrightarrow W_p \cap W_q$. Therefore $t_q \circ t_p^{-1}$ is smooth, and so E acquires a finite-dimensional C^∞ atlas. The uniqueness of this manifold structure follows because, if each function t_p is a diffeomorphism for two manifold structures on E, then id_E is a diffeomorphism between the two manifold structures. It now follows immediately that the map π is smooth because locally it is just $pr_1 \circ t_p$, and it is obviously surjective. The functions t_p therefore become local trivialisations for the bundle (E, π, M).

We may also show that E satisfies the topological conditions which we require of a manifold. First we shall demonstrate the Hausdorff property. So let $a, b \in E$. If $\pi(a) \neq \pi(b)$ then there are open sets $W_a, W_b \subset M$ which separate $\pi(a)$ and $\pi(b)$, so that $\pi^{-1}(W_a)$, $\pi^{-1}(W_b)$ will separate a and b. On the other hand, if $\pi(a) = \pi(b)$ $(= p$, say) then $pr_1(t_p(a)) = pr_1(t_p(b))$ so that $pr_2(t_p(a)) \neq pr_2(t_p(b))$ since t_p is bijective. Then there must be open sets $V_a, V_b \subset F$ which separate $pr_2(t_p(a))$ and $pr_2(t_p(b))$, and therefore open sets $(pr_2 \circ t_p)^{-1}(V_a)$, $(pr_2 \circ t_p)^{-1}(V_b)$ which separate a and b.

Next we shall show that E is second-countable. To do this, we shall first demonstrate that there is a countable family of local trivialisations whose neighbourhoods W_p cover M. So let X_λ be a countable basis for the open sets in M. For each $q \in M$, choose an open set X_{λ_q} such that $q \in X_{\lambda_q} \subset W_q$ and consider the triple

$$\left(X_{\lambda_q}, F, t_q|_{\pi^{-1}(X_{\lambda_q})} \right).$$

Since there are only countably many different open sets X_{λ_q}, we may choose, for each such set, one particular $p \in M$ which gives rise to it and hence obtain the required countable family

$$\left(X_p, F, t_p|_{\pi^{-1}(X_p)} \right).$$

Consequently any open set $O \subset E$ may be written as a countable union

$$O = \bigcup_p (O \cap \pi^{-1}(X_p))$$

where $O \cap \pi^{-1}(X_p)$ is diffeomorphic to an open subset of $X_p \times F$. Since each product manifold $X_p \times F$ has a countable basis of open sets, it follows that E does as well.

Finally we must show that E is connected. This follows from the fact that each map $pr_1 \circ t_p$ is an open map, so that π is an open map which is surjective. ■

EXERCISES

1.1.1 Let E, M be manifolds and let $\pi : E \longrightarrow M$ be a smooth map. Suppose that for each $p \in M$ there is a neighbourhood W_p of p and a map $\phi_p : W_p \longrightarrow E$ satisfying $\pi \circ \phi_p = id_{W_p}$. Show that (E, π, M) is a fibred manifold.

1.1.2 Construct an example of a fibred manifold, all of whose fibres are diffeomorphic to **R**, but which is not locally trivial.

1.1.3 Prove that the Möbius band (regarded as a fibred manifold over the circle) is a bundle but is not trivial. Construct a pair of adapted coordinate systems which together cover the total space of the Möbius band.

1.1.4 Let $\pi : \mathbf{R}^3 - \{0\} \longrightarrow S^2$ be defined by

$$\pi(x) = \frac{x}{\|x\|}.$$

Show that $(\mathbf{R}^3 - \{0\}, \pi, S^2)$ is a trivial bundle, and confirm that spherical polar coordinates $(\theta, \phi; \rho)$ may be used as adapted coordinates in a neighbourhood of $(0,1,0) \in \mathbf{R}^3 - \{0\}$.

1.1.5 Let T^*M denote the cotangent manifold to the m-dimensional manifold M, and let τ^*_M denote the map which associates to each cotangent vector the point of M at which it is located. If $\eta \in T^*_pM$, if the function $f \in C^\infty(M)$ satisfies $df_p = \eta$, and if x^i are coordinate functions on M around p, define a coordinate system (x^i, ∂_i) on T^*M by setting

$$\partial_i(\eta) = \left.\frac{\partial f}{\partial x^i}\right|_p .$$

Show that this coordinate system determines a local trivialisation (τ^*_M, ∂) of (T^*M, τ^*_M, M), and that τ^*_M thereby becomes a bundle with typical fibre $\mathbf{R}^m$.

1.1.6 If (E, π, M) is a bundle, prove that (TE, π_*, TM) is also a bundle.

1.1.7 Let G be a Lie group. Show that the map $t_L : TG \longrightarrow G \times \mathfrak{g}$ given by

$$t_L(\xi) = \left(\tau_G(\xi), L_{(\tau_G(\xi))^{-1}*}(\xi)\right),$$

where $\mathfrak{g}$ is the Lie algebra of G and $L_g : G \longrightarrow G$ is left translation, determines a trivialisation of the tangent bundle (TG, τ_G, G). Is this the same as the corresponding trivialisation determined by right translation?

1.2 Sections

Given a bundle—or, indeed, a general fibred manifold—(E, π, M) we can now return to the idea of a map from M to E as the generalisation of the graph of a function.

Definition 1.2.1 A map $\phi : M \longrightarrow E$ is called a *section of* π if it satisfies the condition $\pi \circ \phi = id_M$. The set of all sections of π will be denoted $\Gamma(\pi)$. ∎

For a trivial bundle given in the form $(M \times F, pr_1, M)$, a section is indeed just the graph of a function from M to F. However, for a general trivial bundle (E, π, M) with typical fibre F the function $M \longrightarrow F$ corresponding to a particular section depends upon the choice of trivialisation, and so for a non-trivial bundle it does not make sense to interpret a section in terms of a function whose codomain is the typical fibre.

A section of a fibred manifold may also be described in terms of coordinates. If $\phi \in \Gamma(\pi)$ and (x^i, u^α) is a family of coordinate functions around $a \in E$ then

$$\begin{aligned} x^i(\phi(a)) &= x^i(\pi(\phi(a))) && \text{(really)} \\ &= x^i(a) && \text{since } \pi \circ \phi = id_M \end{aligned}$$

so that the first m coordinates of $\phi(a)$ are determined by the coordinates of a. Hence only the last n coordinates are of interest in describing ϕ. We may therefore define real-valued functions ϕ^α to represent ϕ in this coordinate system by

$$\phi^\alpha = u^\alpha \circ \phi$$

where in this equation the symbol ϕ actually represents the restriction of the section ϕ to the domain of the appropriate chart in M. This particular abuse of notation will be almost universal when we write equations involving local coordinate representations, and it is to be understood that such equations are meant to hold only on suitably small domains.

Example 1.2.2 A section X of the tangent bundle (TM, τ_M, M) is just a vector field on M, because it associates to each point of M a tangent vector at that point. The set of all vector fields on M will be denoted by $\mathcal{X}(M)$ in preference to $\Gamma(\tau_M)$. Using coordinates $(x^i, \dot{x}^i)$ on TM and defining the real-valued functions X^i by $X^i = \dot{x}^i \circ X$ we can write

$$X = X^i \frac{\partial}{\partial x^i}$$

to represent the relationship between tangent vectors

$$X(p) = X^i(p) \left. \frac{\partial}{\partial x^i} \right|_p$$

for each p in the domain of the coordinate functions x^i. ■

In this last example the symbol $\partial/\partial x^i$ does not, in general, represent a section of τ_M because its domain might not be the whole of M. Indeed, in extreme cases a bundle might not have any sections at all.

Example 1.2.3 Let S^2 be the 2-sphere and let $T^\circ S^2$ be the open subset of TS^2 containing all non-zero tangent vectors. The triple

$$(T^\circ S^2, \tau_{S^2}|_{T^\circ S^2}, S^2)$$

is then a bundle called the *slit tangent bundle of* S^2 with typical fibre $\mathbf{R}^2 - \{0\}$. If ϕ were a section of this bundle then it would define a vector field on S^2 which was never zero, contradicting the Hairy Ball Theorem. ■

Nevertheless, every fibred manifold does have *local* sections: that is, maps defined only on open submanifolds of the base space which satisfy the other conditions for being sections. A section defined on the whole base space is then sometimes referred to as a *global* section for emphasis. Furthermore, if a fibred manifold has any global sections at all then it will have a global section which agrees with any given local section in a neighbourhood of any given point. To prove this assertion, it is convenient to introduce the idea of a *germ*.

Definition 1.2.4 If (E, π, M) is a fibred manifold then a *local section of* π is a map $\phi : W \longrightarrow E$, where W is an open submanifold of M, satisfying the condition $\pi \circ \phi = id_W$. The set of all local sections of π with domain W will be denoted $\Gamma_W(\pi)$, and the set of all local sections of π regardless of domain will be denoted $\Gamma_{loc}(\pi)$. If $p \in M$ then the set of all local sections of π whose domains contain p will be denoted $\Gamma_p(\pi)$. ■

Definition 1.2.5 If $\phi \in \Gamma_p(\pi)$ then the *germ of* ϕ *at* p is the subset of $\Gamma_p(\pi)$ containing those local sections ψ having the property that, for some neighbourhood W of p, $\psi|_W = \phi|_W$. (The neighbourhood W will, of course, depend on ψ.) The germ of ϕ at p will be denoted by $[\phi]_p$. ■

The relation "has the same germ at p" is clearly an equivalence relation on the set $\Gamma_p(\pi)$.

Proposition 1.2.6 *If* $\phi \in \Gamma_p(\pi)$, *and* $\Gamma(\pi)$ *is non-empty, then there is a global section* ψ *satisfying* $[\psi]_p = [\phi]_p$.

Proof Let $\chi \in \Gamma(\pi)$. Both $\phi(p)$ and $\chi(p)$ are in the fibre E_p, which as a manifold is path-connected. Let $\gamma : [0,1] \longrightarrow E$ be a path in this fibre satisfying $\gamma(0) = \chi(p)$, $\gamma(1) = \phi(p)$. Cover $\gamma([0,1])$ (regarded as a subset of E) with the domains of *convex* adapted charts constructed from a local trivialisation around p, and choose a finite subcover $U_1, \ldots, U_n$ where $\chi(p) \in U_1$, $\phi(p) \in U_n$ and $U_r \cap U_{r+1}$ is non-empty for $1 \leq r \leq n-1$. Let the corresponding coordinate systems be $y_r : U_r \longrightarrow \mathbf{R}^m \times \mathbf{R}^n$ and put $x_r : \pi(U_r) \longrightarrow \mathbf{R}^m$ where $x_r \circ \pi = pr_1 \circ u_r$.

Now suppose that, for some r with $1 \leq r \leq n-1$, there is a section $\chi_r \in \Gamma(\pi)$ satisfying $\chi_r(p) \in U_r$. Choose $a_r \in U_r \cap U_{r+1} \cap \pi^{-1}(p)$, and let $W_r \subset M$ be an open subset which satisfies $p \in W_r \subset \chi_r^{-1}(U_r)$ and is sufficiently small that, for every $q \in W_r$,

$$(x_r(q), pr_2(y_r(a_r))) \in y_r(U_r).$$

It is then possible to define the map $\kappa_r : W_r \longrightarrow E$ by

$$\kappa_r(q) = y_r^{-1}(x_r(q), pr_2(y_r(a_r)))$$

which is a local section of π owing to the relationship between x_r and y_r. There is also a compact subset $C_r \subset W_r$ with $p \in C_r$, and a bump function $b_r : W_r \longrightarrow \mathbf{R}$ satisfying $b_r(p) = 1$ and $b_r(q) = 0$ for $q \notin C_r$. We may therefore define a new global section χ_{r+1} by

$$\begin{aligned} \chi_{r+1}(q) &= y_r^{-1}(b_r(q)y_r(\kappa_r(q)) + (1 - b_r(q))y_r(\chi_r(q))) \qquad \text{for } q \in C_r \\ &= \chi_r(q) \qquad \text{otherwise} \end{aligned}$$

which from its method of construction is smooth and which satisfies $\chi_{r+1}(p) = a_r \in U_{r+1}$. Taking χ_1 to be χ, we then obtain a sequence of global sections $\chi_1, \ldots, \chi_n$ where finally $\chi_n(p) \in U_n$. Since the original local section ϕ satisfies $\phi(p) \in U_n$ we may use a similar construction to the one described above with ϕ instead of κ_r and with a bump function which this time equals one in a neighbourhood of p rather than merely at p. The result is a global section ψ satisfying $\psi(q) = \phi(q)$ for q in some neighbourhood of p. ■

As with global sections, a local section ϕ may be represented in coordinates by the functions $\phi^\alpha = u^\alpha \circ \phi$. On the tangent bundle τ_M the symbol $\partial/\partial x^i$ then represents a local section with the particular coordinate representation

$$\left(\dot{x}^j \circ \frac{\partial}{\partial x^i}\right)(p) = \dot{x}^j \left(\left.\frac{\partial}{\partial x^i}\right|_p\right) = \delta^j_i .$$

Finally, we record the useful fact that every local section is actually an embedding.

Proposition 1.2.7 *If $\phi \in \Gamma_W(\pi)$ then $\phi(W)$ is an embedded submanifold of E.*

Proof First, ϕ is an immersion because, for each $p \in W$, $\pi_* \circ \phi_* = id_{T_pM}$ so that $\phi_* : T_pM \longrightarrow T_{\phi(p)}E$ is injective. Secondly, from $\pi \circ \phi = id_W$ it then follows that ϕ is an injective immersion. Finally, from $\phi \circ \pi \circ \phi = \phi$ it follows that $\phi \circ \pi|_{\phi(W)} = id_{\phi(W)}$ so that ϕ is a homeomorphism of W onto $\phi(W)$. ■

EXERCISES

1.2.1 Let (E, π, M) be a fibred manifold and let $a \in E$. Show that there is a local section ϕ of π defined in some neighbourhood of $\pi(a)$ and satisfying $\phi(\pi(a)) = a$.

1.2.2 Let (E, π, M) be a bundle and let the (not necessarily smooth) function $\phi : M \longrightarrow E$ satisfy $\pi \circ \phi = id_M$. Show that ϕ is smooth (and therefore a section of π) if, and only if, for every point $p \in M$ there is an adapted coordinate system around $\phi(p) \in E$ such that the real-valued functions ϕ^α are smooth at p.

1.3 Bundle Morphisms

A morphism from one bundle to another may be described as a pair of maps, one between the total spaces and one between the base spaces. The two maps have to be related by the bundle projections, and indeed the map between the total spaces—if it is able to form part of a bundle morphism at all—determines uniquely the map between the base spaces.

Definition 1.3.1 If (E, π, M) and (H, ρ, N) are bundles then a *bundle morphism from π to ρ* is a pair $(f, \bar{f})$ where $f : E \longrightarrow H$, $\bar{f} : M \longrightarrow N$ and $\rho \circ f = \bar{f} \circ \pi$. The map $\bar{f}$ is called the *projection* of f. ■

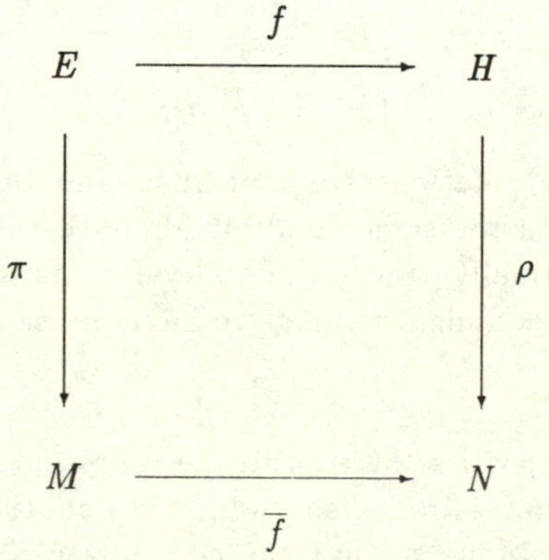

Lemma 1.3.2 *If $f : E \longrightarrow H$ then there is a bundle morphism $(f, \overline{f})$ from π to ρ if, and only if, whenever $p \in M$ and $a, b \in E_p$ then $\rho(f(a)) = \rho(f(b))$. The map $\overline{f}$ is unique.*

Proof If $(f, \overline{f})$ is a bundle morphism then

$$\rho(f(a)) = \overline{f}(\pi(a)) = \overline{f}(\pi(b)) = \rho(f(b)).$$

Conversely, suppose the condition holds. If $p \in M$ then choose $a \in E_p$ and define $\overline{f}(p)$ to equal $\rho(f(a))$, which is independent of the choice of a. It remains to show that $\overline{f}$ is smooth. So let ϕ be any local section of π defined in a neighbourhood W of p. Then $\overline{f}\big|_W = \rho \circ f \circ \phi$, demonstrating that $\overline{f}$ is smooth at p. The map $\overline{f}$ is unique because π is surjective. ∎

A bundle morphism may therefore be described as a map from E to H which maps the fibres of π into the fibres of ρ. For shorthand, a map between the total spaces of two bundles which satisfies this condition will often be called a bundle morphism (although strictly it is the pair of maps which has this description).

Bundle morphisms may be described using local coordinate systems. If $(f, \overline{f})$ is a bundle morphism from (E, π, M) to (H, ρ, N) and if (y^a, v^A) is an adapted coordinate system on H then the real-valued functions f^a, f^A are defined by

$$\begin{aligned} f^a &= y^a \circ f \\ f^A &= v^A \circ f. \end{aligned}$$

This description would apply to any map from E to H. The property that f maps fibres of π to fibres of ρ is reflected in the fact that the functions f^a must be constant on the fibres of π. There must then exist real-valued functions $\overline{f}^a$ defined on an open subset of M and satisfying

$$\begin{aligned} \overline{f}^a &= y^a \circ \overline{f} \\ f^a &= \overline{f}^a \circ \pi \end{aligned}$$

where the y^a in these equations are coordinate functions on N rather than on H. Just as we used the same symbols for both these sets of coordinate functions we shall normally denote the real-valued functions on both M and E by f^a, and with this understanding we have instead

$$f^a = y^a \circ \overline{f}.$$

Note that we normally write the coordinate representation as (f^a, f^A) with the base space coordinates first, even though we denote the bundle morphism itself by $(f, \overline{f})$; this is done so that the coordinate representation matches the order of the coordinate functions (y^a, v^A).

Example 1.3.3 If $(M \times F, pr_1, M)$ and $(N \times K, pr_1, N)$ are trivial bundles and if $f_0 : F \longrightarrow K$, $\overline{f} : M \longrightarrow N$ are maps, then $f = \overline{f} \times f_0 : M \times F \longrightarrow N \times K$ defines a bundle morphism $(f, \overline{f})$. If local coordinates on N and K are y^a, v^A and the coordinate representations of $\overline{f}$ and f_0 are f^a, f^A then the coordinate representation of f is $(f^a \circ pr_1, f^A \circ pr_2)$; as usual, we abbreviate this to (f^a, f^A). ∎

Example 1.3.4 Let $(SL(2,\mathbf{R}), \pi, H)$ be the bundle described in Example 1.1.3, and let $f_\lambda : SL(2,\mathbf{R}) \longrightarrow SL(2,\mathbf{R})$, where $\lambda \in \mathbf{R}$, be the map corresponding to left multiplication by the matrix $\begin{pmatrix} 1 & \lambda \\ 0 & 1 \end{pmatrix}$. Then

$$\begin{aligned} \pi\left(f_\lambda\begin{pmatrix} a & b \\ c & d \end{pmatrix}\right) &= \pi\left(\begin{pmatrix} 1 & \lambda \\ 0 & 1 \end{pmatrix}\begin{pmatrix} a & b \\ c & d \end{pmatrix}\right) \\ &= \pi\begin{pmatrix} a+\lambda c & b+\lambda d \\ c & d \end{pmatrix} \\ &= \frac{(ac+bd)+i}{c^2+d^2} + \lambda \\ &= \pi\begin{pmatrix} a & b \\ c & d \end{pmatrix} + \lambda \end{aligned}$$

so that f_λ determines a map $\overline{f}_\lambda : H \longrightarrow H$ given by $\overline{f}_\lambda(z) = z + \lambda$. It follows that $(f_\lambda, \overline{f}_\lambda)$ is a bundle morphism from π to itself. ∎

Example 1.3.5 If M, N are manifolds and $f : M \longrightarrow N$ then (f_*, f) is a bundle morphism from (TM, τ_M, M) to (TN, τ_N, N), because f_* maps the fibre T_pM to the fibre $T_{f(p)}N$. To find the coordinate representation of (f, f_*), suppose that local coordinates around $\xi \in TM$ are $(x^i, \dot{x}^i)$ and around $f_*(\xi) \in TN$ are $(y^a, \dot{y}^a)$. Then if $p = \tau_M(\xi)$ and

$$\xi = \xi^i \left.\frac{\partial}{\partial x^i}\right|_p$$

we have

$$f_*(\xi) = \xi^i \left.\frac{\partial f^a}{\partial x^i}\right|_p \left.\frac{\partial}{\partial y^a}\right|_{f(p)}$$

so that

$$\begin{aligned} \dot{y}^a(f_*(\xi)) &= \xi^i \left.\frac{\partial f^a}{\partial x^i}\right|_p \\ &= \dot{x}^i(\xi)\left(\frac{\partial f^a}{\partial x^i} \circ \tau_M\right)(\xi). \end{aligned}$$

We may therefore write the coordinate representation of (f_*, f) as

$$\left(f^a, \dot{x}^i \frac{\partial f^a}{\partial x^i}\right).$$

■

Example 1.3.6 If M, N are manifolds and $f : M \longrightarrow N$ is a local diffeomorphism (so that each $f_* : T_pM \longrightarrow T_{f(p)}N$ and its transpose $f^* : T^*_{f(p)}N \longrightarrow T^*_pM$ are isomorphisms) then (f^{*-1}, f) is a bundle morphism from (T^*M, τ^*_M, M) to (T^*N, τ^*_N, N). ■

If $(f, \overline{f})$ is a bundle morphism from π to ρ, and $(g, \overline{g})$ is a bundle morphism from ρ to σ, then $g \circ f$ also maps fibres to fibres and so defines the *composite bundle morphism* $(g \circ f, \overline{g \circ f})$ from ρ to σ. It therefore makes sense to define a bundle isomorphism as a bundle morphism which has a (two-sided) inverse. It should be clear that a bundle which is isomorphic to a trivial bundle is itself trivial, for the isomorphism may be used to pull back the trivialisation from one bundle to the other; indeed a fibred manifold which is isomorphic to a bundle is itself a bundle. In general, since

$$(\overline{g \circ f}) \circ \pi = \sigma \circ g \circ f = \overline{g} \circ \rho \circ f = \overline{g} \circ \overline{f} \circ \pi,$$

it follows that $\overline{g \circ f} = \overline{g} \circ \overline{f}$. If $(f, \overline{f})$ is a bundle isomorphism then both f, $\overline{f}$ are diffeomorphisms, and conversely. Be warned, however, that it is possible for just f (or just $\overline{f}$) to be a diffeomorphism.

Example 1.3.7 The map $f_\lambda : SL(2, \mathbf{R}) \longrightarrow SL(2, \mathbf{R})$ described in Example 1.3.4 defines a bundle isomorphism $(f_\lambda, \overline{f}_\lambda)$; the inverse isomorphism is $(f_{-\lambda}, \overline{f}_{-\lambda})$. ■

Example 1.3.8 If $f : M \longrightarrow N$ is a diffeomorphism then both $(f_*, f) : \tau_M \longrightarrow \tau_N$ and $(f^{*-1}, f) : \tau^*_M \longrightarrow \tau^*_N$ are bundle isomorphisms. ■

Example 1.3.9 The pair (π, id_M) is a bundle morphism from (E, π, M) to (M, id_M, M) but in general π is not a diffeomorphism. Similarly the pair (id_E, π) is a bundle morphism from (E, id_E, E) to (E, π, M). ■

Example 1.3.10 Any section $\phi \in \Gamma(\pi)$ defines a bundle morphism (ϕ, id_M) from (M, id_M, M) to (E, π, M). ■

If (E, π, M) and (H, ρ, N) are bundles then there is an action on the local sections of π by certain bundle morphisms from π to ρ. These are the bundle morphisms whose projections are *diffeomorphisms* from M to N, and they may be used to transport local sections from π to ρ.

Definition 1.3.11 If $(f, \overline{f})$ is a bundle morphism from π to ρ where $\overline{f}$ is a diffeomorphism, and if $\phi \in \Gamma_W(\pi)$ then the local section $\tilde{f}(\phi) \in \Gamma_{\overline{f}(W)}(\pi)$ is defined by

$$\tilde{f}(\phi) = f \circ \phi \circ \overline{f}^{-1}\Big|_{\overline{f}(W)}.$$

■

Example 1.3.12 If $f : M \longrightarrow N$ is a diffeomorphism then (f_*, f) is a bundle isomorphism from τ_M to τ_N, so if $X \in \mathcal{X}(M)$ then $\tilde{f}_*(X) \in \mathcal{X}(N)$ is defined by

$$\tilde{f}_*(X) = f_* \circ X \circ f^{-1}.$$

Usually this vector field is written as $f_*(X)$ rather than $\tilde{f}_*(X)$. ■

Just as one considers local sections as well as global sections, it is often useful to consider local bundle morphisms.

Definition 1.3.13 If $W \subset M$ is an open submanifold and if $f : \pi^{-1}(W) \longrightarrow H$, $\overline{f} : W \longrightarrow N$ then the pair $(f, \overline{f})$ is called a *local bundle morphism from π to ρ* if $\rho \circ f = \overline{f} \circ \pi\Big|_{\pi^{-1}(W)}$. ■

A local bundle morphism which has a (two-sided) inverse is called a local bundle isomorphism. Indeed, it would have been possible to define a bundle as a fibred manifold which was locally isomorphic to a trivial bundle.

A particularly important example of a bundle morphism which will be discussed in more detail in Section 3.2 is given by a pair of π-related vector fields.

Example 1.3.14 A vector field $X \in \mathcal{X}(E)$ which is also a bundle morphism from (E, π, M) to (TE, π_*, TM) is called a *π-projectable vector field*, or sometimes just a *projectable vector field*. The projection of X is a map $\overline{X} : M \longrightarrow TM$ which satisfies

$$\tau_M \circ \overline{X} \circ \pi = \tau_M \circ \pi_* \circ X = \pi \circ \tau_E \circ X = \pi$$

because (π_*, π) is a bundle morphism from τ_E to τ_M. Since π is surjective it follows that $\tau_M \circ \overline{X} = id_M$, so that $\overline{X}$ is a section of τ_M and therefore is a vector field on M.

To find the coordinate representation of X we use coordinate functions $(x^i, \dot{x}^i, u^\alpha, \dot{u}^\alpha)$ on TE and (x^i, u^α) on E. Since X is a section of τ_E the x^i and u^α components of the coordinate representation are fixed, so that X is determined by the real-valued functions X^i, X^α defined by

$$\begin{aligned} X^i &= \dot{x}^i \circ X \\ X^\alpha &= \dot{u}^\alpha \circ X. \end{aligned}$$

As with a general bundle morphism the functions X^i must be constant on the fibres of π, so that there are real-valued functions $\overline{X}^i$ defined on an open subset of M and satisfying

$$\begin{aligned} \overline{X}^i &= \dot{x}^i \circ \overline{X} \\ X^i &= \overline{X}^i \circ \pi \end{aligned}$$

where the $\dot{x}^i$ in these equations are coordinate functions on TM rather than on TE. Once again we usually write X^i instead of $\overline{X}^i$. With this notation we have

$$\begin{aligned} X &= X^i \frac{\partial}{\partial x^i} + X^\alpha \frac{\partial}{\partial u^\alpha} \\ \overline{X} &= X^i \frac{\partial}{\partial x^i}. \end{aligned}$$

■

EXERCISES

1.3.1 If $f : M \longrightarrow N$ is a local diffeomorphism, show that the coordinate representation of the bundle morphism $(f^{*-1}, f) : \tau_M^* \longrightarrow \tau_N^*$ is

$$(f^a, \partial_i \cdot F_a^i)$$

where (x^i, ∂_i) are coordinate functions on T^*M and where F_a^i is the inverse of the matrix of functions $\partial f^a / \partial x^i$.

1.3.2 If M is a manifold then there are two bundles with base space TM and total space TTM, namely τ_{TM} and τ_{M*}. Construct a bundle isomorphism $(f, id_{TM}) : \tau_{TM} \longrightarrow \tau_{M*}$. (Hint: starting with local coordinates x^i on M, what is the effect of τ_{TM} and τ_{M*} in terms of the induced coordinates on TM and TTM? This should suggest the definition of a map whose domain is a chart in TTM, which is smooth and has a smooth inverse. By showing that this map is independent of the choice of coordinate system, deduce the existence of a diffeomorphism of TTM which restricts to this map and which yields the required bundle isomorphism.)

1.4 New Bundles From Old

There are several methods of constructing new bundles from given ones, and most of these methods involve products of some kind. The most general construction is the *product bundle*, which may be formed from an arbitrary pair of bundles.

Definition 1.4.1 If (E,π,M) and (H,ρ,N) are bundles then the *product bundle* is the triple $(E\times H,\pi\times\rho,M\times N)$. ∎

It is straightforward to check that $\pi\times\rho$ really is a bundle whose typical fibre is the Cartesian product of the typical fibres of π and ρ: if (W_p,F,t_p) and (V_q,K,s_q) are local trivialisations of π and ρ around p and q respectively, then

$$t_p\times s_q:(\pi\times\rho)^{-1}(W_p\times V_q)\longrightarrow(W_p\times V_q)\times(F\times K)$$
$$(t_p\times s_q)(a,b)=((\pi(a),\rho(b)),(pr_2\circ t_p(a),pr_2\circ s_q(b)))$$

is a local trivialisation of $\pi\times\rho$ around (a,b).

Given local sections $\phi\in\Gamma_p(\pi)$, $\psi\in\Gamma_q(\rho)$, the *product section* is the map $\phi\times\psi\in\Gamma_{(p,q)}(\pi\times\rho)$ defined by

$$(\phi\times\psi)(r,s)=(\phi(r),\psi(s)).$$

We may also define product coordinates: if (x^i,u^α) is an adapted coordinate system on $U\subset E$ and (y^a,v^A) is an adapted coordinate system on $V\subset H$ then (x^i,y^a,u^α,v^A) is an adapted coordinate system on $U\times V\subset E\times H$.

Example 1.4.2 If $(M\times F,pr_1,M)$ and $(N\times K,pr_1,K)$ are trivial bundles then their product is the trivial bundle $((M\times F)\times(N\times K),pr_1\times pr_1,M\times N)$. ∎

Example 1.4.3 If M and N are manifolds then the product of their tangent bundles τ_M,τ_N is the bundle $(TM\times TN,\tau_M\times\tau_N,M\times N)$. This is isomorphic to the tangent bundle $(T(M\times N),\tau_{M\times N},M\times N)$, and the isomorphism $(f,id_{M\times N}):\tau_{M\times N}\longrightarrow\tau_M\times\tau_N$ may be given explicitly as follows: if $\xi\in T_{(p,q)}(M\times N)$ then

$$f(\xi)=(pr_{1*}(\xi),pr_{2*}(\xi))\in T_pM\times T_qN.$$

In local coordinates x^i on M and y^a on N (and therefore (x^i,y^a) on $M\times N$) we may write

$$\begin{aligned}\xi &= \xi^i\left.\frac{\partial}{\partial x^i}\right|_{(p,q)}+\xi^a\left.\frac{\partial}{\partial y^a}\right|_{(p,q)}\\ f(\xi) &= \left(\xi^i\left.\frac{\partial}{\partial x^i}\right|_p,\xi^a\left.\frac{\partial}{\partial y^a}\right|_q\right)\end{aligned}$$

∎

If there is some relationship between the two bundles π and ρ then other constructions may be performed. In particular, if the base spaces of the two bundles are identical then an important construction called the *fibred product* provides a new bundle over that same base space rather than over the product of the base space with itself. On the other hand, by considering the total space of the fibred product but choosing a different base space we obtain the *pull-back bundle*, which may actually be defined in slightly more general circumstances.

Definition 1.4.4 If (E, π, M) and (H, ρ, M) are bundles over the same base space M then the *fibred product bundle* is the triple $(E \times_M H, \pi \times_M \rho, M)$, where the total space $E \times_M H$ is defined to equal

$$\{(a, b) \in E \times H : \pi(a) = \rho(b)\}$$

and the projection map $\pi \times_M \rho$ is defined by

$$(\pi \times_M \rho)(a, b) = \pi(a) = \rho(b).$$

■

Once again it is necessary to check that $\pi \times_M \rho$ really is a bundle. The first point is that the total space $E \times_M H$ is a manifold because it is the submanifold of $E \times H$ given by $(\pi \times \rho)^{-1}(\Delta_M)$, where $\Delta_M \subset M \times M$ is the diagonal. Having established this, it is again straightforward to see that $\pi \times_M \rho$ has the properties of a bundle whose typical fibre is again the Cartesian product of the typical fibres of π and ρ. The fibred product is also sometimes called the *Whitney sum* and denoted $\pi \oplus \rho$, although this latter notation will be more appropriate when π and ρ are "vector bundles" (to be defined in Section 2.1).

If ϕ and ψ are local sections of π and ρ respectively, we may construct their fibre product provided that the domains of the two sections overlap. If $\phi \in \Gamma_V(\pi)$, $\psi \in \Gamma_W(\pi)$ then $\phi \times_M \psi \in \Gamma_{V \cap W}(\pi \times_M \rho)$ is defined by

$$(\phi \times_M \psi)(q) = (\phi(q), \psi(q)).$$

We may define fibre product coordinates in similar circumstances: if (x^i, u^α) is an adapted coordinate system on $U \subset E$ and (x^i, v^A) is an adapted coordinate system on $V \subset H$ where $\pi(U) \cap \rho(V)$ is non-empty, then we may take (x^i, u^α, v^A) as an adapted coordinate system on

$$(U \cap \pi^{-1}(\rho(V))) \times_M (V \cap \rho^{-1}(\pi(U))) \subset E \times_M H.$$

Although the total space $E \times_M H$ has been given the structure of a bundle over M, there are also maps from this manifold to E and to H given by restricting the Cartesian product projections. However, rather than writing $pr_1|_{E \times_M H} : E \times_M H \longrightarrow E$ we normally denote this map by $\pi^*(\rho)$ and the corresponding map $E \times_M H \longrightarrow H$ by $\rho^*(\pi)$. Both these maps define bundles.

$$\begin{array}{ccc} E \times_M H & \xrightarrow{\pi^*(\rho)} & E \\ \Big\downarrow{\scriptstyle \rho^*(\pi)} & & \Big\downarrow{\scriptstyle \pi} \\ H & \xrightarrow[\rho]{} & M \end{array}$$

In the more general situation where (say) π is a bundle but ρ is merely an arbitrary smooth map, we can still regard $\rho^*(\pi)$ as a bundle. In these circumstances it is conventional to denote the total space $E \times_M H$, using an alternative notation, as $\rho^*(E)$.

Definition 1.4.5 If (E, π, M) is a bundle and $\rho : H \longrightarrow M$ is a map then the *pull-back of* π *by* ρ is the bundle $(\rho^*(E), \rho^*(\pi), H)$, where the total space $\rho^*(E)$ is defined to equal

$$\{(a, b) \in E \times H : \pi(a) = \rho(b)\}$$

and the projection $\rho^*(\pi)$ is defined by

$$\rho^*(\pi)(a, b) = b.$$

■

It is easy to check that the pull-back is a bundle, with typical fibre equal to the typical fibre of π. The pair $(\pi^*(\rho), \rho)$ is then a bundle morphism from $\rho^*(\pi)$ to π. (Indeed, it is often convenient to think of the total space $\rho^*(E)$ as comprising lots of copies of fibres of the form E_p with their base-points transplanted from M to H.) If $\phi \in \Gamma_W(\pi)$ is a local section then its *pull-back section* is the map $\rho^*(\phi) \in \Gamma_{\rho^{-1}(W)}(\rho^*(\pi))$ defined by

$$\rho^*(\phi)(b) = (\phi(\rho(b)), b).$$

Since the composite map $\phi \circ \rho$ may also be written as a pull-back $\rho^*(\phi)$, this notation is reasonably consistent.

Important examples of pull-back bundles arise when a bundle (E, π, M) is given, and tangent or cotangent vectors on M are pulled back to E.

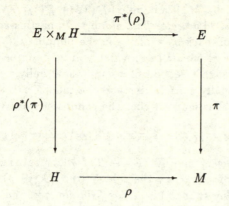

Example 1.4.6 For any map $f : N \longrightarrow M$, the pull-back by f of the tangent bundle (TM, τ_M, M) is a bundle $(f^*(TM), f^*(\tau_M), N)$, and a section X of this bundle is called a *vector field along* f. In particular, if (E, π, M) is a bundle then the pull-back by π of τ_M is the bundle $(\pi^*(TM), \pi^*(\tau_M), E)$ known as the *transverse bundle to* π, which will be described in more detail in Section 3.1. Although a section of $\pi^*(\tau_M)$ is actually a map $X : E \longrightarrow \pi^*(TM)$, we shall usually consider instead the map $\widehat{X} : E \longrightarrow TM$ defined by $\widehat{X} = \tau_M^*(\pi) \circ X$. This map satisfies the condition $\tau_M \circ \widehat{X} = \pi$, because

$$\tau_M \circ \widehat{X} = \tau_M \circ \tau_M^*(\pi) \circ X = \pi \circ \pi^*(\tau_M) \circ X = \pi \circ id_E = \pi.$$

On the other hand, every map $\widehat{X} : E \longrightarrow TM$ which satisfies this condition defines a section X of $\pi^*(\tau_M)$ by the rule $X(p) = (\widehat{X}(p), p)$. We shall often call the map $\widehat{X}$ (rather than X) a *vector field along* π, and we shall denote the set of all such vector fields along π by $\mathcal{X}(\pi)$. ■

Example 1.4.7 If (E, π, M) is a bundle then the pull-back by π of the cotangent bundle (T^*M, τ_M^*, M) is the bundle $(\pi^*(T^*M), \pi^*(\tau_M^*), E)$ known as the *cotangent bundle to E horizontal over* π. This bundle will also be described again in Section 3.1. The elements of the total space $\pi^*(T^*M)$ are pairs (η, a) where $\tau_M^*(\eta) = \pi(a)$.

In this example there is, however, another possible interpretation of the symbol $\pi^*(T^*M)$, and that is as the subset $\{\pi^*(\eta) : \eta \in T^*M\}$ of T^*E. This subset is actually a submanifold, and we obtain a sub-bundle

$$(\pi^*(T^*M), \tau_E^*|_{\pi^*(T^*M)}, E)$$

of τ_E^*. Fortunately the two bundles are isomorphic, and the isomorphism

$$(f, id_E) : \tau_E^*|_{\pi^*(T^*M)} \longrightarrow \pi^*(\tau_M^*)$$

may be given explicitly by $f(\pi^*(\eta)) = (\eta, \tau_E(\pi^*(\eta)))$. We shall usually regard $\pi^*(T^*M)$ as a submanifold of T^*E. ■

In the particular case when the pull-back map is an embedding then the pull-back bundle is called the restricted bundle.

Definition 1.4.8 If (E, π, M) is a bundle and $\iota_W : W \longrightarrow M$ is an embedding then the *restriction of* π *to* W is the bundle $\iota_W^*(\pi)$, which is usually denoted $\pi|_W$. ■

Technically, the total space $\iota_W^*(E)$ of the restricted bundle consists of pairs $(a, b) \in E \times W$ where $\pi(a) = b$. However this is clearly diffeomorphic to the submanifold $\pi^{-1}(W)$ of E, and with this identification the restricted bundle may be regarded as a sub-bundle of π.

Definition 1.4.9 If (E, π, M) is a bundle and $E' \subset E$ is a submanifold such that the triple $(E', \pi|_{E'}, \pi(E'))$ is itself a bundle, then $\pi|_{E'}$ is called a *sub-bundle* of π. ∎

(The reason for the qualification in this definition is that there may be submanifolds $E' \subset E$ where $\pi|_{E'}$ is not a bundle, or even a fibred manifold.) A restricted bundle may therefore be considered as a sub-bundle with the particular property that $\pi^{-1}(\pi(E')) = E'$. It will be clear from the context whether the notation $\pi|_{E'}$ is meant to denote a restricted bundle or a sub-bundle, for this just depends upon whether E' is a submanifold of M or of E.

Example 1.4.10 If W is an *open* submanifold of M then whenever $p \in M$ the vector spaces T_pW and T_pM are isomorphic. Correspondingly the restricted bundle $(\iota_W^*(TM), \tau_M|_W, W)$ and the tangent bundle (TW, τ_W, W) are isomorphic, and the isomorphism $(f, id_W) : \tau_M|_W \longrightarrow \tau_W$ may be given explicitly by $f(\xi, p) = \xi$. If, however, W is not open then $\iota_W^*(TM)$ will contain elements (ξ, p) where the vector $f(\xi, p) \in T_pM$ is *not* tangent to W but instead "points out into the surrounding space". ∎

Example 1.4.11 Considering now the bundle (E, π, M), the subset

$$\{\xi \in TE : \pi_*(\xi) = 0 \in T_{\tau_M(\pi_*(\xi))}M\}$$

is called the set of *vectors vertical to* π. It is denoted $V\pi$ and is a submanifold of TE. The triple $(V\pi, \tau_E|_{V\pi}, E)$ is a sub-bundle of τ_E which is called the *vertical bundle to* π. This bundle will be described again in Section 3.1. ∎

EXERCISES

1.4.1 Construct an example of a bundle (E, π, M) and a submanifold $E' \subset E$ where $\pi|_{E'}$ is *not* a fibred manifold.

1.4.2 Let (E, π, M) and (H, ρ, M) be bundles. Show that the fibred product $\pi \times_M \rho$ is locally trivial (and is therefore a bundle).

1.4.3 Let M be a manifold and denote by T^2M the subset of elements $\xi \in TTM$ which satisfy $\tau_{TM}(\xi) = \tau_{M*}(\xi)$; denote by $\tau_M^{2,1}$ the map $\tau_{TM}|_{T^2M}$. Show that T^2M is a submanifold of TTM and that $(T^2M, \tau_M^{2,1}, TM)$ is a sub-bundle both of (TTM, τ_{TM}, TM) and of (TTM, τ_{M*}, TM).

1.4.4 From the previous exercise, a section X of $\tau_M^{2,1}$ is also a section of τ_{TM} and is therefore a vector field on TM of the particular type called a *second-order vector field.* Show, using coordinate functions $(x^i, \dot{x}^i)$ on TM, that the most general coordinate expression for such a vector field X is

$$X = \dot{x}^i \frac{\partial}{\partial x^i} + X^i \frac{\partial}{\partial \dot{x}^i}$$

where X^i are functions defined locally on TM.

REMARKS

The description we have given of a "bundle" is not the most general possible; in particular, the base and total spaces need only have a topology rather than a differentiable structure. This is the point of view adopted in [8], where the most general bundle is just a continuous map between topological spaces which need be neither surjective nor locally trivial.

The most common type of bundle found in applications is the *fibre bundle.* This involves, as well as the triple (E, π, M), a group of transformations of the fibres: a Lie group for the differentiable case, or more generally a topological group for the continuous case. Details of the properties of fibre bundles may be found in [8] or [16].

The properties of local sections of a bundle may be expressed in the language of *sheaf theory*, where the collection of germs of local sections may be considered as a sheaf. An introduction to the ideas of sheaf theory may be found in [18].

Finally, it is worth pointing out that the collection of bundles and bundle morphisms forms a *category*, and that (for example) the correspondence which associates to a manifold M its tangent bundle (TM, τ_M, M), and to a map $f : M \longrightarrow N$ its derivative $f_* : TM \longrightarrow TN$, is a covariant functor. We shall need to use a few of the ideas of category theory in a specialised context in Chapter 7, and we shall give the necessary definitions there. An introduction to category theory may be found in [13].

Chapter 2

Linear Bundles

In this chapter, we introduce bundles which have additional structure on their fibres: the fibres of vector bundles and affine bundles are, respectively, vector spaces and affine spaces. Although most of our discussion in subsequent chapters will start with general bundles, we shall have a great deal to say about various vector bundles associated with a general bundle. These will be constructed from the tangent and cotangent bundles of two different manifolds, namely the base and total spaces of the original bundle.

Affine bundles are less familiar objects; their importance lies in the fact that a first-order jet, being just a first-order Taylor polynomial in disguise, is the "best linear inhomogeneous approximation" to a section, and so is naturally represented as an element of an affine space.

2.1 Vector Bundles

A vector bundle is a mathematical object which combines two different types of structure. As such, its definition falls naturally into *three* parts, where the third part is a consistency condition relating the two structures. For a vector bundle, this consistency condition requires the existence of a family of local trivialisations, each of which has $\mathbf{R}^n$ as its typical fibre, and each of which is linear on every fibre.

Definition 2.1.1 A *vector bundle* is a quintuple (E, π, M, σ, μ) where:

1. (E, π, M) is a bundle;
2. (a) $\sigma : E \times_M E \longrightarrow E$ satisfies, for each $p \in M$, $\sigma(E_p \times E_p) \subset E_p$;

   (b) $\mu : \mathbf{R} \times E \longrightarrow E$ satisfies, for each $p \in M$, $\mu(\mathbf{R} \times E_p) \subset E_p$;

   (c) for each $p \in M$, $(E_p, \sigma|_{E_p \times E_p}, \mu|_{\mathbf{R} \times E_p})$ is a real vector space;
3. for each $p \in M$ there is a local trivialisation $(W_p, \mathbf{R}^n, t_p)$, called a *linear local trivialisation*, satisfying the condition that, for $q \in W_p$,

the composite of

$$t_p|_{E_q} : E_q \longrightarrow \{q\} \times \mathbf{R}^n$$

with $pr_2 : \{q\} \times \mathbf{R}^n \longrightarrow \mathbf{R}^n$ is a linear isomorphism.

■

In a vector bundle, the typical fibre and the actual fibres are all isomorphic vector spaces, and it is possible to select a family of local trivialisations which provide the isomorphisms. However, in general there is no canonical isomorphism between the typical fibre and any particular fibre. Notice also that the maps σ and μ are automatically smooth, because under the linear local trivialisation $(W, \mathbf{R}^n, t)$ they correspond locally to $id_W \times s$ and $id_W \times m$ where s and m are addition and scalar multiplication on $\mathbf{R}^n$.

To avoid our notation for vector bundles getting too ridiculous we shall usually refer to (E, π, M, σ, μ) as (E, π, M) or even sometimes as π; we do not normally consider two different vector bundle structures on the same underlying bundle. Indeed, most of the vector bundles we shall consider will be constructed in some way from the following three basic examples.

Example 2.1.2 For any manifold M, the *trivial line bundle* is the vector bundle $(M \times \mathbf{R}, pr_1, M)$. The *trivial n-plane bundle* is the vector bundle $(M \times \mathbf{R}^n, pr_1, M)$. ■

Example 2.1.3 The tangent bundle (TM, τ_M, M) is a vector bundle. Each fibre T_pM may be given the structure of a vector space, and if (W, x) is a coordinate system around $p \in M$ then the map

$$(\tau_M|_{\tau_M^{-1}(W)}, \dot{x}) : \tau_M^{-1}(W) \longrightarrow W \times \mathbf{R}^m$$

is a local trivialisation around p. Since $\dot{x} : \tau_M^{-1}(W) \longrightarrow \mathbf{R}^m$ is linear on each fibre, the linearity condition is automatically satisfied. ■

Example 2.1.4 The cotangent bundle (T^*M, τ_M^*, M) is also a vector bundle, where now the linear local trivialisations are

$$(\tau_M^*|_{\tau_M^{*-1}(W)}, \partial) : \tau_M^{*-1}(W) \longrightarrow W \times \mathbf{R}^m$$

■

Example 2.1.5 The bundle $(T^2M, \tau_M^{2,1}, TM)$ described in Exercise 1.4.3 is *not* a vector bundle in any natural way, even though its typical fibre is $\mathbf{R}^m$. The reason for this is that the fibres of $\tau_M^{2,1}$ do not have a natural vector space structure. Although T^2M is a submanifold of TTM and $\tau_M^{2,1} = \tau_{TM}|_{T^2M}$, and although the fibres of τ_{TM} are vector spaces, the corresponding fibres of $\tau_M^{2,1}$ are not vector subspaces. For instance, if $\xi \in TM$ and if $(x^i, \dot{x}^i)$ are coordinates on TM around ξ, then an arbitrary element $\lambda \in T_\xi TM$ has the form

$$\lambda = \lambda^i \left.\frac{\partial}{\partial x^i}\right|_\xi + \mu^i \left.\frac{\partial}{\partial \dot{x}^i}\right|_\xi .$$

If $\lambda \in T^2M$ then, from the results in Exercise 1.4.4, we must have

$$\lambda^i = \dot{x}^i(\xi);$$

it follows that scalar multiples of λ are generally not elements of T^2M. ∎

As with more general bundles, we may show that the manifold structure on the total space of a vector bundle may be deduced from the manifold structures on its base space and its typical fibre, although the linearity allows us to express the result slightly differently.

Proposition 2.1.6 *Let M be a manifold, E a set, and $\pi : E \longrightarrow M$ a function such that, for each $p \in M$, $\pi^{-1}(p)$ has the structure of an n-dimensional real vector space. Suppose also that, for each $p \in M$, there is a neighbourhood W_p of p and a bijection $t_p : \pi^{-1}(W_p) \longrightarrow W_p \times \mathbf{R}^n$ satisfying:*

1. $pr_1 \circ t_p = \pi|_{\pi^{-1}(W_p)}$;
2. *for each $q \in W_p$, $pr_2 \circ t_p|_{\pi^{-1}(q)} : \pi^{-1}(q) \longrightarrow \mathbf{R}^n$ is a linear isomorphism.*

Then E may be given a unique structure as a manifold such that π becomes a vector bundle and the maps t_p become linear local trivialisations.

Proof By Proposition 1.1.14, π is a bundle with typical fibre $\mathbf{R}^n$. The vector space structure on the fibres and the linearity of the maps $pr_2 \circ t_p|_{\pi^{-1}(q)}$ then show that π is actually a vector bundle. ∎

Example 2.1.7 The real projective plane RP^2 may be defined as the quotient space $S^2/\{\pm\}$, where the equivalence relation $\pm$ on S^2 identifies p and $-p$. The manifold structure on RP^2 is the one induced from S^2; in particular, each coordinate system defined on a suitably small domain in S^2 yields a coordinate system on RP^2.

If S^2 is regarded as a subset of $\mathbf{R}^3$, then we may identify RP^2 with the set of lines (in this context, one-dimensional vector subspaces) in $\mathbf{R}^3$. We may therefore attempt to define a vector bundle with base space RP^2 and typical fibre $\mathbf{R}$, by attaching to each point $\pm p \in RP^2$ the line $\{\lambda p : \lambda \in \mathbf{R}\} \subset \mathbf{R}^3$ with which it is identified, and by letting $E = \{(\pm p, \lambda p)\} \subset RP^2 \times \mathbf{R}^3$; the result (E, h, RP^2) is called the *tautological bundle on* RP^2. To show that this is indeed a vector bundle, let W' be a neighbourhood of $p \in S^2$ which is sufficiently small that $W' \cap (-W') = \emptyset$, and let W be the corresponding neighbourhood of $\pm p \in RP^2$, so that the equivalence relation $\pm$ defines a diffeomorphism between W' and W. We may then specify the function $t : h^{-1}(W) \longrightarrow W \times \mathbf{R}$ by

$$t(\pm p, \lambda p) = (\pm p, \lambda),$$

using the diffeomorphism to select the sign of λ. The map then satisfies the conditions of Proposition 2.1.6. ∎

If (E, π, M) is a vector bundle and (x^i, u^α) are adapted coordinate systems on E, then in general there is no reason why the maps u^α should be linear on the fibres of π. However, there are always adapted coordinate systems which *are* linear on the fibres; these may be constructed from coordinates on the base space and the linear local trivialisations. The domain of such a coordinate system is then of the form $\pi^{-1}(W)$ where $W \subset M$ is the domain of the base coordinates, and the coordinate system is called a ***vector bundle coordinate system***. When dealing with vector bundles we invariably use vector bundle coordinate systems.

Corresponding to each vector bundle coordinate system there is a family of local sections which may be regarded as dual to the coordinates. If (x^i, u^α) is the coordinate system and $W \subset M$ is the domain of the coordinates x^i, then we may define the family of local sections $e_\beta \in \Gamma_W(\pi)$ by $u^\alpha(e_\beta(p)) = \delta^\alpha_\beta$ for every $p \in W$. This family has the property that every local section $\phi \in \Gamma_W(\pi)$ may be written as a linear combination $\phi = \phi^\alpha e_\alpha$, where $\phi^\alpha = u^\alpha \circ \phi|_W \in C^\infty(W)$ and the linear operations on sections are defined pointwise. In fact, when referring to the coordinate representation of a section of a *vector* bundle, we shall invariably write $\phi^\alpha e_\alpha$ and so refer to these local sections explicitly.

Example 2.1.8 On the tangent bundle (TM, τ_M, M) the standard coordinates $(x^i, \dot{x}^i)$ are vector bundle coordinates, and the corresponding local sections are the vector fields $\partial/\partial x^i$. ∎

Example 2.1.9 On the cotangent bundle (T^*M, τ^*_M, M) the standard coordinates (x^i, ∂_i) are vector bundle coordinates, and the corresponding local sections are the 1-forms dx^i. ∎

We may also consider the sets of sections of a vector bundle. The set of global sections forms a vector space, and many of the constructions which may be performed with vector bundles may similarly be performed with their spaces of sections; for some of these constructions, however, it is more appropriate to regard the sections as forming a module over the ring of functions on the base manifold. The existence of partitions of unity on the base manifold will be an important tool for establishing the global validity of certain results.

Lemma 2.1.10 *If (E, π, M) is a vector bundle then $\Gamma(\pi)$ is a vector space under pointwise operations. If $W \subset M$ is an open submanifold then $\Gamma_W(\pi)$ is a vector space in the same way.*

Proof First, $\Gamma(\pi)$ is non-empty because the zero section (which maps $p \in M$ to $0 \in E_p$) is smooth. If $\phi, \psi \in \Gamma(\pi)$ and $\lambda \in \mathbf{R}$ then we must check that $\phi + \psi$ and $\lambda\phi$ are smooth. So let (x^i, u^α) be vector bundle coordinates on E; then $x^i \circ (\phi + \psi)(p) = x^i(p)$ and $u^\alpha \circ (\phi + \psi)(p) = (u^\alpha \circ \phi + u^\alpha \circ \psi)(p)$. Since addition is a linear map from $\mathbf{R}^2$ to $\mathbf{R}$ and hence smooth, it follows that the component functions of $\phi + \psi$ are all smooth so that $\phi + \psi$ is smooth. The result for $\lambda\phi$ follows similarly. There is no essential difference when considering $\Gamma_W(\pi)$. ■

Note that the vector space $\Gamma(\pi)$ is infinite-dimensional (provided that the fibre dimension of E is non-zero). Although the choice of a topology for $\Gamma(\pi)$ is an important question in the study of global analysis, it is not one which we shall consider here.

Example 2.1.11 For any manifold M, the space of sections of the tangent bundle (TM, τ_M, M) will be denoted $\mathcal{X}(M)$; elements of $\mathcal{X}(M)$ are just vector fields on M. ■

Example 2.1.12 For any manifold M, the space of sections of the cotangent bundle (T^*M, τ^*_M, M) will be denoted $\bigwedge^1 M$; elements of $\bigwedge^1 M$ are just 1-forms on M. We shall also use the notation $\bigwedge^r M$ for the space of r-forms on M; this will be the space of sections of the vector bundle $\bigwedge^r \tau^*_M$ whose fibres are the vector spaces $\bigwedge^r T^*_p M$ (where $p \in M$). ■

Example 2.1.13 If π is the trivial line bundle $(M \times \mathbf{R}, pr_1, M)$ then $\Gamma(\pi)$ is a vector space which is canonically isomorphic to $C^\infty(M)$: each section ϕ corresponds to the function $pr_1 \circ \phi$. This is just a special case of the relationship between functions and graphs which introduced our discussion of bundles at the very beginning of Chapter 1. For this bundle π we shall choose not to distinguish between $\Gamma(\pi)$ and $C^\infty(M)$, although conceptually they are different objects. Furthermore, for any real vector space V, there is a canonical isomorphism from the zeroth alternating product space $\bigwedge^0 V$ to $\mathbf{R}$; applying this to the fibres of the vector bundle $\bigwedge^0 \tau^*_M$ we obtain a global trivialisation $\bigwedge^0 T^*M \cong M \times \mathbf{R}$, so that the space of 0-forms $\bigwedge^0 M$ may also be identified with $C^\infty(M)$. ■

Of course, $C^\infty(M)$ is more than a vector space: pointwise multiplication makes it a commutative ring, and we may use it to define a module structure on the space of sections of a vector bundle over M.

Proposition 2.1.14 *If (E, π, M) is a vector bundle then $\Gamma(\pi)$ is a module over $C^\infty(M)$ under pointwise operations. If $W \subset M$ is an open submanifold then $\Gamma_W(\pi)$ is similarly a module over $C^\infty(W)$. Furthermore, the module $\Gamma(\pi)$ is locally finitely generated, in the sense that for each $p \in M$ there is an open submanifold W containing p and a finite family $\phi_\mu \in \Gamma(\pi)$ such that $\phi_\mu|_W$ generate $\Gamma_W(\pi)$.*

Proof If $\phi \in \Gamma(\pi)$ and $f \in C^\infty(M)$ then $f\phi$ is defined by $f\phi(p) = f(p)\phi(p)$, and the smoothness of this new section is proved in the same way as for $\lambda\phi$. If $p \in M$ then take a vector bundle coordinate system (U, u) on E with $p \in \pi(U)$, and let $e_\alpha \in \Gamma(\pi)$ be zero outside $\pi(U)$ and satisfy $u^\beta(e_\alpha(q)) = \delta^\beta_\alpha$ for all q in some neighbourhood $W \subset \pi(U)$ of p: the global sections e_α may be constructed using bump functions from the local sections of $\pi|_U$ dual to the coordinate system. ∎

In general, the methods of constructing new bundles from old ones may be applied to vector bundles, with results which, in most cases, are themselves vector bundles. However not every sub-bundle of a vector bundle is itself a vector bundle, and so in this case an amended definition is needed.

Definition 2.1.15 If (E, π, M) is a vector bundle and $E' \subset E$ is a submanifold such that $(E', \pi|_{E'}, \pi(E'))$ is itself a vector bundle under the restriction of the fibre-linear operations σ, μ to E', then $\pi|_{E'}$ is termed a *vector sub-bundle of* π. ∎

A vector sub-bundle is therefore a sub-bundle which is a vector bundle in its own right under the induced operations. In most cases we deal with vector sub-bundles where $\pi(E') = M$, although this is not a requirement of our definition. Notice, however, that the mere specification of a linear subspace of each fibre of π does not necessarily create a vector sub-bundle: although Proposition 2.1.6 implies that the result of such a specification always yields a vector bundle provided that the subspaces all have the same dimension, the definition of a sub-bundle also requires that the union of the subspaces be a submanifold of the original total space.

Example 2.1.16 Let π be the trivial bundle $(\mathbf{R} \times \mathbf{R}^2, pr_1, \mathbf{R})$. Define $E' \subset \mathbf{R} \times \mathbf{R}^2$ by

$$E' = (-\infty, 0) \times (\mathbf{R} \times \{0\}) \cup [0, \infty) \times (\{0\} \times \mathbf{R}),$$

so that $\pi(E') = \mathbf{R}$. Although $\pi|_{E'}$ may be given the structure of a vector bundle isomorphic to $(\mathbf{R}\times\mathbf{R}, pr_1, \mathbf{R})$ using Proposition 2.1.6, the result is *not* a vector sub-bundle of π because the subspace topology on E' is different from the usual topology on $\mathbf{R}\times\mathbf{R}$. ■

By contrast, if subspaces of the fibres of a vector bundle are defined by the span of a family of sections then they *do* create a vector sub-bundle. We shall prove this result by first establishing a local result. This generalises to vector bundles the idea of creating a basis of a vector space by extending an arbitrary basis of a subspace.

Proposition 2.1.17 *Let (E, π, M) be a vector bundle and let $(E', \pi|_{E'}, M)$ be a vector sub-bundle. If $a \in E'$ and (x^i, u'^α), $1 \leq \alpha \leq k = \dim E' - \dim M < n$ is a vector bundle coordinate system on E' on the neighbourhood U of a, then there is a neighbourhood W of $\pi(a) \in M$ and a vector bundle coordinate system (x^i, u^α) on $\pi^{-1}(W)$ such that*

$$\begin{aligned} u^\alpha|_{\pi^{-1}(W)\cap U} &= u'^\alpha|_{\pi^{-1}(W)\cap U} \qquad 1 \leq \alpha \leq k \\ &= 0 \qquad \textit{otherwise.} \end{aligned}$$

Proof Let e_α, $1 \leq \alpha \leq k$ be the local sections of $\pi|_{E'}$ dual to the vector bundle coordinates (x^i, u^α); they are also smooth local sections of π because the inclusion map $E' \longrightarrow E$ is smooth. Choose also an arbitrary vector bundle coordinate system on E around a, and let f_α, $1 \leq \alpha \leq n$, be the corresponding local sections.

Let $p = \pi(a)$. At least one of the elements $f_\alpha(p)$, say $f_\beta(p)$, will be linearly independent of $\{e_1(p), \ldots, e_k(p)\}$ in E_p, and by continuity a similar property will be true of $f_\beta(q)$ for all q in some neighbourhood of p. We may therefore restrict the definition of $e_1, \ldots, e_k$ to this neighbourhood and define e_{k+1} to equal f_β there. This procedure may clearly be continued to yield a neighbourhood W of p and local sections $e_1, \ldots, e_n \in \Gamma_W(\pi)$. The vector bundle coordinates (x^i, u^α) defined by

$$u^\alpha(\lambda^\beta e_\beta(q)) = \lambda^\alpha$$

and dual to the local sections e_α then satisfy the conditions of the proposition. ■

Proposition 2.1.18 *Let $\phi_\mu \in \Gamma(\pi)$ be a family of sections, and for each $p \in M$ let the subspace $\Delta_p \subset E_p$ be the linear span of the elements $\phi_\mu(p)$. If $k = \dim \Delta_p$ is independent of p then $E' = \bigcup_{p\in M} \Delta_p$ is a submanifold of E and $(E', \pi|_{E'}, M)$ is a vector sub-bundle of π.*

Proof Choose $p \in M$, and for $1 \leq \alpha \leq k$ select a section ϕ_{μ_α} such that $\{\phi_{\mu_\alpha}(p)\}$ forms a basis of Δ_p. The elements $\phi_{\mu_\alpha}(q)$ will then be linearly independent for q in some neighbourhood W' of p, and will also span Δ_q since $\dim \Delta_q = k$; we may also choose W' sufficiently small that it lies within the domain of coordinate functions x^i around p. Denote the local sections $\phi_{\mu_\alpha}|_{W'}$ by e'_α, and extend these as in Proposition 2.1.17 to form a basis $\{e_\alpha : 1 \leq \alpha \leq n\}$ of the local sections on some smaller neighbourhood W of p. Let (x^i, u^α) be vector bundle coordinates on $\pi^{-1}(W) \subset E$ which are dual to the local sections e_α. Then the subset $E' \subset E$ will be defined locally by the equations $u^{k+1} = \ldots = u^n = 0$, and so will be a submanifold of E; furthermore, the map $(\pi|_{\pi^{-1}(W)}, u) : \pi^{-1}(W) \longrightarrow W \times \mathbf{R}^n$ will be a linear local trivialisation around p. ■

EXERCISES

2.1.1 Let $(\mathbf{R}^3 - \{0\}, \pi, S^2)$ be the bundle defined in Exercise 1.1.4. Show that π may be given the structure of a vector bundle, but that the spherical polar coordinates $(\theta, \phi; \rho)$ around $(0,1,0) \in \mathbf{R}^3 - \{0\}$ are *not* vector bundle coordinates. Confirm that $(\theta, \phi; \log \circ \rho)$ may be used instead as vector bundle coordinates.

2.1.2 If (E, π, M) and (H, ρ, N) are vector bundles with typical fibres $\mathbf{R}^n$ and $\mathbf{R}^k$ respectively, show that the product bundle $(E \times H, \pi \times \rho, M \times N)$ may be given the structure of a vector bundle with typical fibre $\mathbf{R}^{n+k}$. (The resulting vector bundle is called the *product vector bundle.*)

2.1.3 If (E, π, M) and (H, ρ, M) are vector bundles over the same base manifold M with typical fibres $\mathbf{R}^n$ and $\mathbf{R}^k$ respectively, show that the fibre product bundle $(E \times_M H, \pi \times_M \rho, M)$ may also be given the structure of a vector bundle with typical fibre $\mathbf{R}^{n+k}$. (The resulting vector bundle is called the *direct sum* or *Whitney sum* vector bundle and is denoted $(E \oplus H, \pi \oplus \rho, M)$.)

2.1.4 If (E, π, M) is a vector bundle with typical fibre $\mathbf{R}^n$ and $\rho : H \longrightarrow M$ is a map, show that the pull-back bundle $(\rho^*(E), \rho^*(\pi), H)$ may be given the structure of a vector bundle with typical fibre $\mathbf{R}^n$. (The resulting vector bundle is called the *pull-back vector bundle.*)

2.1.5 Let (E, π, M) be a bundle and suppose that, for each $p \in M$,

$$(W_p, \mathbf{R}^n, t_p)$$

is a local trivialisation around p which has the property that for each $r \in W_p \cap W_q$,

$$t_q \circ t_p^{-1}\Big|_{\{r\}\times\mathbf{R}^n} : \{r\} \times \mathbf{R}^n \longrightarrow \{r\} \times \mathbf{R}^n$$

defines a linear automorphism of $\mathbf{R}^n$. Show that π may then be given the structure of a vector bundle such that each $(W_p, \mathbf{R}^n, t_p)$ is a linear local trivialisation.

2.1.6 Show that the tautological bundle on RP^2 defined in Example 2.1.7 is not trivial. (Hint: if $M \times \mathbf{R}$ is the total space of a trivial line bundle, and if $\gamma : [0,1] \longrightarrow M \times \mathbf{R}$ is a curve satisfying $\gamma(0) = (p, -1)$, $\gamma(1) = (p, 1)$ for some $p \in M$, then there is a $t \in (0,1)$ satisfying $\gamma(t) = (q, 0)$ for some $q \in M$.)

2.2 Vector Bundle Morphisms

If (E, π, M) and (H, ρ, N) are two vector bundles, then a bundle morphism $(f, \bar{f})$ from π to ρ may respect the additional structure by being linear on each fibre. Such vector bundle morphisms have many of the properties of linear maps between vector spaces. We may define the kernel of a vector bundle morphism as the set of all elements in the domain total space which map to the zero elements of the codomain total space, and in certain circumstances this is a vector sub-bundle of the domain bundle. We are also able to introduce the idea of an "exact sequence" of vector bundles, and there is one particular exact sequence which will be of great importance in our study of the partial differential equations associated with a connection. Finally, we shall show how morphisms of vector bundles give rise to morphisms of their modules of sections.

Definition 2.2.1 A *vector bundle morphism from π to ρ* is a bundle morphism $(f, \bar{f})$ which has the property that, for each $p \in M$, $f|_{E_p} : E_p \longrightarrow H_{\bar{f}(p)}$ is a linear map. ∎

As a bundle morphism, $(f, \bar{f})$ may be represented in coordinates. Let (x^i, u^α) be vector bundle coordinates around $a \in E$, and let (y^a, v^A) be vector bundle coordinates around $f(a) \in H$. The coordinates of f are then $f^A = v^A \circ f$, $f^a = y^a \circ f$. However, if f is a *vector* bundle morphism then we may express it in matrix terms: whereas a linear map between vector spaces may be represented by a matrix of numbers, a vector bundle morphism is represented by a matrix of *functions* defined locally on the base space M. To see how this matrix arises, let e_α be the local sections of π dual to the coordinates u^α, and let h_A be the local sections of ρ dual to the coordinates v^A. Put $p = \pi(a) \in M$, and let $\lambda^\alpha = u^\alpha(a)$ (so that $a = \lambda^\alpha e_\alpha(p)$). Then

$$f(a) = \lambda^\alpha f(e_\alpha(p)) \in H_{\overline{f}(p)}$$

and so we may write

$$f(a) = \lambda^\alpha f_\alpha^A(p) h_A(\overline{f}(p))$$

where the real numbers $f_\alpha^A(p)$ are the coordinates of $f(e_\alpha(p))$ with respect to the basis vectors $h_A(\overline{f}(p))$ of $H_{\overline{f}(p)}$; another way of saying this is that $f_\alpha^A(p) = v^A(f(e_\alpha(p)))$. It is then immediate that the functions $f_\alpha^A = f^A \circ e_\alpha$ are smooth, and the resulting matrix of smooth functions defined near p is called the *local matrix representation* of the vector bundle morphism f. (Of course the functions f_α^A may be defined for any map $f : E \longrightarrow H$, whether or not it is linear on the fibres, but these functions may only be used to reconstruct f in the manner we have described when f is a vector bundle morphism.)

Example 2.2.2 If $f : M \longrightarrow N$ then (f_*, f) is a vector bundle morphism from τ_M to τ_N. If $(x^i, \dot{x}^i)$ are coordinates on TM and $y^a, \dot{y}^a$ are coordinates on TN then the local matrix representation of f_* is $\partial f^a / \partial x^i$. If in addition f is a local diffeomorphism (so that $\partial f^a / \partial x^i$ is a square non-singular matrix at each point), then $(f^{*-1}, f) : \tau_M^* \longrightarrow \tau_N^*$ is a vector bundle morphism whose local matrix representation F_a^i satisfies

$$F_a^i(p) \left. \frac{\partial f^a}{\partial x^j} \right|_p = \delta_j^i.$$

■

Example 2.2.3 Let (E, π, M) be a vector bundle, and let $(E', \pi|_{E'}, M)$ be a vector sub-bundle. It follows from Proposition 2.1.17 that it is always possible to find vector bundle coordinates on E and E' such that locally the inclusion map $E' \longrightarrow E$ may be represented in matrix form by the constant matrix

$$\left(\frac{I}{O} \right).$$

■

Since each map $f|_{E_p}$ obtained from a vector bundle morphism f is linear, we may define its rank in the usual manner as $\dim f(E_p)$. We therefore arrive at the definition of the rank of a vector bundle morphism as a function defined on the base space M; it should be clear that the rank of f at a point p is equal to the rank of its local matrix representation $f_\alpha^A(p)$ in any pair of coordinate systems. If the rank of f does not depend upon the particular point $p \in M$ then we sat that f is of *constant rank*; this is the condition which will ensure that the kernel and image of f define vector sub-bundles of π and ρ respectively.

Definition 2.2.4 If (E, π, M) and (H, ρ, N) are vector bundles, and if $(f, \overline{f}) : \pi \longrightarrow \rho$ is a vector bundle morphism, then the *kernel of* f is the subset

$$\ker f = \{a \in E : f(a) = 0 \in F_{\overline{f}(\pi(a))}\}.$$

■

Proposition 2.2.5 *If f has constant rank then the kernel of f defines a vector sub-bundle*

$$(\ker f, \pi|_{\ker f}, M)$$

of π, and the image of f defines a vector sub-bundle $(\operatorname{im} f, \rho|_{\operatorname{im} f}, N)$ *of ρ.*

Proof We shall show first that $\ker f$ defines a vector sub-bundle of π.

It is clear that $\ker f$ is a closed submanifold of E. Since f has constant rank, the subspaces $(\ker f)_p \subset E_p$ all have the same dimension. To prove that $\pi|_{\ker f}$ is locally trivial, choose $p \in M$ and let

$$\begin{aligned} t : \pi^{-1}(\overline{f}^{-1}(W)) &\longrightarrow \overline{f}^{-1}(W) \times \mathbf{R}^n \\ s : \rho^{-1}(W) &\longrightarrow W \times \mathbf{R}^k \end{aligned}$$

be linear local trivialisations of π, ρ around p, $\overline{f}(p)$ respectively. For each $q \in \overline{f}^{-1}(W)$, define the linear map $f_q : \mathbf{R}^n \longrightarrow \mathbf{R}^k$ by

$$f_q(v) = pr_2(s(f(t^{-1}(q, v)))),$$

so that each f_q has the same rank. Put $F_1 = \ker f_p$, $K_1 = \operatorname{im} f_p$ and let F_2, K_2 satisfy $F_1 \oplus F_2 = \mathbf{R}^n$, $K_1 \oplus K_2 = \mathbf{R}^k$. Put $F = \mathbf{R}^n \oplus K_2$ and $K = \mathbf{R}^k \oplus F_1$, so that $\dim F = \dim K$.

For each $q \in \overline{f}^{-1}(W)$, define the linear map $g_q : F \longrightarrow K$ by

$$g_q(u, v, w) = (f_q(u + v) + w, u) \in \mathbf{R}^k \oplus F_1,$$

where $(u, v, w) \in F_1 \oplus F_2 \oplus K_2 = F$. Since f_p is an isomorphism, it follows that g_p is injective and hence (by dimensionality) also an isomorphism; therefore g_q is also an isomorphism for each q in some neighbourhood $\overline{f}^{-1}(W')$ of p. Now observe that, if $(u, v) \in F_1 \oplus F_2$, then $(u, v) \in \ker f_q$ exactly when $g_q(u, v, 0) = (0, u) \in \mathbf{R}^k \oplus F_1$, so that $\ker f_q = g_q^{-1}(F_1)$. We may therefore define a diffeomorphism

$$\left(\pi|_{\ker f}\right)^{-1}(\overline{f}^{-1}(W')) \longrightarrow \overline{f}^{-1}(W') \times F_1$$

by

$$a \longmapsto (\pi(a), pr_{F_1}(g_{\pi(a)}(pr_2(t(a)), 0)))$$

where pr_{F_1} denotes the projection $K \longrightarrow F_1$, and so $\ker f$ is locally trivial.

We shall now show that $\operatorname{im} f$ defines a vector sub-bundle of ρ. Given $p \in M$, the vector space E_p is spanned by $\{\phi(p) : \phi \in \Gamma(\pi)\}$, and so $\operatorname{im} f_p$ is spanned by $\{f(\phi(p)) : \phi \in \Gamma(\pi)\}$. It follows that $\operatorname{im} f$ is spanned by the sections $\tilde{f}(\phi)$ of ρ, and so defines a vector sub-bundle by Proposition 2.1.18. ∎

Example 2.2.6 The set $V\pi$ of vertical vectors defined in Example 1.4.11 is the kernel of the vector bundle morphism $(\pi_*, \pi) : \tau_E \longrightarrow \tau_M$. Since π_* has constant rank (because π is a submersion), it follows from Proposition 2.2.5 that $(V\pi, \tau_E|_{V\pi}, E)$ is a vector sub-bundle of τ_E. Note that, for this example, we may also see directly that $\tau_E|_{V\pi}$ is locally trivial: for if (x^i, u^α) are adapted coordinates on $U \subset E$ then $(x^i, u^\alpha; \dot{x}^i, \dot{u}^\alpha)$ are vector bundle coordinates on TE with dual local sections $\partial/\partial x^i$, $\partial/\partial u^\alpha$, and since $\pi_* \circ \partial/\partial x^i = \partial/\partial x^i$, $\pi_* \circ \partial/\partial u^\alpha = 0$ we may take coordinates on $V\pi$ as $(x^i, u^\alpha; \dot{u}^\alpha)$. From the local trivialisation

$$(\tau_E|_{\tau_E^{-1}(U)}, (\dot{x}, \dot{u})) : \tau_E^{-1}(U) \longrightarrow U \times (\mathbf{R}^m \times \mathbf{R}^n)$$

of τ_E we then obtain the local trivialisation

$$\left(\tau_E|_{\tau_E^{-1}(U)\cap V\pi}, \dot{u}\right) : (\tau_E|_{V\pi})^{-1}(U) \longrightarrow U \times \mathbf{R}^n$$

of $\tau_E|_{V\pi}$. ∎

Since the kernel and image of a vector bundle morphism are vector sub-bundles, it makes sense to consider exact sequences of vector bundle morphisms. First, however, we shall recall the corresponding definition for ordinary linear maps. If V_1, V_2, V_3 are vector spaces (or modules over some commutative ring) and $f_{12} : V_1 \longrightarrow V_2$, $f_{23} : V_2 \longrightarrow V_3$ are linear maps, we say that the sequence of maps

$$V_1 \xrightarrow{f_{12}} V_2 \xrightarrow{f_{23}} V_3$$

is *exact* (at V_2) if $\operatorname{im} f_{12} = \ker f_{23}$. A longer sequence is called exact if it is exact at each vector space in the sequence (apart from the initial and final spaces). A very common form of exact sequence involves five spaces,

$$0 \longrightarrow V_1 \xrightarrow{f_{12}} V_2 \xrightarrow{f_{23}} V_3 \longrightarrow 0,$$

where exactness at V_1 means that f_{12} is injective, and exactness at V_3 means that f_{23} is surjective; such a sequence is commonly called a *short exact sequence*. If there is another map $s : V_3 \longrightarrow V_2$ such that $f_{23} \circ s = id_{V_3}$ then we say that the short exact sequence *splits*; when this happens, we may regard V_2 as the direct sum $V_1 \oplus V_3$ and the maps f_{12}, s as the canonical inclusion maps. The standard example of a short exact sequence is

$$0 \longrightarrow \ker f \xrightarrow{\iota} V \xrightarrow{f} \operatorname{im} f \longrightarrow 0$$

where $f : V \longrightarrow W$ is an arbitrary linear map and ι is the inclusion. If this sequence splits then we may write $V = \ker f \oplus \operatorname{im} f$, so that any element $\xi \in V$ has a unique representative $\xi - s(f(\xi))$ in the kernel of f.

A similar idea may be applied to vector bundles over the same base space M.

Definition 2.2.7 Let (E_1, π_1, M), (E_2, π_2, M), (E_3, π_3, M) be vector bundles, and let $(f_{12}, id_M) : \pi_1 \longrightarrow \pi_2$, $(f_{23}, id_M) : \pi_2 \longrightarrow \pi_3$ be vector bundle morphisms of constant rank. The sequence

$$E_1 \xrightarrow{f_{12}} E_2 \xrightarrow{f_{23}} E_3$$

is said to be *exact* at E_2 (or π_2) if $\operatorname{im} f_{12} = \ker f_{23}$. ■

If we let 0 denote the trivial vector bundle $(M \times 0, pr_1, M)$ then we can define a short exact sequence of vector bundles to be one of the form

$$0 \longrightarrow E_1 \xrightarrow{f_{12}} E_2 \xrightarrow{f_{23}} E_3 \longrightarrow 0$$

so that f_{12} is injective and f_{23} surjective. (Note that the latter assertions depend on the fact the we have required $\overline{f}_{12} = \overline{f}_{23} = id_M$.) We say that the sequence *splits* if there is another vector bundle morphism $(s, id_M) : \pi_3 \longrightarrow \pi_2$, necessarily of constant rank, such that $f_{23} \circ s = id_{E_3}$, and then we may regard π_2 as the direct sum bundle $\pi_1 \oplus \pi_3$. The standard example of a short exact sequence of vector bundles is

$$0 \longrightarrow \ker f \xrightarrow{\iota} E \xrightarrow{f} \operatorname{im} f \longrightarrow 0$$

where (E, π, M) and (F, ρ, M) are bundles, and where $(f, id_M) : \pi \longrightarrow \rho$ is a vector bundle morphism and (ι, id_M) is the inclusion.

Example 2.2.8 If (E, π, M) is an arbitrary bundle then (TE, τ_E, E) and (TM, τ_M, M) do not have the same base space. However, we may instead consider the pull-back vector bundle $(\pi^*(TM), \pi^*(\tau_M), E)$ and we may define a new map $\overline{\pi}_* : TE \longrightarrow \pi^*(TM)$ by

$$\overline{\pi}_*(\xi) = (\pi_*(\xi), \tau_E(\xi)).$$

Then $\ker \overline{\pi}_* = \ker \pi_* = V\pi$ and $\operatorname{im} \overline{\pi}_* = \pi^*(TM)$, so we may construct a short exact sequence

$$0 \longrightarrow V\pi \stackrel{\iota}{\longrightarrow} TE \stackrel{\overline{\pi}_*}{\longrightarrow} \pi^*(TM) \longrightarrow 0.$$

Further properties of this particular sequence will be described in Chapter 3.

∎

In the final part of this section, we shall show that each exact sequence of vector bundles gives rise to an exact sequence of modules of sections.

Lemma 2.2.9 *If (E, π, M) and (H, ρ, N) are vector bundles and $(f, \overline{f}) : \pi \longrightarrow \rho$ is a vector bundle morphism such that $\overline{f}$ is a diffeomorphism, then $\tilde{f} : \Gamma(\pi) \longrightarrow \Gamma(\rho)$ is a module homomorphism over the ring isomorphism $\overline{f}^{-1*} : C^\infty(M) \longrightarrow C^\infty(N)$.*

Proof If $\phi, \psi \in \Gamma(\pi)$ and $g \in C^\infty(M)$ then

$$\begin{aligned} \tilde{f}(\phi + \psi) &= f \circ (\phi + \psi) \circ \overline{f}^{-1} \\ &= (f \circ \phi \circ \overline{f}^{-1}) + (f \circ \psi \circ \overline{f}^{-1}) \\ &= \tilde{f}(\phi) + \tilde{f}(\psi) \end{aligned}$$

and

$$\begin{aligned} \tilde{f}(g\phi) &= f \circ (g\phi) \circ \overline{f}^{-1} \\ &= f \circ (g \circ \overline{f}^{-1})(\phi \circ \overline{f}^{-1}) \\ &= (g \circ \overline{f}^{-1})(f \circ \phi \circ \overline{f}^{-1}) \\ &= \overline{f}^{-1*}(g)\tilde{f}(\phi) \end{aligned}$$

because f is a linear map from the fibres of π to the fibres of ρ. ∎

Proposition 2.2.10 *If (E_1, π_1, M), (E_2, π_2, M) and (E_3, π_3, M) are vector bundles and if $(f_{12}, id_M) : \pi_1 \longrightarrow \pi_2$, $(f_{23}, id_M) : \pi_2 \longrightarrow \pi_3$ are vector bundle morphisms of constant rank such that the sequence*

$$E_1 \stackrel{f_{12}}{\longrightarrow} E_2 \stackrel{f_{23}}{\longrightarrow} E_3$$

is exact, then the sequence of $C^\infty(M)$-module morphisms

$$\Gamma(\pi_1) \stackrel{\tilde{f}_{12}}{\longrightarrow} \Gamma(\pi_2) \stackrel{\tilde{f}_{23}}{\longrightarrow} \Gamma(\pi_3)$$

is also exact.

Proof To see that $\operatorname{im} \tilde{f}_{12} \subset \ker \tilde{f}_{23}$, let $\phi \in \Gamma(\pi)$ and $p \in M$; then

$$\tilde{f}_{23}(\tilde{f}_{12}(\phi))(p) = f_{23} \circ f_{12} \circ \phi(p) \in f_{23} \circ f_{12}(E)_p = \{0\}_p$$

so that $\tilde{f}_{12}(\phi) \in \ker \tilde{f}_{23}$. For the converse inclusion, let $\psi \in \ker \tilde{f}_{23}$; then $\operatorname{im} \psi \subset \ker f_{23}$ so that $\psi \in \Gamma(\rho|_{\ker f_{23}}) = \Gamma(\rho|_{\operatorname{im} f_{12}})$. Let $p \in M$ and let $e_\alpha \in \Gamma_p(\pi)$ be a family of linearly independent local sections of π such that $e_1, \ldots, e_k$ span $\ker f_{12}$ in a neighbourhood of p (as in the proof of Proposition 2.1.17). Then $\tilde{f}_{12}(e_{k+1}), \ldots, \tilde{f}_{12}(e_n)$ are linearly independent and span $\operatorname{im} f_{12}$ in a neighbourhood of p, so that locally $\psi = \psi^\alpha \tilde{f}_{12}(e_\alpha) = \tilde{f}_{12}(\psi^\alpha e_\alpha)$ where $k + 1 \leq \alpha \leq n$. By using a partition of unity, it is then possible to construct from the local sections $\psi^\alpha e_\alpha$ a global section $\phi \in \Gamma(\pi)$ such that $\psi = \tilde{f}_{12}(\phi)$. ∎

Corollary 2.2.11 *If $(f, id_M) : \pi \longrightarrow \rho$ is a vector bundle morphism then $\Gamma(\ker f) = \ker \tilde{f}$ and $\Gamma(\operatorname{im} f) = \operatorname{im} \tilde{f}$.* ∎

The standard example of a short exact sequence of vector bundle morphisms

$$0 \longrightarrow \ker f \longrightarrow E \longrightarrow \operatorname{im} f \longrightarrow 0$$

therefore gives rise to a short exact sequence

$$0 \longrightarrow \ker \tilde{f} \longrightarrow \Gamma(\pi) \longrightarrow \operatorname{im} \tilde{f} \longrightarrow 0$$

of modules over $C^\infty(M)$.

EXERCISES

2.2.1 If (E, π, M) is a vector bundle, construct a canonical isomorphism between the vector bundles $(V\pi, \tau_E|_{V\pi}, E)$ and $(\pi^*(E), \pi^*(\pi), E)$. Deduce the existence of a canonical map $V\pi \longrightarrow E$ which in general is *not* the restriction of the tangent bundle projection τ_E. (Hint: the isomorphism is a generalisation, to vector bundles, of the standard isomorphism $T_pV \cong V$ where V is a vector space and $p \in V$.

2.2.2 Use the result of Exercise 2.2.1 to define a "vertical lift" operation on the tangent bundle (TM, τ_M, M), whereby a tangent vector $\xi \in T_pM$ may be lifted to a tangent vector $\xi^{\mathrm{v}} \in V_\eta \tau_M$ whenever $\eta \in T_pM$. Show that, if the coordinate representation of ξ is

$$\xi = \xi^i \left.\frac{\partial}{\partial x^i}\right|_p,$$

then the coordinate representation of ξ^{v} is

$$\xi^{\mathrm{v}} = \xi^i \left.\frac{\partial}{\partial \dot{x}^i}\right|_{\eta}.$$

Use this construction to give an intrinsic definition of a vector field $\Delta \in \mathcal{X}(TM)$ whose coordinate representation is

$$\Delta = \dot{x}^i \frac{\partial}{\partial \dot{x}^i}.$$

2.2.3 Let M be a closed embedded submanifold of the manifold H. Show that the inclusion map $\iota : M \longrightarrow H$ defines a vector bundle morphism of constant rank $\iota_* : \tau_M \longrightarrow \tau_H$. For each $p \in M$, let the quotient space T_pH/T_pM be denoted by N_pM, and let $N_HM = \bigcup_{p\in M} N_pM$. Let the map $\nu : N_HM \longrightarrow M$ be defined by $\nu[\xi] = p$ if $[\xi] \in N_pM$. Show that (N_HM, ν, M) becomes a vector bundle (called the *normal bundle of M in H*), and that there is an exact sequence

$$0 \longrightarrow \tau_M \longrightarrow \iota^*(\tau_H) \longrightarrow \nu \longrightarrow 0$$

of vector bundles over M.

2.3 Duality and Tensor Products

There are certain ways of constructing new vector bundles from old ones which make essential use of the linearity of the fibres, and so do not correspond to any constructions applicable in the more general case. In one of these, the fibres are the dual spaces to the fibres of the original bundle. We shall normally apply this construction in the case where the original bundle is a tangent bundle, or a sub-bundle of a tangent bundle (such as a bundle of vertical vectors).

Definition 2.3.1 Let (E, π, M) be a vector bundle with fibres E_p. The *dual bundle* is the vector bundle with fibres E^*_p, and is denoted (E^*, π^*, M). ∎

In order to apply Proposition 2.1.6 to show that π^* is indeed a bundle, we must define suitable maps $t^*_p : \pi^{*-1}(W_p) \longrightarrow W_p \times \mathbf{R}^n$ which will become the linear local trivialisations of π^*. To do this, suppose that $t_p : \pi^{-1}(W_p) \longrightarrow W_p \times \mathbf{R}^n$ is a linear local trivialisation of π around p. If $q \in W_p$, $pr_2 \circ t_p|_{\pi^{-1}(q)} : E_q \longrightarrow \mathbf{R}^n$ is a linear isomorphism, so that the inverse of its transpose is a linear isomorphism $t^*_{pq} : E^*_q \longrightarrow \mathbf{R}^n$ (where we have identified $\mathbf{R}^{n*}$ with $\mathbf{R}^n$). We may therefore define $t^*_p : \pi^{*-1}(W_p) \longrightarrow W_p \times \mathbf{R}^n$ by

$$t^*_p(a) = (\pi^*(a), t^*_{p\pi^*(a)}(a))$$

and this map clearly satisfies the conditions of Proposition 2.1.6.

If (x^i, u^α) is a system of vector bundle coordinates with domain $\pi^{-1}(W) \subset E$ then we may define the dual coordinates (x^i, u_α) on $\pi^{*-1}(W) \subset E^*$ by taking the fibre coordinates on E^* to be the inverse transpose of the fibre coordinates on E. Explicitly, let $e_\beta \in \Gamma_W(\pi)$ be the local sections dual to u^α, so that $u^\alpha(e_\beta(p)) = \delta^\alpha_\beta$ for each $p \in W$. Now $e_\beta(p) \in E_p \cong E_p^{**}$, so that we may regard each $e_\beta(p)$ as a linear map from E_p^* to $\mathbf{R}$. We may therefore define coordinate functions $u_\beta : \pi^{*-1}(W) \longrightarrow \mathbf{R}$ on E^* by

$$u_\beta|_{E_p^*} = e_\beta(p).$$

If $e^\alpha \in \Gamma_W(\pi^*)$ are the local sections dual to the coordinates u_β then we also have

$$u^\alpha|_{E_p} = e^\alpha(p).$$

Example 2.3.2 The bundle dual to the tangent bundle (TM, τ_M, M) is the cotangent bundle (T^*M, τ_M^*, M), and the coordinates dual to $(x^i, \dot{x}^i)$ are (x^i, ∂_i). The fibre coordinates $\dot{x}^i$ on TM therefore correspond to the local sections dx^i of T^*M, in that

$$\dot{x}^i\Big|_{T_pM} = dx^i\Big|_p \in T_p^*M.$$

Similarly the fibre coordinates ∂_i on T^*M correspond to the local sections $\partial/\partial x^i$ of TM, in that

$$\partial_i|_{T_p^*M} = \left.\frac{\partial}{\partial x^i}\right|_p .$$

(In mechanics, the coordinates on TM are often denoted $(q^i, \dot{q}^i)$ rather than $(x^i, \dot{x}^i)$, and then the dual coordinates on T^*M are denoted (q^i, p_i). We shall sometimes use this alternative labelling convention in later chapters.) ■

The other vector bundle construction which we shall need to use is that of the tensor product. If V, W are finite-dimensional vector spaces, then their tensor product $V \otimes W$ may be defined to be the space of bilinear maps from $V^* \times W^*$ to $\mathbf{R}$. We may therefore apply this definition to two vector bundles over the same base space M.

Definition 2.3.3 Let (E, π, M) and (F, ρ, M) be vector bundles with fibres E_p, F_p respectively. The *tensor product of* π *and* ρ is the vector bundle with fibres $E_p \otimes F_p$ and is denoted $(E \otimes F, \pi \otimes \rho, M)$. ■

This construction clearly generalises to the tensor product of a finite number of vector bundles. It may be considered "associative" in the same way that the tensor product of vector spaces may be considered associative.

Example 2.3.4 An element of the total space of the tensor product bundle $(TM \otimes TM, \tau_M \otimes \tau_M, M)$ is a type $(2,0)$ tensor. In local coordinates, such an element would be written

$$A = A^{ij}\left(\frac{\partial}{\partial x^i} \otimes \frac{\partial}{\partial x^j}\right)_p .$$

(We shall generally distinguish between a *tensor field*, which is a local section of a bundle such as this, and a *tensor*, which is an element of the total space. A tensor field evaluated at a point gives a tensor.) ∎

Example 2.3.5 An element of the total space of the tensor product bundle $(T^*M \otimes TM, \tau_M^* \otimes \tau_M, M)$ is a type (1,1) tensor, and may be written in coordinates as

$$A = A_i^j\left(dx^i \otimes \frac{\partial}{\partial x^j}\right)_p .$$

However, since the fibre $T_p^*M \otimes T_pM$ is canonically isomorphic to the space $\mathrm{L}(T_pM, T_pM)$ of endomorphisms of T_pM, we may also regard $\tau_M^* \otimes \tau_M$ as a bundle of endomorphisms of TM. A section of this bundle is also called a *vector-valued 1-form* on M. ∎

We are also interested in certain sub-bundles of tensor product bundles, where the tensors are either completely alternating or completely symmetric. We use the symbols $\bigwedge^r V$ and $S^r V$ to denote the subspaces of $V \otimes \ldots \otimes V$ containing, respectively, the alternating and the symmetric r-linear maps from $V^* \times \ldots \times V^*$ to $\mathbf{R}$.

Definition 2.3.6 Let (E, π, M) be a vector bundle with fibres E_p. The *r-fold alternating product of* π is the vector bundle with fibres $\bigwedge^r E_p$ and is denoted $(\bigwedge^r E, \bigwedge^r \pi, M)$. The *$r$-fold symmetric product of* π is the vector bundle with fibres $S^r E_p$ and is denoted $(S^r E, S^r \pi, M)$. ∎

Example 2.3.7 An element of the total space of $(\bigwedge^2 T^*M, \bigwedge^2 \tau_M^*, M)$ is a 2-covector, with local coordinate representation

$$\omega = \omega_{ij}(dx^i \wedge dx^j)_p$$

where $\omega_{ij} = \omega_{ji}$. A local section of this bundle is called a 2-form. ∎

We shall now show that these $\mathbf{R}$-linear constructions involving vector bundles may be matched by corresponding $C^\infty(M)$-linear constructions involving the corresponding modules of sections.

Proposition 2.3.8 *If (E^*, π^*, M) is the vector bundle dual to (E, π, M), then $\Gamma(\pi^*)$ is isomorphic to the $C^\infty(M)$-module $(\Gamma(\pi))^*$ dual to $\Gamma(\pi)$.*

Proof If $\psi \in \Gamma(\pi^*)$, define the element $\widehat{\psi}$ in $(\Gamma(\pi))^*$ by

$$\widehat{\psi}(\phi)(p) = \psi(p)(\phi(p))$$

where $\phi \in \Gamma(\pi)$ and $p \in M$. It is clear that $\widehat{\psi}(\phi)$ is smooth and hence an element of $C^\infty(M)$, because in coordinates $\widehat{\psi}(\phi) = \psi_\alpha \phi^\alpha$; it is then straightforward to check that $\psi \longmapsto \widehat{\psi}$ is a $C^\infty(M)$-module homomorphism. To see that it is injective, suppose $\psi_1 \neq \psi_2$ and choose $p \in M$ such that $\psi_1(p) \neq \psi_2(p)$. Choose $a \in E_p$ such that $\psi_1(p)(a) \neq \psi_2(p)(a)$, and let ϕ be a section of π satisfying $\phi(p) = a$. Then $\widehat{\psi}_1(\phi(p)) \neq \widehat{\psi}_2(\phi(p))$ so that $\widehat{\psi}_1 \neq \widehat{\psi}_2$.

Finally, to show that the correspondence is surjective, we shall employ a local argument involving coordinates and then use a partition of unity. So suppose χ is an element of $(\Gamma(\pi))^*$. Let (x^i, u^α) be vector bundle coordinates on E (where x^i are coordinates around $p \in M$); let e_β be the local sections dual to u^α, and let $\overline{e}_\beta$ be global sections of π which equal e_β in a neighbourhood W of p (using Proposition 1.2.6). Define the functions $\overline{\chi}_\beta \in C^\infty(M)$ by $\overline{\chi}_\beta = \chi(\overline{e}_\beta)$, and put $\chi_\beta = \overline{\chi}_\beta\big|_W$. Now define the local sections $\psi_W \in \Gamma_W(\pi^*)$ by $\psi_W = \chi_\beta e^\beta$, where e^β are the local sections of π^* defined by the coordinates u^β on E. These local sections ψ_W may be combined using a partition of unity on M, to give a global section $\psi \in \Gamma(\pi^*)$.

We may then see that $\widehat{\psi} = \chi$ by the following argument. Let $\phi \in \Gamma(\pi)$ with coordinates $\phi^\alpha = u^\alpha \circ \phi$, and extend these coordinates to smooth functions $\overline{\phi}^\alpha \in C^\infty(M)$ where $\overline{\phi}^\alpha(q) = \phi^\alpha(q)$ for each q in a neighbourhood $\overline{W} \subset W$ of p. The global section $\phi^0 = \phi - \overline{\phi}^\alpha \overline{e}_\alpha$ is then zero on $\overline{W}$, and so we may write $\phi^0 = z\phi^0$, where $z \in C^\infty(M)$ is a bump function which satisfies $z(p) = 0$, $z(q) = 1$ for $q \in M - \overline{W}$. Then

$$\begin{aligned}\chi(\phi)(p) &= \chi(\overline{\phi}^\alpha \overline{e}_\alpha + z\phi^0)(p) \\ &= \overline{\phi}^\alpha(p)\chi(\overline{e}_\alpha)(p) \\ &= \phi^\alpha(p)\chi_\alpha(p)\end{aligned}$$

whereas

$$\begin{aligned}\widehat{\psi}(\phi)(p) &= \psi(p)(\phi(p)) \\ &= \chi_\beta(p)e^\beta(p)(\phi^\alpha(p)e_\alpha(p)) \\ &= \chi_\alpha(p)\phi^\alpha(p)\end{aligned}$$

so that, for each $\phi \in \Gamma(\pi)$ and for each $p \in M$,

$$\chi(\phi)(p) = \widehat{\psi}(\phi)(p)$$

which establishes the result. ■

Example 2.3.9 For an arbitrary manifold M, the module of 1-forms $\bigwedge^1 M$ is isomorphic to the module dual to $\mathcal{X}(M)$. We shall normally denote the pairing of a vector field $X \in \mathcal{X}(M)$ and a 1-form $\omega \in \bigwedge^1 M$ by

$$X \lrcorner \omega \in C^\infty(M)$$

rather than $\omega(X)$. We shall use a similar notation for local vector fields and differential forms, so that we may write (for example)

$$X^i \frac{\partial}{\partial x^i} \lrcorner \omega_j dx^j = X^i \omega_i \in C^\infty(W)$$

if x^i are coordinate functions defined on $W \subset M$. We shall also extend this notation to r-forms, and if $\theta \in \bigwedge^r M$ then we shall write $X \lrcorner \theta$ for the element of $\bigwedge^{r-1} M$ defined by

$$(X \lrcorner \theta)(Y_1, \ldots, Y_{r-1}) = \theta(X, Y_1, \ldots, X_{r-1}).$$

∎

There are similar results for the module of sections of a tensor product bundle.

Proposition 2.3.10 *If (E, π, M) and (H, ρ, M) are vector bundles then*

$$\Gamma(\pi \otimes \rho) = \Gamma(\pi) \otimes_{C^\infty(M)} \Gamma(\rho),$$

where the tensor product of modules is indicated explicitly.

Proof We shall construct a map

$$\Psi : \Gamma(\pi) \otimes_{C^\infty(M)} \Gamma(\rho) \longrightarrow \Gamma(\pi \otimes \rho)$$

by using the fact that $\Gamma(\pi) \otimes_{C^\infty(M)} \Gamma(\rho)$ is generated over $C^\infty(M)$ by elements of the form $\phi \otimes \psi$, where $\phi \in \Gamma(\pi)$ and $\psi \in \Gamma(\rho)$. We may therefore define $\Psi(\phi \otimes \psi)$ by the rule that, for each $p \in M$,

$$\Psi(\phi \otimes \psi)(p) = \phi(p) \otimes \psi(p) \in E_p \otimes H_p.$$

The resulting map, extended to the whole of $\Gamma(\pi) \otimes_{C^\infty(M)} \Gamma(\rho)$, is then $C^\infty(M)$-linear by construction. It is injective, for if

$$\Psi(\phi \otimes \psi) = 0 \in \Gamma(\pi \otimes \rho)$$

then, for every $p \in M$,

$$\phi(p) \otimes \psi(p) = \Psi(\phi \otimes \psi)(p) = 0,$$

so that both $\phi(p) = 0$ and $\psi(p) = 0$; it follows that both $\phi = 0$ and $\psi = 0$, so that $\phi \otimes \psi = 0 \in \Gamma(\pi) \otimes_{C^\infty(M)} \Gamma(\rho)$.

We must now show that Ψ is also surjective. So let $\chi \in \Gamma(\pi \otimes \rho)$, let W be a coordinate neighbourhood of M, and let e_α, f_A be bases of $\Gamma_W(\pi)$, $\Gamma_W(\rho)$ respectively. If $p \in W$ then

$$\chi(p) = \chi^{\alpha A}(p) e_\alpha(p) \otimes f_A(p),$$

so that the restriction $\chi|_W$ yields an element

$$\chi^{\alpha A} e_\alpha \otimes f_A \in \Gamma(\pi|_W) \otimes_{C^\infty(W)} \Gamma(\rho|_W).$$

We may now use a partition of unity λ_W to obtain

$$\sum_W \lambda_W \overline{\chi}^{\alpha A} \overline{e}_\alpha \otimes \overline{f}_A \in \Gamma(\pi) \otimes_{C^\infty(M)} \Gamma(\rho),$$

where $\overline{e}_\alpha$, $\overline{f}_A$ and $\overline{\chi}^{\alpha A}$ are e_α, f_A and $\chi^{\alpha A}$ extended to the whole of M. By construction, at each $p \in M$,

$$\chi(p) = \sum_W \lambda_W(p) \overline{\chi}^{\alpha A}(p) \overline{e}_\alpha(p) \otimes \overline{f}_A(p),$$

so that

$$\chi = \Psi \left(\sum_W \lambda_W \overline{\chi}^{\alpha A} \overline{e}_\alpha \otimes \overline{f}_A \right).$$

■

Example 2.3.11 For any manifold M, the module of vector-valued 1-forms $\Gamma(\tau_M^* \otimes \tau_M)$ is equal to the tensor product $\bigwedge^1 M \otimes_{C^\infty(M)} \mathcal{X}(M)$. A typical element A of this module may be written in local coordinates as

$$A = A_i^j \, dx^i \otimes \frac{\partial}{\partial x^j}.$$

■

The preceding result may of course be extended to arbitrary finite tensor product bundles. It is then straightforward to see that $\Gamma(\bigwedge^r \pi) = \bigwedge^r \Gamma(\pi)$ and $\Gamma(S^r \pi) = S^r \Gamma(\pi)$.

EXERCISES

2.3.1 If (E, π, M) and (H, ρ, M) are vector bundles, construct explicit linear local trivialisations for the tensor product $\pi \otimes \rho$, and hence confirm that $\pi \otimes \rho$ is a vector bundle.

2.3.2 Let (E, π, M) be a vector bundle, and let (f, id_M) be a vector bundle morphism from π to itself. Define the section $A \in \Gamma(\pi^* \otimes \pi)$ by, for $p \in M$,

$$A(p) = f|_{E_p} \in \mathrm{L}(E_p, E_p) \cong E_p^* \otimes E_p.$$

Let (x^i, u^α) and (x^i, u_α) be dual vector bundle coordinates on E and E^*, and suppose that the local matrix representation of f is f^α_β and that the coordinate representation of A is A^α_β. Show that $f^\alpha_\beta = A^\alpha_\beta$.

2.3.3 Let $L \in C^\infty(E)$ be a function on the total space of a vector bundle (E, π, M). If $\iota_p : E_p \longrightarrow E$ is the inclusion, show that the map

$$a \longmapsto d(L \circ \iota_{\pi(a)}) \in T_a^* E_{\pi(a)} \cong E_{\pi(a)}^*$$

defines a bundle morphism $(\mathcal{F}L, id_M) : \pi \longrightarrow \pi^*$ called the *fibre derivative* of L. Show further that if each $L \circ \iota_p$ is a quadratic function on the vector space E_p (so that L is derived from a *fibre metric* on π), then $\mathcal{F}L$ is actually a vector bundle morphism.

2.4 Affine Bundles

By analogy with vector bundles, we may describe affine bundles as bundles whose fibres are affine spaces, and where there are the local trivialisations which are affine maps on each fibre. Now every vector space has a distinguished point, namely its origin, and any linear transformation of vector spaces maps one origin to another. In an affine space, however, there is no distinguished point: the definition of an affine space is intended to retain those linear properties of a vector space which may be described without reference to its origin, and the morphisms between affine spaces may be described as inhomogeneous linear transformations.

Definition 2.4.1 If V is a vector space, A is a set and $\alpha : A \times V \longrightarrow A$ is a function, then the triple (A, V, α) is called an *affine space* if:

1. for each $x \in A$, $\alpha(x, 0) = x$;
2. for each $x \in A$ and each $v, w \in V$, $\alpha(\alpha(x, v), w) = \alpha(x, v + w)$;
3. if $x, y \in A$ then there is a unique $v \in V$ satisfying $\alpha(x, v) = y$.

■

We may regard the function α as expressing a displacement of the point x by a vector v. The conditions in the definition may be made to seem more familiar by writing α as addition, so that $x + 0 = x$ and $(x + v) + w =$

$x + (v + w)$, although in the latter equation the symbol $+$ is used in two different senses. In the third condition, we may (suggestively) write the unique element v as $x - y$. We usually say that A (rather than the triple (A, V, α)) is an affine space, with an underlying vector space V, and we often say that A is *modelled* on V. Every vector space V may be regarded as an affine space modelled on itself, where the map α is just addition in V.

The fact that an affine space encapsulates all those properties of a vector space which remain after ignoring the origin may be demonstrated by selecting a distinguished point $p \in A$ to serve as an "origin". The choice of p then allows the vector space structure of V to be transported to A, in such a way that the zero vector corresponds to p: $v \in V$ corresponds to $\alpha(p, v) \in A$, and $x \in A$ corresponds to $x - p \in V$. If $x, y \in A$ and $\lambda \in \mathbf{R}$, we may set $x + y$ to equal $\alpha(p, (x - p) + (y - p))$ and λx to equal $\alpha(p, \lambda(x - p))$. We shall also define the *dimension* of an affine space to equal the dimension of its underlying vector space; we shall assume that V is finite-dimensional.

In a finite-dimensional affine space, we may introduce coordinates.

Definition 2.4.2 Let A be an affine space modelled on the vector space V, let (e_i) be a basis of V, and let $p \in A$. If $x \in A$ then the *affine coordinates of* x with respect to (p, e_i) are the real numbers x^i satisfying $x = \alpha(p, x^i e_i)$. ∎

In other words, x^i are the coordinates of $x - p$ with respect to the basis of V.

For each point x, different coordinates may be obtained by choosing a different basis of V, or a different origin in A. If (f_j) is another basis of V such that $e_i = T_i^j f_j$, then $x = \alpha(p, x^i T_i^j f_j)$, so that the coordinates of x with respect to (p, f_j) are $x^i T_i^j$. If instead we choose $q \in A$ to be the origin, and if q has affine coordinates q^i with respect to (p, e_i), then

$$\begin{aligned} x &= \alpha(p, x^i e_i) \\ &= \alpha(p, q^i e_i + (x^i - q^i) e_i) \\ &= \alpha(\alpha(p, q^i e_i), (x^i - q^i) e_i) \\ &= \alpha(q, (x^i - q^i) e_i) \end{aligned}$$

so that the coordinates of x with respect to (q, e_i) are $x^i - q^i$: this is just a "translation of coordinates". The most general rule for a coordinate transformation is then obtained from

$$x = \alpha(q, (x^i - q^i) T_i^j f_j),$$

so that the new coordinates of x with respect to (q, f_j) are $(x^i - q^i)T_i^j$ and are clearly related in an inhomogeneous linear manner to the original coordinates.

As with vector spaces, coordinate transformations may be related to the morphisms of affine spaces.

Definition 2.4.3 Let A, B be affine spaces modelled on V, W by the maps α, β respectively. The function $T : A \longrightarrow B$ is called an *affine morphism* if there is a linear map $\overline{T} : V \longrightarrow W$ such that, whenever $x \in A$ and $v \in V$,

$$T(\alpha(x, v)) = \beta(T(x), \overline{T}(v)).$$

■

It may be seen that $\overline{T}$ is completely determined by T, as follows: if $v \in V$, $p, q \in A$ and we write $\overline{T}_p$, $\overline{T}_q$ for the linear maps $V \longrightarrow W$ defined by

$$\begin{aligned} \overline{T}_p(v) &= T(\alpha(p, v)) - T(p) \\ \overline{T}_q(v) &= T(\alpha(q, v)) - T(q) \end{aligned}$$

then

$$\begin{aligned} \beta(T(q), \overline{T}_q(v)) &= T(\alpha(q, v)) \\ &= T(\alpha(p, (q - p) + v)) \\ &= \beta(T(p), \overline{T}_p((q - p) + v)) \\ &= \beta(T(p), \overline{T}_p(q - p) + \overline{T}_p(v)) \\ &= \beta(T(q), \overline{T}_p(v)) \end{aligned}$$

so that $\overline{T}_p(v) = \overline{T}_q(v)$ by uniqueness. The map $\overline{T}$ is called the *linear part* of T. An affine morphism which is invertible is called an *affine isomorphism* because its inverse is also an affine morphism: in fact T is an affine isomorphism if, and only if, its linear part $\overline{T}$ is a vector space isomorphism, and then $\overline{(T^{-1})} = \overline{T}^{-1}$. An affine morphism from A to itself whose linear part is id_V is called a *translation*.

To find the coordinate description of the affine morphism $T : A \longrightarrow B$, suppose that we use (p, e_i) to give coordinates on A, and (q, f_A) to give coordinates on B. Then $T(p)$ and $T(\alpha(p, e_i))$ are all elements of B. Suppose the coordinates of $T(p)$ are p^A, and those of $T(\alpha(p, e_i)) = \beta(T(p), \overline{T}(e_i))$ are

T_i^A. Then if $x \in A$ has coordinates x^i, we have

$$\begin{aligned} T(x) &= T(\alpha(p, x^i e_i)) \\ &= \beta(T(p), \overline{T}(x^i e_i)) \\ &= \beta(q, (T(p) - q) + \overline{T}(x^i e_i)) \\ &= \beta(q, p^A f_A + x^i T_i^A f_A) \end{aligned}$$

so that the coordinates of $T(x)$ are $T_i^A x^i + p^A$. The numbers T_i^A are the components of the matrix of the linear transformation $\overline{T}$. As we might have expected, the transformation rule for affine coordinates is just the reverse of the coordinate representation of an affine isomorphism.

With this machinery at our disposal, we can give a definition of an affine bundle.

Definition 2.4.4 Let (E, π, M) be a vector bundle. An *affine bundle modelled on π* is a quadruple (A, ρ, M, α) such that:

1. (A, ρ, M) is a bundle;
2. (a) $\alpha : A \times_M E \longrightarrow A$ satisfies, for each $p \in M$, $\alpha(A_p \times E_p) \subset A_p$;
   (b) for each $p \in M$, $(A_p, E_p, \alpha|_{A_p \times E_p})$ is an affine space;
3. for each $p \in M$ there is a local trivialisation $(W_p, \mathbf{R}^n, t_p)$, called an *affine local trivialisation*, satisfying the condition that, for $q \in W_p$, the composite of
$$t_p|_{A_q} : A_q \longrightarrow \{q\} \times \mathbf{R}^n$$
with $pr_2 \times \mathbf{R}^n \longrightarrow \mathbf{R}^n$ is an affine isomorphism.

■

We sometimes say that α is an action of the vector bundle π on the affine bundle ρ. Although we have written α as a right action, it is also a left action because the addition of vectors is commutative.

Example 2.4.5 If (E, π, M) and (H, ρ, N) are vector bundles, $(f, \overline{f}) : \pi \longrightarrow \rho$ is a vector bundle morphism of constant rank, and $\psi \in \Gamma(\rho)$ is an arbitrary section, then

$$\left(f^{-1}(\operatorname{im}\psi), \pi|_{f^{-1}(\operatorname{im}\psi)}, M, \alpha\right)$$

is an affine bundle modelled on the vector bundle $\pi|_{\ker f}$, where the action α is simply addition in the fibres of π. Each fibre $f^{-1}(\operatorname{im}\psi)_p$ is a coset of $(\ker f)_p$ in E_p. ■

Example 2.4.6 As a special case of the previous example, if $Y \in \mathcal{X}(M)$ then the set of *vectors projecting to* Y is defined to be

$$T_Y\pi = \{\xi \in TE : \pi_*(\xi) = Y_{\pi(\tau_E(\xi))}\}$$

and then $\tau_E|_{T_Y\pi}$ becomes an affine bundle modelled on the vector bundle $\tau_E|_{V\pi}$. ∎

Example 2.4.7 Every vector bundle (E, π, M) yields an affine bundle

$$(E, \pi, M, \sigma)$$

where $\sigma : E \times_M E \longrightarrow E$ is addition in the fibres of π. ∎

Lemma 2.4.8 *Let (A, ρ, M, α) be an affine bundle modelled on the vector bundle (E, π, M). Let $z \in \Gamma(\rho)$; then the section z (known as the* zero section*) determines a vector bundle structure on (A, ρ, M).*

Proof Let $p \in M$. If $x, y \in A_p$ then there is a unique $v = x - z(p) \in E_p$ such that $\alpha(z(p), v) = x$, so define $x + y$ to equal $\alpha(y, v)$; similarly if $\lambda \in \mathbf{R}$, define λx to equal $\alpha(z(p), \lambda v)$. The fibre A_p then becomes a vector space. If $t_p : \rho^{-1}(W_p) \longrightarrow W_p \times \mathbf{R}^n$ is an affine local trivialisation, then the map $\bar{t}_p : \rho^{-1}(W_p) \longrightarrow W_p \times \mathbf{R}^n$ defined by

$$\bar{t}_p(a) = (\rho(a), pr_2(t_p(a - z(\rho(a)))))$$

is a linear local trivialisation. ∎

Just as for vector bundles, there are special coordinate systems (x^i, a^α) appropriate to the total space of an affine bundle. These are called *affine bundle coordinate systems*, and they have the property that a^α is an affine map on each fibre, whose linear part is the corresponding vector bundle coordinate map u^α. These coordinates may be derived from the affine local trivialisations. In each fibre A_p there is a point b such that each $a^\alpha(b) = 0$ and so every affine bundle coordinate system determines a local zero section of ρ; conversely, given a basis of local sections $e_\beta \in \Gamma_W(\pi)$ and a local section $z \in \Gamma_W(\rho)$, we may define affine coordinates a^α on the fibres of ρ.

Definition 2.4.9 Let (A, ρ, M, α) be an affine bundle modelled on the vector bundle (E, π, M). If $(E', \pi|_{E'}, \pi(E'))$ is a vector sub-bundle of π and $A' \subset A$ is a submanifold such that $\rho(A') = \pi(E')$ and such that

$$(A', \rho|_{A'}, \rho(A'), \alpha|_{A' \times E'})$$

is an affine bundle modelled on $\pi|_{E'}$, then $\rho|_{A'}$ is termed an *affine sub-bundle of* ρ. ∎

Example 2.4.10 The bundle $\pi|_{f^{-1}(\mathrm{im}\,\psi)}$ described in Example 2.4.5 is an affine sub-bundle of π (where the latter vector bundle is regarded as an affine bundle modelled on itself). ∎

Finally in this section we shall describe affine bundle morphisms: as might be expected, these are bundle morphisms which, when restricted to each fibre, are morphisms of affine spaces.

Definition 2.4.11 Let (A, ρ, M, α) and (B, σ, N, β) be affine bundles. A bundle morphism $(f, \bar{f}) : \rho \longrightarrow \sigma$ is called an *affine bundle morphism* if, for each $p \in M$, $f|_{A_p} : A_p \longrightarrow B_{\bar{f}(p)}$ is an affine morphism. ∎

Just as each vector bundle morphism has a local matrix representation obtained from vector bundle coordinates on its domain and codomain, there is a similar local representation for each affine bundle morphism. Indeed, suppose that (x^i, a^α) and (y^a, b^A) are affine bundle coordinates on A and B respectively; suppose also that a^α correspond to the local sections e_β of the vector bundle π on which ρ is modelled, and a local section z of ρ. Put $f^A_\alpha = b^A \circ f \circ \alpha \circ (z, e_\alpha)$ and put $f^A = b^A \circ f \circ z$; then if $c \in A_p$ has coordinates $c^\alpha = a^\alpha(c)$, the coordinates of $f(c) \in B_{\bar{f}(p)}$ are

$$b^A(f(c)) = f^A_\alpha(p)c^\alpha + f^A(p).$$

The matrix of functions f^A_α is also the local matrix representation of the linear part of the affine bundle morphism f.

EXERCISES

2.4.1 Every affine space A modelled on the vector space V is automatically a differentiable manifold with a global coordinate system. If $p \in A$, show that there is a canonical isomorphism $T_p A \cong V$.

2.4.2 Deduce from the results of Exercises 2.2.1 and 2.4.1 that, if (A, ρ, M) is an affine bundle modelled on the vector bundle (E, π, M) then there is a canonical isomorphism between the vector bundles $(V\rho, \tau_A|_{V\rho}, A)$ and $(\rho^*(E), \rho^*(\pi), A)$.

2.4.3 Show that the bundle $(T^2M, \tau^{2,1}_M, TM)$ described in Example 2.1.5 is an affine bundle modelled on the vector bundle $(V\tau_M, \tau_{TM}|_{V\tau_M}, TM)$.

REMARKS

Many of the ideas introduced in this chapter may also be described in the language of fibre bundles. A vector bundle may be considered as a particular type of fibre bundle with structure group $GL(n, \mathbf{R})$, and similarly an affine bundle may be regarded as a fibre bundle whose structure group is the group of affine transformations of $\mathbf{R}^n$. The relationship between fibre bundles and vector bundles is described in [8]. More information about affine spaces, and their relationship with vector spaces, may be found in [3].

Vector bundles are also of importance in topology, where they may be used to classify global properties of manifolds (or, indeed, of more general topological spaces). We have already seen that the tangent bundle to the sphere S^2 is not globally trivial, whereas it may be shown that (for example) every vector bundle with base space $\mathbf{R}^n$ is trivial. The study of topological properties of spaces using isomorphism classes of vector bundles is known as *K-theory*, and accounts of this theory may be found in [1] or [8].

Chapter 3

Linear Operations on General Bundles

In this chapter, we return to the study of a general bundle (E, π, M) and the various vector bundles and modules of sections associated with it. These vector bundles are constructed from the tangent and cotangent bundles to E and to M. The main theme of the chapter comes from isolating those tangent vectors and vector fields on E which are tangent to the fibres of π, and those cotangent vectors and differential forms which annihilate them. Many of the definitions in the early part of the chapter are restatements of examples from the previous chapter, but are examined here in more detail.

3.1 Tangent and Cotangent Vectors

Definition 3.1.1 If (E, π, M) is a bundle, then the *vertical bundle to* π is the vector sub-bundle $(V\pi, \tau_E|_{V\pi}, E)$ of the tangent bundle τ_E whose total space $V\pi$ is defined by

$$V\pi = \{\xi \in TE : \pi_*(\xi) = 0 \in T_{\tau_M(\pi_*(\xi))}M\}$$

(see Example 1.4.11). The fibre of $V\pi$ over $a \in E$ is usually denoted $V_a\pi$ rather than $(V\pi)_a$. ■

The total space of the vertical bundle may also be considered as the collection of those vectors which are "tangent to the fibres of π".

Lemma 3.1.2 *If $a \in E$ then $T_a(E_{\pi(a)}) \cong V_a\pi$.*

Proof Let $\iota_{\pi(a)} : E_{\pi(a)} \longrightarrow E$ be the inclusion, so that $\pi \circ \iota_{\pi(a)} : E_{\pi(a)} \longrightarrow M$ is a constant map and that therefore that $\pi_* \circ \iota_{\pi(a)*} : T_a(E_{\pi(a)}) \longrightarrow T_{\pi(a)}M$ is the zero map. Since $V_a\pi$ is the kernel of $\pi_*|_{T_aE}$, it follows that $\operatorname{im}(\iota_{\pi(a)*}) \subset V_a\pi$. But since $\iota_{\pi(a)*}$ is an injection it also follows that

$$\dim \operatorname{im}(\iota^*_{\pi(a)}) \quad = \quad \dim T_a(E_{\pi(a)})$$

$$
\begin{aligned}
&= \dim E_{\pi(a)} \\
&= \dim E - \dim M \\
&= \dim T_a E - \dim T_{\pi(a)} M \\
&= \dim \ker \pi_*|_{T_a E} \\
&= \dim V_a \pi
\end{aligned}
$$

so that $\iota_{\pi(a)*}$ is the required isomorphism. ■

One important use of the vertical bundle is in demonstrating when a map between the total spaces of two bundles gives rise to a bundle morphism: such a map must take vertical vectors to vertical vectors, and is characterised by this property.

Proposition 3.1.3 *Let (E, π, M) and (H, ρ, N) be bundles, and let $f : E \longrightarrow H$. Then f defines a bundle morphism $\pi \longrightarrow \rho$ if, and only if, $f_*(V\pi) \subset V\pi$.*

Proof Suppose first that $(f, \overline{f})$ is a bundle morphism, and let $\xi \in V_a\pi$ so that $\pi_*(\xi) = 0 \in T_{\pi(a)}M$. Then

$$
\begin{aligned}
\rho_*(f_*(\xi)) &= \overline{f}_*(\pi_*(\xi)) \\
&= 0 \in T_{\rho(f(a))}N
\end{aligned}
$$

so that $f_*(\xi) \in V_{f(a)}\rho$.

To prove the converse, we shall use coordinates (x^i, u^α) on E and (y^b, v^B) on H. Let $\xi \in V_a\pi$ have the coordinate representation

$$\xi = \xi^\alpha \left.\frac{\partial}{\partial u^\alpha}\right|_a$$

so that $f_*(\xi)$ has coordinate representation

$$f_*(\xi) = \xi^\alpha \left(\left.\frac{\partial f^b}{\partial u^\alpha}\right|_a \left.\frac{\partial}{\partial y^b}\right|_{f(a)} + \left.\frac{\partial f^B}{\partial u^\alpha}\right|_a \left.\frac{\partial}{\partial v^B}\right|_{f(a)} \right).$$

But $f_*(\xi) \in V_{f(a)}\rho$, so that the coefficient of $\partial/\partial y^b$ must vanish. By choosing vectors ξ with a single non-zero coordinate ξ^α we may deduce that, for each α and each b,

$$\left.\frac{\partial f^b}{\partial u^\alpha}\right|_a = 0.$$

Since this must be true for each point a in the domain of the coordinate system, it follows that $y^b \circ f$ is constant on the fibres of π in a neighbourhood of each point of a. It then follows from the connectedness of the fibres that f is constant on each complete fibre of π, so that it defines a bundle morphism. ■

The complementary entity to the vertical bundle is called the transverse bundle, and may be thought of as containing "horizontal" vectors. It is not, however, a sub-bundle of τ_E.

Definition 3.1.4 The *transverse bundle to* π is the pull-back vector bundle $(\pi^*(TM), \pi^*(\tau_M), E)$ (see Example 1.4.6). ■

These two bundles, and the tangent bundle τ_E, are related by a short exact sequence of vector bundle morphisms projecting to the identity on E, as described in Example 2.2.8, where the map $V\pi \longrightarrow TE$ is the inclusion and the map $TE \longrightarrow \pi^*(TM)$ is given by $\xi \longmapsto (\pi_*(\xi), \tau_E(\xi))$.

Lemma 3.1.5 *The sequence of vector bundle morphisms*

$$0 \longrightarrow V\pi \longrightarrow TE \longrightarrow \pi^*(TM) \longrightarrow 0$$

is exact.

Proof The sequence is exact at $V\pi$ since the map $V\pi \longrightarrow TE$ is an inclusion and so injective. It is exact at TE since, given $\xi \in TE$, then $\xi \in V\pi$ if, and only if, $\pi_*(\xi) = 0 \in T_{\pi(\tau_E(\xi))}M$, and this corresponds to $(\pi_*(\xi), \tau_E(\xi)) = 0 \in \pi^*(TM)_{\tau_E(\xi)}$. Finally, the surjectivity of the map $TE \longrightarrow \pi^*(TM)$ may be seen in local coordinates: if $(\eta, a) \in \pi^*(TM)$ with $a \in E$ and $\eta \in T_{\pi(a)}M$, and if

$$\eta = \eta^i \left.\frac{\partial}{\partial x^i}\right|_{\pi(a)}$$

then put

$$\xi = \eta^i \left.\frac{\partial}{\partial x^i}\right|_a \in TE$$

so that $\xi \longmapsto (\eta, a)$. ■

In general this short exact sequence does not split: there is no distinguished sub-bundle of τ_E which complements the vertical bundle. The choice of such a sub-bundle is precisely the choice of a *connection* on π, and this will be examined in Section 3.5.

A manifestation of this phenomenon may be seen in coordinates. Using adapted local coordinates (x^i, u^α) on E, the induced coordinates on TE are $(x^i, u^\alpha; \dot{x}^i, \dot{u}^\alpha)$. An arbitrary element $\xi \in TE$ may be written as

$$\xi^i \left.\frac{\partial}{\partial x^i}\right|_a + \xi^\alpha \left.\frac{\partial}{\partial u^\alpha}\right|_a$$

where $a = \tau_E(\xi) \in E$, so that $\xi^i = \dot{x}^i(\xi)$, $\xi^\alpha = \dot{u}^\alpha(\xi)$. However, $\xi \in V\pi$ precisely when ξ may be written as

$$\xi^\alpha \left.\frac{\partial}{\partial u^\alpha}\right|_a$$

so that $\dot{x}^i(\xi) = 0$, and in fact $(x^i, u^\alpha; \dot{u}^\alpha)$ may be used as adapted coordinates on $V\pi$. Changing to a different adapted coordinate system (y^j, v^β) on E will not introduce any terms in $\partial/\partial y^j$ into the coordinate description of elements of $V\pi$, because the x^i depend only on the y^j and not on the v^β. However, a tangent vector of the form

$$\xi^i \left.\frac{\partial}{\partial x^i}\right|_a$$

will in general, when written in the new coordinate system, have terms in both $\partial/\partial y^j$ and $\partial/\partial v^\beta$. Although $(x^i, u^\alpha, \dot{x}^i)$ may be used as coordinates on $\pi^*(TM)$ and a general element of this manifold may be written as

$$\left(\xi^i \left.\frac{\partial}{\partial x^i}\right|_{\pi(a)}, a\right),$$

there is no canonical injection $\pi^*(TM) \longrightarrow TE$.

Example 3.1.6 Let $(SL(2,\mathbf{R}), \pi, H)$ be the bundle described in Example 1.1.3, and let $(\xi, \eta; \theta)$ be the coordinates in a neighbourhood of the identity $I \in SL(2,\mathbf{R})$ defined by

$$\begin{aligned} \xi(A) &= \frac{ac+bd}{c^2+d^2} &&= \operatorname{Re}(\pi(A)) \\ \eta(A) &= \frac{1}{c^2+d^2} &&= \operatorname{im}(\pi(A)) \\ \theta(A) &= \tan^{-1}\frac{c}{a} \end{aligned}$$

where

$$A = \begin{pmatrix} a & b \\ c & d \end{pmatrix};$$

these coordinates correspond to the trivialisation ρ_1 of Example 1.1.7. Now the tangent space to $SL(2,\mathbf{R})$ at the identity I may be represented by the vector space $\mathbf{sl}(2,\mathbf{R})$ of 2×2 matrices with zero trace (we shall not need to use the Lie algebra structure of $\mathbf{sl}(2,\mathbf{R})$ in this example). A short calculation then gives

$$\begin{aligned} \left.\frac{\partial}{\partial \xi}\right|_I &= \begin{pmatrix} 0 & 1 \\ 0 & 0 \end{pmatrix} \\ \left.\frac{\partial}{\partial \eta}\right|_I &= \tfrac{1}{2}\begin{pmatrix} 1 & 0 \\ 0 & -1 \end{pmatrix} \\ \left.\frac{\partial}{\partial \theta}\right|_I &= \begin{pmatrix} 0 & -1 \\ 1 & 0 \end{pmatrix} \end{aligned}$$

so that the subspace of $\mathbf{sl}(2,\mathbf{R})$ spanned by $\partial/\partial\xi$ and $\partial/\partial\eta$ contains matrices of the form

$$\begin{pmatrix} \lambda & \mu \\ 0 & -\lambda \end{pmatrix}.$$

However, we may choose instead the coordinates $(\xi', \eta'; \theta')$ where $\xi' = \xi$, $\eta' = \eta$ and

$$\theta'(A) = \tan^{-1}\frac{b}{d};$$

these coordinates correspond to the trivialisation ρ_2. We now find that

$$\left.\frac{\partial}{\partial\xi}\right|_I = \begin{pmatrix} 0 & 0 \\ 1 & 0 \end{pmatrix}$$
$$\left.\frac{\partial}{\partial\eta}\right|_I = \tfrac{1}{2}\begin{pmatrix} 1 & 0 \\ 0 & -1 \end{pmatrix}$$
$$\left.\frac{\partial}{\partial\theta}\right|_I = \begin{pmatrix} 0 & 1 \\ -1 & 0 \end{pmatrix}$$

so that the subspace of $\mathbf{sl}(2,\mathbf{R})$ spanned by $\partial/\partial\xi'$ and $\partial/\partial\eta'$ contains matrices of the form

$$\begin{pmatrix} \lambda & 0 \\ \mu & -\lambda \end{pmatrix}.$$

Of course the vertical subspace of $\mathbf{sl}(2,\mathbf{R})$, containing matrices of the form

$$\begin{pmatrix} 0 & \nu \\ -\nu & 0 \end{pmatrix},$$

is spanned by both $\partial/\partial\theta$ and $\partial/\partial\theta'$. ■

We may carry out a similar analysis of the bundles of cotangent vectors which are associated with E. Once again we may define "vertical" and "horizontal" cotangent vectors, although this time it is the bundle of horizontal cotangent vectors which may be considered as a sub-bundle of τ_E^*.

Definition 3.1.7 The *vertical cotangent bundle to* π is defined to be the vector bundle dual to $(V\pi, \tau_E|_{V\pi}, E)$ and is denoted $(V^*\pi, (\tau_E|_{V\pi})^*, E)$. ■

We may call an element of the total space $V^*\pi$ a "vertical cotangent vector"; it is not, however, a cotangent vector in the usual sense of the word, and the bundle $(\tau_E|_{V\pi})^*$ is not normally the pull-back of a bundle over some other manifold.

Definition 3.1.8 The *cotangent bundle to E horizontal over π* is defined to be the pull-back vector bundle $(\pi^*(T^*M), \pi^*(\tau_M^*), E)$, which is identified with the sub-bundle $(\pi^*(T^*M), \tau_E^*|_{\pi^*(T^*M)}, E)$ of τ_E^* (see Example 1.4.7). ∎

Lemma 3.1.9 *The bundle $(\pi^*(T^*M), \pi^*(\tau_M^*), E)$ is isomorphic to the vector bundle dual to $(\pi^*(TM), \pi^*(\tau_M), E)$.*

Proof Let $a \in E$; we shall show that the fibres $\pi^*(T^*M)_a$ and $\pi^*(TM)_a$ may be regarded as dual vector spaces. So suppose $(\eta, a) \in \pi^*(T^*M)_a$ and $(\xi, a) \in \pi^*(TM)_a$. Then $\eta \in T^*_{\pi(a)}M$ and $\xi \in T_{\pi(a)}M$, so that the duality relationship may be obtained from the obvious isomorphisms $\pi^*(T^*M)_a \cong T^*_{\pi(a)}M$ and $\pi^*(TM)_a \cong T_{\pi(a)}M$. ∎

We may now relate these bundles to the cotangent bundle τ_E^* using another short exact sequence of vector bundle morphisms projecting to the identity on E. The map $\pi^*(T^*M) \longrightarrow T^*E$ will be given by $(\eta, a) \longmapsto \pi^*\eta \in T_a^*E$ (or equivalently will be the inclusion, using the identification mentioned in Definition 3.1.8), and the map $T^*E \longrightarrow V^*\pi$ will be the transpose of the inclusion $V\pi \longrightarrow TE$.

Lemma 3.1.10 *The sequence of vector bundle morphisms*

$$0 \longrightarrow \pi^*(T^*M) \longrightarrow T^*E \longrightarrow V^*\pi \longrightarrow 0$$

is exact.

Proof By duality from Lemma 3.1.5, using the fact that $(\eta, a) \longmapsto \pi^*\eta$ is the transpose of $\xi \longmapsto (\pi_*(\xi), \tau_E(\xi))$. ∎

Lemma 3.1.11 *The vector sub-bundle $(\pi^*(T^*M), \tau_E^*|_{\pi^*(T^*M)}, E)$ is the annihilator in τ_E^* of $\tau_E|_{V\pi}$; the vector sub-bundle $(V\pi, \tau_E|_{V\pi}, E)$ is the annihilator in τ_E of $\tau_E^*|_{\pi^*(T^*M)}$.*

Proof If $\eta \in \pi^*(T^*M)_a$ then $\eta = \pi^*\zeta$ for some $\zeta \in T^*_{\pi(a)}M$, so if $\xi \in V_a\pi$ then

$$\eta(\xi) = (\pi^*\zeta)(\xi) = \zeta(\pi_*\xi) = 0;$$

consequently $(V_a\pi)^\circ \subset \pi^*(T^*M)_a$. However, $\dim V_a\pi = \dim E_{\pi(a)} = n$ because $V_a\pi \cong T_a(E_{\pi(a)})$ and $\dim \pi^*(T^*M)_a = \dim T^*_{\pi(a)}M = m$ because π^* is injective so that $(V_a\pi)^\circ$ actually equals $\pi^*(T^*M)_a$. The other half of the result is obtained by duality. ∎

Using adapted local coordinates (x^i, u^α) on E, an arbitrary element $\eta \in T^*E$ may be written as

$$\eta_i \, dx^i\Big|_a + \eta_\alpha \, du^\alpha|_a$$

where $a = \tau_E^*(\eta) \in E$, whereas an arbitrary element of $\pi^*(T^*M)$ may be written as

$$\eta_i \, dx^i\Big|_a .$$

Once again, changing to a different adapted coordinate system (y^j, v^β) on E will not introduce any terms in du^β into the coordinate description of elements of $\pi^*(T^*M)$. However, a cotangent vector of the form

$$\eta_\alpha \, du^\alpha|_a$$

will in general, when written in the new coordinate system, have terms in both dy^j and dv^β. Although an element of $V^*\pi$ may be written in coordinates in this form, a better description would be as the coset

$$\eta_\alpha \left(du^\alpha|_a + (\pi^*(T^*M)_a) \right),$$

for there is no canonical injection $V^*\pi \longrightarrow T^*E$.

Example 3.1.12 Let $(\xi, \eta; \theta)$ be coordinates on the total space of the bundle $(SL(2, \mathbf{R}), \pi, H)$ described in Example 3.1.6. If (e, f, h) is the basis of $\mathbf{sl}(2, \mathbf{R})^*$ dual to the basis

$$\begin{pmatrix} 0 & 1 \\ 0 & 0 \end{pmatrix}, \begin{pmatrix} 0 & 0 \\ 1 & 0 \end{pmatrix}, \begin{pmatrix} 1 & 0 \\ 0 & -1 \end{pmatrix}$$

of $\mathbf{sl}(2, \mathbf{R})$, then a short calculation gives

$$\begin{aligned} d\xi|_I &= e + f \\ d\eta|_I &= 2h \\ d\theta|_I &= f. \end{aligned}$$

On the other hand, if we use the coordinates $(\xi', \eta'; \theta')$ then

$$\begin{aligned} d\xi'|_I &= e + f \\ d\eta'|_I &= 2h \end{aligned}$$

as before, but

$$d\theta'|_I = e,$$

so by choosing two different coordinate systems we obtain two different complements to the subspace of horizontal cotangent vectors at the identity. ∎

There is, however, one way in which the arrangement for cotangent vectors differs from that for tangent vectors, and this occurs when the base manifold M is orientable; in these circumstances, it is possible to construct isomorphisms between the vertical cotangent bundle and a certain exterior power of ordinary cotangent bundles. The particular bundle of interest here is the sub-bundle

$$(T^*E \wedge \textstyle\bigwedge^m \pi^*(T^*M), \tau_E^* \wedge \bigwedge^m(\tau_E^*|_{\pi^*(T^*M)}), E)$$

of $\bigwedge^{m+1}\tau_E^*$, where $m = \dim M$, containing those $(m+1)$-covectors $\eta \in \bigwedge^{m+1}T_a^*E$ where

$$\eta(\xi_0, \xi_1, \ldots, \xi_m) = 0$$

whenever two or more of the vectors $\xi_i \in T_aE$ are vertical vectors.

Proposition 3.1.13 *If (E, π, M) is a bundle and M is orientable then each volume form Ω on M determines a vector bundle isomorphism between $\tau_E^* \wedge \bigwedge^m(\tau_E^*|_{\pi^*(T^*M)})$ and $(\tau_E|_{V\pi})^*$.*

Proof Suppose $\eta \in (T^*E \wedge \bigwedge^m\pi^*(T^*M))_a$, and denote the pull-back of the m-form Ω to E also by Ω. Then for each $\xi \in V_a\pi$ the m-linear map $\xi \lrcorner \eta : T_aE \times \ldots \times T_aE \longrightarrow \mathbf{R}$ defined by

$$(\xi \lrcorner \eta)(\xi_1, \ldots, \xi_m) = \eta(\xi, \xi_1, \ldots, \xi_m)$$

is an element of $(\bigwedge^m\pi^*(T^*M))_a$, and so $\xi \lrcorner \eta = \lambda_{\xi,\eta}\Omega_a$ for some $\lambda_{\xi,\eta} \in \mathbf{R}$. We may therefore define a function $\overline{\eta} : V_a\pi \longrightarrow \mathbf{R}$ by $\overline{\eta}(\xi) = \lambda_{\xi,\eta}$, and since $\overline{\eta}$ is obviously linear it is an element of $V_a^*\pi$.

The correspondence $\overline{\Omega} : (T^*E \wedge \bigwedge^m\pi^*(T^*M))_a \longrightarrow V_a^*\pi$ given by $\overline{\Omega}(\eta) = \overline{\eta}$ is linear on each fibre. It is surjective, for starting with an element $\overline{\eta} \in V_a^*\pi$ there is certainly a cotangent vector $\sigma \in T_a^*E$ satisfying $\sigma(\xi) = \overline{\eta}(\xi)$ for all $\xi \in V_a\pi \subset T_aE$, so that we may define $\eta = \sigma \wedge \Omega_a$.

On the other hand, starting with η, define $\sigma \in T_a^*E$ using coordinates by

$$\sigma = (-1)^{\frac{1}{2}m(m+1)} \left.\frac{\partial}{\partial x^1}\right|_a \lrcorner \ldots \lrcorner \left.\frac{\partial}{\partial x^m}\right|_a \lrcorner\, \eta$$

where the coordinate functions x^i around $\pi(a) \in M$ are chosen so that $\Omega = dx^1 \wedge \ldots \wedge dx^m$. (Of course, the cotangent vector σ obtained in this way will depend upon the coordinate system used.) Then $\eta = \sigma \wedge \Omega_a$ and, for $\xi \in V_a\pi$,

$$\lambda_{\xi,\eta}\Omega_a = \xi \lrcorner \eta = \sigma(\xi)\Omega_a$$

so that $\sigma(\xi) = \lambda_{\xi,\eta} = \overline{\eta}(\xi)$. If $\overline{\eta}_1 = \overline{\eta}_2$ then, for all $\xi \in V_a\pi$, $\sigma_1(\xi) = \sigma_2(\xi)$ so that $\sigma_1 - \sigma_2 \in \pi^*(T^*M)_a$ and hence $\eta_1 - \eta_2 = (\sigma_1 - \sigma_2) \wedge \Omega_a = 0$, demonstrating that $\overline{\Omega}$ is also injective. It follows that $\overline{\Omega}$ is a linear isomorphism between the fibres $(T^*E \wedge \bigwedge^m \pi^*(T^*M))_a$ and $V_a^*\pi$. To see that the collection of these isomorphisms defines a smooth map between the two total spaces (and hence a vector bundle isomorphism projecting to the identity on E), observe that in coordinates this correspondence is simply

$$\eta_\alpha (du^\alpha \wedge \Omega)_a \longmapsto \eta_\alpha \left(du^\alpha|_a + \pi^*(T^*M)_a \right).$$

■

Example 3.1.14 On the bundle $(SL(2, \mathbf{R}), \pi, H)$ a vertical cotangent vector at the identity may be therefore be written in coordinates as

$$\lambda \, d\theta \wedge (d\xi \wedge d\eta)|_I .$$

Notice that, although $d\theta'|_I = d\xi|_I - d\theta|_I$, taking the wedge product with the volume form $(d\xi \wedge d\eta)|_I$ absorbs the term $d\xi|_I$ and so the vertical cotangent vector may also be written as

$$- \lambda \, d\theta' \wedge (d\xi \wedge d\eta)|_I .$$

■

EXERCISES

3.1.1 Let π be the trivial bundle $(M \times H, pr_1, M)$. Show that, in this case, the vertical cotangent bundle $(V^*\pi, (\tau_{M\times H}|_{V\pi})^*, M \times H)$ *is* isomorphic to a pull-back bundle (namely the pull-back of τ_H^* to $M \times H$).

3.1.2 Let (E, π, M) be a bundle, and let ϕ be a global section of π (so that $\phi(M)$ is a closed embedded submanifold of E). Show that the normal bundle $(N_E\phi(M), \nu, \phi(M))$ defined in Exercise 2.2.3 is isomorphic (as a vector bundle) to the restricted bundle

$$\left(V\pi|_{\phi(M)}, (\tau_E|_{V\pi})|_{\phi(M)}, \phi(M) \right).$$

3.2 Vector Fields

In this section, we shall describe some special types of vector field which are particularly important in the theory of bundles. Some of these, the vertical vector fields and the vector fields along the bundle projection π, may be obtained by taking sections of bundles which have already been introduced. Others, notably the projectable vector fields, do not in general have a pointwise description, and may be defined instead as vector bundle morphisms.

Definition 3.2.1 A section of the bundle $(V\pi, \tau_E|_{V\pi}, E)$ is called a *vertical vector field* on E; the space of vertical vector fields will be denoted $\mathcal{V}(\pi)$. ■

So a vertical vector field is just a vector field on E which is π-related to zero. In adapted local coordinates, a vertical vector field appears as

$$X = X^\alpha \frac{\partial}{\partial u^\alpha}$$

so that the coefficients of $\partial/\partial x^i$ are all zero.

Example 3.2.2 If π is itself a vector bundle, then multiplication by real numbers on the fibres gives a well-defined mapping $\mathbf{R} \times E \longrightarrow E$. For each $a \in E$ there is a canonical vertical tangent vector $[t \longmapsto e^t a] \in T_a E$, and the vector field $\Delta \in \mathcal{V}(\pi)$ defined by

$$\Delta_a = [t \longmapsto e^t a]$$

is called the *dilation field* of π. In coordinates,

$$\Delta = u^\alpha \frac{\partial}{\partial u^\alpha}.$$

■

A simple characterisation of vertical vector fields is given by the following lemma, which describes a condition on the Lie derivative action.

Lemma 3.2.3 *$X \in \mathcal{X}(E)$ is vertical if, and only if, for each $f \in C^\infty(M)$, $\mathcal{L}_X(\pi^*(f)) = 0$.*

Proof This is obtained directly from the definitions. At each $a \in E$,

$$\mathcal{L}_X(\pi^*(f))(a) = X_a(\pi^*(f)) = (\pi_*(X_a))(f).$$

If X is vertical then $\mathcal{L}_X(\pi^*(f))(a) = 0$ for each $a \in E$, giving the condition of the lemma. Conversely, if the condition holds then $(\pi_*(X_a))(f) = 0$ for every $f \in C^\infty(M)$, so that $\pi_*(X_a) = 0$. ■

Lemma 3.2.4 *The space of vertical vector fields forms an (infinite-dimensional) Lie algebra.*

Proof This uses the result from elementary differential geometry involving the Lie bracket and π-related vector fields. If $X, Y \in \mathcal{V}(\pi)$ then both X and Y are π-related to zero, and so $[X, Y]$ is π-related to $[0, 0]$. ■

To obtain vector fields along π, we shall start with the exact sequence of vector bundles over E

$$0 \longrightarrow V\pi \longrightarrow TE \longrightarrow \pi^*(TM) \longrightarrow 0$$

and use Proposition 2.2.10 to construct an exact sequence

$$0 \longrightarrow \mathcal{V}(\pi) \longrightarrow \mathcal{X}(E) \longrightarrow \mathcal{X}(\pi) \longrightarrow 0$$

of modules of sections.

Definition 3.2.5 A *vector field along* π is an element of $\mathcal{X}(\pi)$, the space of sections of the transverse bundle described in Example 1.4.6. ∎

In local coordinates, a vector field along π is written as

$$X = X^i \frac{\partial}{\partial x^i}$$

where X^i are functions on the total space E, but $\partial/\partial x^i$ are supposed to be local vector fields on M. Where confusion is possible a coordinate description like this will be written explicitly as

$$X_a = X^i(a) \left. \frac{\partial}{\partial x^i} \right|_{\pi(a)} \in T_{\pi(a)}M.$$

It is not in general possible to define the contraction of a vector field along π with a differential form on the total space E. However, it *is* possible to define its contraction with a differential form on the base space M.

Definition 3.2.6 If $X \in \mathcal{X}(\pi)$ and $\sigma \in \bigwedge^1 M$, define $X \lrcorner \sigma \in C^\infty(E)$ by, for $a \in E$,

$$(X \lrcorner \sigma)(a) = \sigma_{\pi(a)}(X_a).$$

∎

In local coordinates, if $\sigma = \sigma_j dx^j$ where σ_j are functions defined locally on M, then

$$X \lrcorner \sigma = X^i \sigma_i$$

where the resulting function is defined locally on E. If $\theta \in \bigwedge^r M$ then $X \lrcorner \theta$ is defined in a similar way, and results in an "$(r-1)$-form on M with coefficients on E"; differential forms of this type will be described in Section 3.3.

The coordinate expression for a vector field along π suggests that it would also be possible to define $\mathcal{X}(\pi)$ in terms of derivations, in an analogous way to the standard definition given for vector fields on manifolds; this is indeed the case.

Proposition 3.2.7 *Let $\mathcal{D}(\pi)$ be the space of linear maps $\widetilde{X} : C^\infty(M) \longrightarrow C^\infty(E)$ satisfying*

$$\widetilde{X}(fg) = \pi^*(f)\widetilde{X}(g) + \pi^*(g)\widetilde{X}(f).$$

Then $\mathcal{D}(\pi) \cong \mathcal{X}(\pi)$.

Proof This is just a variation on the standard proof which applies to vector fields and derivations on manifolds, rather than along maps. Given a function $\widetilde{X} : C^\infty(M) \longrightarrow C^\infty(E)$ and a point $a \in E$, define $X_a : C^\infty(M) \longrightarrow \mathbf{R}$ by $X_a(f) = (\widetilde{X}(f))(a)$. If $\widetilde{X}$ is linear and satisfies the derivation property above, then X_a is clearly linear and satisfies

$$\begin{aligned} X_a(fg) &= \widetilde{X}(fg)(a) \\ &= \left(\pi^*(f)\widetilde{X}(g) + \pi^*(g)\widetilde{X}(f)\right)(a) \\ &= f(\pi(a))X_a(g) + g(\pi(a))X_a(f) \end{aligned}$$

for $f, g \in C^\infty(M)$. Since M is a finite-dimensional C^∞ manifold, this property is sufficient to show that $X_a \in T_{\pi(a)}M$. The map $X : a \longmapsto X_a$ is smooth because in coordinates

$$X = (\widetilde{X}(x^i))\frac{\partial}{\partial x^i}$$

where x^i and therefore $\widetilde{X}(x^i)$ are smooth. The reverse implication, showing that a vector field along π is a derivation, is straightforward. ■

If $X \in \mathcal{X}(\pi)$, we may also use the notation $\mathcal{L}_X$ for this action of X on functions in $C^\infty(M)$, and indeed $\mathcal{L}_X f = X \lrcorner\, df$ as specified in Definition 3.2.6. In fact, the Lie derivative action of a vector field along π may be combined with the surjectivity of the map $\widetilde{(\pi_*, \tau_E)} : \mathcal{X}(E) \longrightarrow \mathcal{X}(\pi)$.

Lemma 3.2.8 *If $Y \in \mathcal{X}(\pi)$ then $\mathcal{L}_Y = \mathcal{L}_X \circ \pi^*$ for some $X \in \mathcal{X}(E)$.*

Proof Let $X \in \mathcal{X}(E)$ satisfy $\widetilde{(\pi_*, \tau_E)}(X) = Y$. Then for $a \in E$,

$$Y_a = \widetilde{(\pi_*, \tau_E)}(X(a)) = (\pi_*(X(a)), a)$$

so that, if $f \in C^\infty(M)$,

$$Y_a(f) = (\pi_*(X_a))(f) = X_a(\pi^*(f))$$

and hence $\mathcal{L}_Y f = \mathcal{L}_X(\pi^*(f))$. ■

The final type of vector field which we shall consider is the projectable vector field. This is a particular type of vector field on E which can give rise, not merely to a vector field $\pi_* \circ X$ along π, but also to a vector field $\overline{X}$ on the base space M. To examine projectable vector fields, we shall first consider the affine bundle $\tau_E|_{T_Y\pi}$ described in Example 2.4.6.

Definition 3.2.9 A section of $\tau_E|_{T_Y\pi}$ is called a *vector field projecting to* Y; the affine space of all vector fields projecting to Y will be denoted $\mathcal{X}_Y(\pi)$. ∎

So "X projects to Y" is just another way of saying that X and Y are π-related. The property of being a vector field projecting to Y is of course a pointwise property; however, there is normally no reason to choose a particular vector field on M (unless it is the zero field), and we usually consider the space of all vector fields on E which project to *some* vector field on M.

Definition 3.2.10 A vector field X on E is called *projectable on* M if it defines a bundle morphism from π to π_*:

$$\begin{array}{ccc} E & \xrightarrow{\;X\;} & TE \\ \Big\downarrow{\scriptstyle \pi} & & \Big\downarrow{\scriptstyle \pi_*} \\ M & \longrightarrow & TM \end{array}$$

The set of projectable vector fields will be denoted $\mathcal{X}_{\mathrm{proj}}(\pi)$. ∎

Lemma 3.2.11 *The map* $\overline{X} : M \longrightarrow TM$ *which satisfies* $\overline{X} \circ \pi = \pi_* \circ X$ *is a vector field on* M, *such that* $X \in \mathcal{X}_{\overline{X}}(\pi)$.

Proof From $\tau_M \circ \overline{X} \circ \pi = \tau_M \circ \pi_* \circ X = \pi \circ \tau_E \circ X = \pi$ it follows, since π is surjective, that $\tau_M \circ \overline{X} = id_M$. ∎

Of course the general property of being projectable, unlike the property of projecting to a particular vector field, is *not* a pointwise property; the requirement is that the tangent vectors at all points of a given fibre must project to the same (but otherwise arbitrary) tangent vector on M, rather than to a pre-assigned tangent vector. In fact,

$$\mathcal{X}_{\text{proj}}(\pi) = \bigcup_{Y \in \mathcal{X}(M)} \mathcal{X}_Y(\pi).$$

In coordinates, a projectable vector field may be written

$$X = X^i \frac{\partial}{\partial x^i} + X^\alpha \frac{\partial}{\partial u^\alpha}$$

where the functions X^i have all been pulled back from the base space M; the vector field $\overline{X}$ then has the coordinate representation

$$\overline{X} = X^i \frac{\partial}{\partial x^i}.$$

Some elementary properties of projectable vector fields are given in the following three lemmas.

Lemma 3.2.12 *The vector fields on E which are projectable on M form an (infinite-dimensional) Lie algebra.*

Proof The projectable vector fields form a vector space; indeed if $X \in \mathcal{X}_{\overline{X}}(\pi)$, $Y \in \mathcal{X}_{\overline{Y}}(\pi)$ and $\lambda, \mu \in \mathbf{R}$ then $\lambda X + \mu Y \in \mathcal{X}_{\lambda\overline{X}+\mu\overline{Y}}(\pi)$. If X, $\overline{X}$ are π-related and Y, $\overline{Y}$ are π-related then so are $[X, Y]$ and $[\overline{X}, \overline{Y}]$. ∎

Lemma 3.2.13 *If X is projectable and Y is vertical then $[X, Y]$ is vertical.*

Proof The Lie bracket $[X, Y]$ is π-related to $[\overline{X}, 0]$ by the same argument as above, and so projects to zero. ∎

Lemma 3.2.14 *If X is projectable then $\mathcal{L}_X \circ \pi^* = \pi^* \circ \mathcal{L}_{\overline{X}}$.*

Proof The proof of this is just a sequence of basic manipulations, using the definition of $\overline{X}$. If $f \in C^\infty(M)$, $a \in E$ then

$$\begin{aligned}
\mathcal{L}_X(\pi^*(f))(a) &= X_a(\pi^*(f)) \\
&= (\pi_*(X_a))(f) \\
&= \overline{X}_{\pi(a)}(f) \\
&= (\mathcal{L}_{\overline{X}} f)(\pi(a)) \\
&= \pi^*(\mathcal{L}_{\overline{X}}(f))(a).
\end{aligned}$$

∎

A rather more substantial property of projectable vector fields is that their flows define bundle morphisms, and that this property actually characterises those vector fields on E which are projectable.

Proposition 3.2.15 *If $X \in \mathcal{X}(E)$ is a complete vector field with flow ψ then X is projectable to $\overline{X}$ if, and only if, for each $t \in \mathbf{R}$ the diffeomorphism ψ_t defines a bundle isomorphism $(\psi_t, \overline{\psi}_t)$ from π to itself, where $\overline{\psi}$ is the flow of $\overline{X}$.*

Proof Suppose first that each ψ_t gives rise to a bundle isomorphism $(\psi_t, \overline{\psi}_t)$; the proof that X is projectable then just uses the definitions. For each $a \in E$,

$$\begin{aligned} \pi_*(X_a) &= \pi_*[t \longmapsto \psi_t(a)] \\ &= [t \longmapsto \pi(\psi_t(a))] \\ &= [t \longmapsto \overline{\psi}_t(\pi(a))] \end{aligned}$$

so that the tangent vector $\pi_*(X_a)$ depends only on the image $\pi(a) \in M$ rather than $a \in E$, and hence X defines a bundle morphism from π to π_*. The projection of the vector field X to a vector field $\overline{X} : M \longrightarrow TM$ then satisfies $\overline{X}_{\pi(a)} = [t \longmapsto \overline{\psi}_t(\pi(a))]$, so that $\overline{\psi}$ is the flow of $\overline{X}$.

The proof of the converse assertion relies on the uniqueness of integral curves. Suppose that X is projectable to $\overline{X}$. Given $a \in E$, the integral curve of X through a is $t \longmapsto \psi_t(a)$ and so the integral curve of $\overline{X}$ through $\pi(a)$ is $t \longmapsto \pi(\psi_t(a))$; consequently

$$\pi(\psi_t(a)) = \overline{\psi}_t(\pi(a))$$

by uniqueness. It follows that, for each t, $(\psi_t, \overline{\psi}_t)$ is a bundle morphism. It is a bundle isomorphism because both ψ_t and $\overline{\psi}_t$ are diffeomorphisms. ∎

A similar result holds when the vector field X is not complete; however the domain and image of the bundle isomorphism are then only sub-bundles of π.

Finally in this section, we shall see how a vector field on E may act on a section ϕ of π to give a vector field along ϕ (which may be regarded as a section of $\pi \circ \tau_E$). Of course the composite $X \circ \phi$ is always a section of $\pi \circ \tau_E$; we shall, however, be interested in constructing sections of $\pi \circ \tau_E$ which are vertical (that is, which take their values in $V\pi$).

Definition 3.2.16 The action of $\mathcal{X}(E)$ on $\Gamma_{loc}(\pi)$ is the map $(X, \phi) \longmapsto \widetilde{X}(\phi)$ given pointwise by

$$(\widetilde{X}(\phi))_p = [t \longmapsto \psi_t(\phi(\pi(\psi_{-t}(\phi(p)))))] \in T_{\phi(p)}E$$

where ψ_t is the flow of X in a neighbourhood of $\phi(p) \in E$. ∎

We may check that, with this definition, $\pi_*(\tilde{X}(\phi))_p = 0$, so that $(\tilde{X}(\phi))_p$ is indeed vertical.

If the original vector field X is projectable then the slightly lengthy expression in the definition above may be simplified.

Lemma 3.2.17 *If $X \in \mathcal{X}(E)$ projects to $\overline{X} \in \mathcal{X}(M)$ and $\overline{\psi}_t$ is the flow of $\overline{X}$ in a neighbourhood of $p \in M$ then*

$$(\tilde{X}(\phi))_p = [t \longmapsto (\tilde{\psi}_t(\phi))(p)].$$

Proof From the property of $(\psi_t, \overline{\psi}_t)$ as a bundle morphism and the definition of $\tilde{\psi}_t$,

$$\begin{aligned} \psi_t \circ \phi \circ \pi \circ \psi_{-t} \circ \phi &= \psi_t \circ \phi \circ \overline{\psi}_{-t} \circ \pi \circ \phi \\ &= \tilde{\psi}_t(\phi). \end{aligned}$$

■

If the vector field X is itself vertical then there is a further simplification.

Lemma 3.2.18 *If $X \in \mathcal{V}(\pi)$ then $\tilde{X}(\phi) = X \circ \phi$.*

Proof Directly from the definitions:

$$\tilde{X}(\phi)_p = [t \longmapsto (\tilde{\psi}_t(\phi))(p)] = [t \longmapsto \psi_t(\phi(p))] = X_{\phi(p)}.$$

■

Another interpretation of the construction of $\tilde{X}(\phi))$ is that it defines a vector field on the submanifold $\operatorname{im}\phi \subset E$; according to Definition 3.2.16 this should technically be denoted $\tilde{X}(\phi) \circ \pi|_{\operatorname{im}\phi}$. It is evident from Lemma 3.2.18 that if X is vertical then $\tilde{X}(\phi) \circ \pi$ is just the restriction of X to $\operatorname{im}\phi$. Furthermore, we may use Definition 3.2.16 to write

$$(\tilde{X}(\phi))_p = X_{\phi(p)} - \phi_*(\pi_*(X_{\phi(p)}));$$

in coordinates, if

$$X = X^i \frac{\partial}{\partial x^i} + X^\alpha \frac{\partial}{\partial u^\alpha}$$

then

$$\tilde{X}(\phi) = \left(X^\alpha - X^i \frac{\partial \phi^\alpha}{\partial x^i} \right) \frac{\partial}{\partial u^\alpha}.$$

One might therefore ask whether it would be possible to obtain a vertical vector field from an arbitrary vector field X on E by mapping each tangent vector $X_a \in T_a E$ to $(\tilde{X}(\phi))_{\pi(a)}$, where ϕ is a local section satisfying $\phi(p) =$

a. The trouble with this idea is that such a mapping of tangent vectors involves ϕ_* and therefore depends, not just on the value of ϕ at $\pi(a)$, but also on its first derivatives at that point. In fact, this is the same difficulty as we found when considering the question of a complement to the vector bundle $(V\pi, \tau_E|_{V\pi}, E)$ in (TE, τ_E, E), and its resolution requires the use of a connection on π.

EXERCISES

3.2.1 If $X \in \mathcal{X}(E)$ and $\phi \in \Gamma_{loc}(\pi)$, confirm by an argument using coordinates that the tangent vector $(\tilde{X}(\phi))_p$ specified in Definition 3.2.16 does indeed satisfy the condition $\pi_*(\tilde{X}(\phi))_p = 0$.

3.2.2 For an arbitrary manifold M, define a map $\tilde{T} : C^\infty(M) \longrightarrow C^\infty(TM)$ by

$$(\tilde{T}(f))(\xi) = \xi(f) \in \mathbf{R}$$

where $f \in C^\infty(M)$ and $\xi \in TM$. Show that this map is a derivation, and that the corresponding vector field along the tangent bundle projection τ_M may be represented in coordinates by

$$T = \dot{q}^i \frac{\partial}{\partial q^i}$$

using coordinates $(q^i, \dot{q}^i)$ on TM. (This vector field along τ_M is called the *total time derivative* on M.)

3.3 Differential Forms

In the same way as for vector fields, there are certain differential forms on E which are distinguished by the bundle projection π.

Definition 3.3.1 A section σ of the bundle $(\pi^*(T^*M), \pi^*(\tau_M^*), E)$ is called a *1-form on E horizontal over M*. The set of all such 1-forms will be denoted $\bigwedge_0^1\pi$. ∎

Another name for a horizontal 1-form is a *semi-basic* 1-form. The reason for the notation $\bigwedge_0^1\pi$ will become evident when we consider k-forms which are horizontal (or partly horizontal) over M.

It is clear that $\bigwedge_0^1\pi$ is a vector space, and that it is the module generated over $C^\infty(E)$ by $\{\pi^*(\sigma) : \sigma \in \bigwedge^1 M\}$; the following lemma shows that it is the annihilator of $\mathcal{V}(\pi)$ under the operation of contraction.

Lemma 3.3.2 *If $\sigma \in \bigwedge^1 E$ then $\sigma \in \bigwedge^1_0 \pi$ if, and only if, for every vertical vector field $X \in \mathcal{V}(\pi)$,*

$$X \lrcorner \sigma = 0.$$

Proof The structure of this proof is similar to that of the proof in basic differential geometry that the module dual to $\mathcal{X}(E)$ is $\bigwedge^1 E$. If $\sigma \in \bigwedge^1_0 \pi$ then, for each $a \in E$, $\sigma_a = \pi^*(\eta)$ for some $\eta \in T^*_{\pi(a)}M$. Then if $X \in \mathcal{V}(\pi)$,

$$\begin{aligned} (X \lrcorner \sigma)_a &= \sigma_a(X_a) \\ &= \pi^*(\eta)(X_a) \\ &= \eta(\pi_*(X_a)) \\ &= 0. \end{aligned}$$

Conversely, suppose $\sigma \in \bigwedge^1 E$ and that $X \lrcorner \sigma = 0$ for every $X \in \mathcal{V}(\pi)$. Let $a \in E$. For each $\zeta \in V_a\pi$ there is a vertical vector field $X \in \mathcal{V}(\pi)$ such that $X_a = \zeta$; X may be constructed by, for example, writing ζ in local coordinates, choosing smooth functions whose values at a are those coordinates, and then extending the local vertical vector field so defined to the whole of E by using a bump function. Then

$$\sigma_a(\zeta) = \sigma_a(X_a) = (X \lrcorner \sigma)_a = 0.$$

Define a cotangent vector $\eta \in T^*_{\pi(a)}M$ by, for $\bar{\xi} \in T_{\pi(a)}M$,

$$\eta(\bar{\xi}) = \sigma_a(\xi)$$

where $\xi \in T_a E$ satisfies $\pi_*(\xi) = \bar{\xi}$; if $\pi_*(\xi_1) = \pi_*(\xi_2) = \bar{\xi}$ then $\pi_*(\xi_1 - \xi_2) = 0$, so that $\xi_1 - \xi_2 \in V_a\pi$ and therefore $\sigma_a(\xi_1) = \sigma_a(\xi_2)$. It follows that, for any $\xi \in T_a E$,

$$\sigma_a(\xi) = \eta(\pi_*(\xi)) = \pi^*(\eta)(\xi)$$

so that $\sigma_a = \pi^*(\eta)$ and therefore $\sigma \in \bigwedge^1_0 \pi$. ∎

In local coordinates, an element $\sigma \in \bigwedge^1 E$ may be written

$$\sigma = \sigma_i dx^i + \sigma_\alpha du^\alpha.$$

If $\sigma \in \bigwedge^1_0 \pi$ then

$$\sigma = \sigma_i dx^i$$

so that there are no terms in du^α; however the functions σ^i are elements of $C^\infty(E)$.

A similar definition may be used for r-forms.

Definition 3.3.3 A section of the bundle $(\bigwedge^r \pi^*(T^*M), \bigwedge^r \pi^*(\tau_M^*), E)$ is called an r-form on E *horizontal over* M. The space of horizontal r-forms will be denoted by $\bigwedge_0^r \pi$ and the algebra of all horizontal forms by $\bigwedge_0 \pi$. ∎

Lemma 3.3.4 *If* $\theta \in \bigwedge^r E$ *then* $\theta \in \bigwedge_0^r \pi$ *if, and only if, for every* $X \in \mathcal{V}(\pi)$, $X \lrcorner \theta = 0$.

Proof Similar to the proof of Lemma 3.3.2. ∎

In local coordinates a horizontal r-form is written

$$\theta = \theta_{i_1 \ldots i_r} dx^{i_1} \wedge \ldots \wedge dx^{i_r}$$

where the set of functions $\theta_{i_1 \ldots i_r}$ is completely skew-symmetric, so that again there are no terms involving du^α. Note that this feature of horizontal forms allows the definition of their contraction with vector fields on M and vector fields along π, as well as with vector fields on E.

Definition 3.3.5 If $X \in \mathcal{X}(M)$ and $\sigma \in \bigwedge_0^1 \pi$, define the contraction $X \lrcorner \sigma \in C^\infty(E)$ by, for $a \in E$,

$$(X \lrcorner \sigma)(a) = \eta(X_{\pi(a)})$$

where $\eta \in T^*_{\pi(a)} M$ satisfies $\pi^*(\eta) = \sigma_a$. ∎

Definition 3.3.6 If $X \in \mathcal{X}(\pi)$ and $\sigma \in \bigwedge_0^1 \pi$, define the contraction $X \lrcorner \sigma \in C^\infty(E)$ by, for $a \in E$,

$$(X \lrcorner \sigma)(a) = \eta(X_a)$$

where $\eta \in T^*_{\pi(a)} M$ satisfies $\pi^*(\eta) = \sigma_a$. ∎

The notation $\bigwedge_0^r \pi$is useful because it may be generalised to "partly horizontal" r-forms, where $\bigwedge_s^r \pi$ denotes the space of $(r-s)$-horizontal r-forms.

Definition 3.3.7 A section of the bundle

$$(\bigwedge^s T^*E \wedge \bigwedge^{r-s} \pi^*(T^*M), \bigwedge^s \tau_E^* \wedge \bigwedge^{r-s} \pi^*(\tau_M^*), E),$$

$(1 \leq s \leq r-1)$ is an r-form on E which is called $(r-s)$*-horizontal over* M. The space of all $(r-s)$-horizontal r-forms on E is denoted $\bigwedge_s^r \pi$. ∎

Lemma 3.3.8 *If* $\theta \in \bigwedge^r E$ *then* $\theta \in \bigwedge_s^r \pi$, $(1 \leq s \leq r-1)$ *if, and only if, for every* $X \in \mathcal{V}(\pi)$, $X \lrcorner \theta \in \bigwedge_{s-1}^{r-1} \pi$.

Proof Again similar to the proof of Lemma 3.3.2, but this time using multilinear algebra to demonstrate that $\theta_a \in \bigwedge^s T_a^* E \wedge \bigwedge_s^r \pi^*(T^*_{\pi(a)} M)$. ∎

The specification of partly horizontal r-forms defines a filtration on the space of r-forms on E,

$$\textstyle\bigwedge_0^r \pi \subset \bigwedge_1^r \pi \subset \ldots \subset \bigwedge_{r-1}^r \pi \subset \bigwedge^r E,$$

where if $s < r - \dim M$ then $\bigwedge_s^r \pi = \{0\}$. In local coordinates, a form $\theta \in \bigwedge_s^r \pi$ may be written

$$\theta = \theta_{\alpha_1 \ldots \alpha_k i_{k+1} \ldots i_r} du^{\alpha_1} \wedge \ldots \wedge du^{\alpha_k} \wedge dx^{i_{k+1}} \wedge \ldots \wedge dx^{i_r},$$

$0 \leq k \leq s$, where the set of functions $\theta_{\alpha_1 \ldots \alpha_k i_{k+1} \ldots i_r}$ is skew-symmetric in the α indices and the i indices separately, and so a form in $\bigwedge_s^r \pi$ contains s (or fewer) du^α's in each term of its coordinate expression. Note that, without the additional structure of a connection, there is no distinguished complement of $\bigwedge_s^r \pi$ in $\bigwedge_{s+1}^r \pi$.

Lemma 3.3.9 *The $C^\infty(E)$-module $\bigwedge_0^1 \pi$ is isomorphic to the module dual to $\mathcal{X}(\pi)$.*

Proof The pairing $(X, \sigma) \longmapsto X \lrcorner \sigma$ clearly gives an isomorphism of $\bigwedge_0^1 \pi$ with a submodule of the dual of $\mathcal{X}(\pi)$; the fact that it is the whole of this module follows from an argument similar to that used when showing that $\bigwedge^1 M$ is the dual of $\mathcal{X}(M)$. ∎

We shall also mention briefly the space of vertical 1-forms. These are not in fact 1-forms on any manifold, but may be regarded as cosets (just as vertical cotangent vectors are cosets).

Definition 3.3.10 A section of the vector bundle $(V^*\pi, (\tau_E|_{V\pi})^*, E)$ is called a *vertical 1-form*. The space of all vertical 1-forms is denoted $\mathcal{V}^*(\pi)$. ∎

It follows from Proposition 2.3.8 that $\mathcal{V}^*(\pi)$ is the $C^\infty(E)$-module dual to $\mathcal{V}(\pi)$. In certain circumstances we also have the following realisation of $\mathcal{V}^*(\pi)$.

Proposition 3.3.11 *If M is orientable then each volume form Ω determines an isomorphism between $\bigwedge_1^{m+1} \pi$ and $\mathcal{V}^*(\pi)$.*

Proof From Proposition 3.1.13, Ω determines a vector bundle isomorphism between $\tau_E^* \wedge \bigwedge^m (\tau_E^*|_{\pi^*(T^*M)})$ and $(\tau_E|_{V\pi})^*$. ∎

In the final part of this section we shall consider vector-valued forms defined in the context of bundles. We have defined three different bundles of tangent vectors over E, namely $V\pi$, TE and $\pi^*(TM)$, and similarly three different bundles of cotangent vectors. We may therefore construct nine different types of vector-valued 1-form, and a correspondingly larger number of different types of vector-valued r-form. We shall, however, restrict attention to just two kinds of vector-valued form, depending roughly on whether the vector field part or the differential form part is projected along π. These will be important in our later consideration of connections and derivations respectively.

Definition 3.3.12 A *vector-valued r-form on E horizontal over M* is a section of the tensor product bundle $(\bigwedge^r \pi^*(T^*M) \otimes TE, \bigwedge^r \pi^*(\tau_M^*) \otimes \tau_E, E)$. ∎

One may check that if R is a vector-valued r-form on E then R is horizontal over M if, and only if, for every $X \in \mathcal{V}(\pi)$, $X \lrcorner R = 0$. The module of all horizontal vector-valued r-forms is then $\bigwedge_0^r \pi \otimes \mathcal{X}(E)$. Just as with ordinary horizontal forms, we may define an operation of contraction with vector fields on M and vector fields along π, as well as with vector fields on E.

Definition 3.3.13 If $X \in \mathcal{X}(M)$ and $R \in \bigwedge_0^1 \pi \otimes \mathcal{X}(E)$, define the contraction $X \lrcorner R \in \mathcal{X}(E)$ by, for $a \in E$,

$$(X \lrcorner R)_a = R_a(X_{\pi(a)})$$

where $R_a \in (\pi^*(T^*M))_a \otimes T_a E$ is regarded as a linear map $T^*_{\pi(a)} M \longrightarrow T_a E$. ∎

Definition 3.3.14 If $X \in \mathcal{X}(\pi)$ and $R \in \bigwedge_0^1 \pi \otimes \mathcal{X}(E)$, define the contraction $X \lrcorner R \in \mathcal{X}(E)$ by, for $a \in E$,

$$(X \lrcorner R)_a = R_a(X_a).$$

∎

A similar definition may be used if R is a horizontal vector-valued r-form rather than a 1-form.

On the other hand, if the vector field part of the vector-valued form is projected along π then the result is *not* a vector-valued form on the manifold E.

Definition 3.3.15 A *vector-valued r-form along π* is a section of the tensor product bundle $(\bigwedge^r T^*E \otimes \pi^*(TM), \bigwedge^r \tau_E^* \otimes \pi^*(\tau_M), E)$. ∎

The module of all vector-valued r-forms along π is $\bigwedge^r E \otimes \mathcal{X}(\pi)$, so that such a form may be identified either with a $C^\infty(E)$-linear map from $\bigwedge_0^1 \pi$ to $\bigwedge^r E$, or alternatively as an alternating $C^\infty(E)$-multilinear map from $\mathcal{X}(E) \times \ldots \times \mathcal{X}(E)$ to $\mathcal{X}(\pi)$. These forms will be used in the construction of derivations on the bundle π.

Example 3.3.16 There is a natural vector-valued 1-form along π corresponding to the inclusion map $\iota : \bigwedge_0^1 \pi \longrightarrow \bigwedge^1 E$ (or, equivalently, to its transpose $\pi_* : \mathcal{X}(E) \longrightarrow \mathcal{X}(\pi)$) which will be denoted by I. In coordinates $I = dx^i \otimes \partial/\partial x^i$, or to be more precise

$$I_a = (\pi^*(dx^i))_a \otimes \left.\frac{\partial}{\partial x^i}\right|_{\pi(a)}.$$

■

EXERCISES

3.3.1 If M is any manifold, show that the map which assigns to any $\eta \in T^*M$ the cotangent vector $\tau^*_{T^*M}(\eta) \in T^*_\eta T^*M$ defines a horizontal 1-form $\theta \in \bigwedge_0^1 \tau_M^*$; if (q^i, p_i) are coordinates on T^*M, show that this 1-form has coordinate representation

$$\theta = p_i dq^i.$$

(This is known as the *canonical 1-form* on T^*M, and its differential $d\theta \in \bigwedge_1^2 \tau_M^*$ is called the *canonical symplectic form* on T^*M.)

3.3.2 If $L \in C^\infty(M)$, show that the fibre derivative $\mathcal{F}L : TM \longrightarrow T^*M$ specified in Exercise 2.3.3 may be used to define a horizontal 1-form $\theta_L = (\mathcal{F}L)^*(\theta) \in \bigwedge_0^1 \tau_M$ with coordinate representation

$$\theta_L = \frac{\partial L}{\partial \dot{q}^i} dq^i.$$

Show further that $\mathcal{F}L$ itself may be regarded as a section of the pull-back bundle

$$((\tau_M)^*(T^*M), (\tau_M)^*(\tau_M^*), TM),$$

where we distinguish between the cotangent bundle projection τ_M^* and the action $(\tau_M)^*$ of pull-back by the tangent bundle projection τ_M, and that with this interpretation we may actually identify $\mathcal{F}L$ and θ_L. (In mechanics, the mapping $\theta \longrightarrow \theta_L$ is called the *Legendre transformation*, and the 1-form θ_L is called the *Cartan 1-form* of L.)

3.4 Derivations

In the context of differential forms, a derivation is an operation D which is $\mathbf{R}$-linear, maps s-forms to $(r+s)$-forms for some fixed integer r, and satisfies the following version of Leibniz' rule,

$$D(\alpha \wedge \beta) = D\alpha \wedge \beta \pm \alpha \wedge D\beta,$$

where the choice of sign depends on circumstances. The integer r is called the *degree* of the derivation: for example, d, $\mathcal{L}_X$ and i_X are derivations of degree 1, 0 and -1 respectively, where $i_X\theta$ is an alternative notation for the contraction $X \lrcorner\, \theta$. It is worth mentioning here that the choice of numerical factor in the definition of the wedge product affects the statement of Leibniz' rule. If we had adopted the alternative convention then (for example) if α, β were 1-forms and X, Y were vector fields, we would have

$$(X,Y) \lrcorner\, (\alpha \wedge \beta) = \tfrac{1}{2}((X \lrcorner\, \alpha)(Y \lrcorner\, \beta) - (Y \lrcorner\, \alpha)(X \lrcorner\, \beta))$$

yielding

$$i_X(\alpha \wedge \beta) = \tfrac{1}{2}((i_X\alpha)\beta - (i_X\beta)\alpha)$$

so that Leibniz' rule would require a numerical factor which would depend on the degree of the forms involved. The convention we have adopted has the merit of giving simpler formulæ in many of our applications.

In the context of bundles, we shall be interested in derivations mapping differential forms on M to differential forms on E, and we shall call them *derivations along* π.

Definition 3.4.1 A *derivation along* π *of degree* r is an $\mathbf{R}$-linear map $D : \bigwedge M \longrightarrow \bigwedge E$ satisfying the properties

1. if $\theta \in \bigwedge^s M$ then $D\theta \in \bigwedge^{r+s} E$;
2. if $\theta_1 \in \bigwedge^{s_1} M$ and $\theta_2 \in \bigwedge^{s_2} M$ then
$D(\theta_1 \wedge \theta_2) = D\theta_1 \wedge \pi^*(\theta_2) + (-1)^{rs_1}\pi^*(\theta_1) \wedge D\theta_2.$

■

We may distinguish two particular types of derivation, which we shall call derivations of type i_* and of type d_*. The model for derivations of type i_* is contraction with a vector field, and for those of type d_* is the Lie derivative. The importance of these two special types of derivation is that every derivation may be decomposed into derivations of these two types.

Definition 3.4.2 A derivation D along π is of type i_* if, for every $f \in C^\infty(M) \cong \bigwedge^0 M$, $Df = 0$. ■

Definition 3.4.3 A derivation D along π of degree r is of type d_* if

$$D \circ d = (-1)^r d \circ D$$

where d on the left-hand side of this equation is exterior derivative on M, and on the right-hand side is exterior derivative on E. ■

To construct derivations of type i_*, we shall generalise the contraction operation between vector fields along π and 1-forms on M specified in Definition 3.2.6. This operation may be extended to s-forms to give a derivation of degree -1; by using a vector-valued r-form instead of a vector field we may obtain a derivation of degree $r-1$.

Proposition 3.4.4 *If R is a vector-valued r-form along π then R determines a derivation along π of type i_* and degree $(r-1)$, denoted i_R, by the following rules: if $f \in C^\infty(M)$ then $i_R f = 0$; if $\theta \in \bigwedge^s M$ $(s \geq 1)$ and $X_1, \ldots, X_{r+s+1} \in \mathcal{X}(E)$ then*

$$(X_1, \ldots, X_{r+s-1}) \lrcorner\, i_R\theta =$$

$$\sum_{\sigma \in S_{r,s-1}} \varepsilon_\sigma((X_{\sigma(1)}, \ldots, X_{\sigma(r)}) \lrcorner\, R, \pi_* \circ X_{\sigma(r+1)}, \ldots, \pi_* \circ X_{\sigma(r+s-1)}) \lrcorner\, \theta,$$

where $S_{r,s-1}$ is the subgroup of the permutation group S_{r+s-1} containing those permutations σ which satisfy $\sigma(1) < \ldots < \sigma(r)$ and $\sigma(r+1) < \ldots < \sigma(r+s-1)$. Furthermore, every derivation along π of type i_* and degree $(r-1)$ is determined in this way by a unique vector-valued r-form along π.

Proof The map i_R is clearly $\mathbf{R}$-linear and of degree $(r-1)$; a (not very illuminating) combinatorial argument shows that it satisfies Leibniz' rule. Since $i_R f = 0$ it is therefore a derivation of type i_*.

Conversely, suppose D is a derivation along π of type i_* and degree $(r-1)$. Define the mapping $\check{D}$ from 1-forms on M to r-forms on E by $\check{D} = D|_{\bigwedge^1 M}$. Then if $\omega \in \bigwedge^1 M$, $f \in C^\infty(M)$,

$$\check{D}(f\omega) = (Df)\pi^*\omega + (\pi^*(f))\check{D}\omega = (\pi^*(f))\check{D}\omega$$

since $Df = 0$; consequently $\check{D}$ is $C^\infty(M)$-linear.

Now the map $\check{D}$ defines a vector-valued r-form along π, for given $a \in E$ let $D_a : T^*_{\pi(a)}M \longrightarrow \bigwedge^r T^*_a E$ by the rule

$$D_a(\omega_{\pi(a)}) = (\check{D}\omega)_a;$$

this does not depend on the particular 1-form ω used to define the cotangent vector $\omega_{\pi(a)}$. The map D_a then defines an element of the tensor product space $\bigwedge^r T^*_a E \otimes T_{\pi(a)}M$, and so the correspondence $\overline{D} : a \longmapsto D_a$ yields a section of the bundle $(\bigwedge^r T^*E \otimes \pi^*(TM), \bigwedge^r \tau^*_E \otimes \pi^*(\tau_M), E)$.

The final part of the proof relies on the fact that any derivation of differential forms is characterised by its action on functions and 1-forms, because its action on s-forms may be deduced from Leibniz' rule. Since $Df = i_{\overline{D}}f = 0$ and $D\omega = i_{\overline{D}}\omega$ for $\omega \in \bigwedge^1 M$ by construction, $D = i_{\overline{D}}$; if $D = i_R$ for some other vector-valued r-form R then clearly $R = \overline{D}$. ∎

Example 3.4.5 Let $E = M$ and $\pi = id_M$. Then if X is a vector field, i_X is just contraction with X, so the notation is consistent. If I is the identity vector-valued 1-form then $i_I\theta = s\theta$ for $\theta \in \bigwedge^s M$. ■

Example 3.4.6 For general bundles (E, π, M), if X is a vector field along π then i_X is contraction with X as specified in Definition 3.3.6. If I is the vector-valued 1-form along π defined by the inclusion $\bigwedge_0^1 \pi \longrightarrow \bigwedge^1 E$ then again $i_I\theta = s\pi^*(\theta)$ for $\theta \in \bigwedge^s M$. ■

Proposition 3.4.7 *If i_R is a derivation along π of type i_* and degree $(r-1)$, then i_R determines a derivation along π of type d_* and degree r, denoted d_R, by the rule*

$$d_R = i_R \circ d + (-1)^r d \circ i_R.$$

Furthermore, every derivation along π of type d_* and degree r is determined in this way by a unique derivation of type i_*.

Proof The map d_R is certainly $\mathbf{R}$-linear and of degree r. In addition, a straightforward calculation shows that d_R satisfies Leibniz' rule, and so is a derivation along π. Clearly $d_R \circ d = (-1)^r d \circ d_R$.

Conversely, suppose D is a derivation along π of type d_* and degree r. For each $a \in E$, define $D_a : T^*_{\pi(a)}M \longrightarrow \bigwedge^r T^*_a E$ by $D_a(df_{\pi(a)}) = (Df)_a$ where $f \in C^\infty(M)$; once again this does not depend on the particular choice of f used to define the cotangent vector $df_{\pi(a)}$. Linearity of D_a follows from $\mathbf{R}$-linearity of D, so as in Proposition 3.4.4 we may obtain a vector-valued r-form along π, denoted $\overline{D}$, satisfying $Df = i_{\overline{D}}df$; since $i_{\overline{D}}f = 0$ this gives $Df = d_{\overline{D}}f$. The commutation relation with d then shows that any derivation of type d_* is completely determined by its action on functions, and hence $D = d_{\overline{D}}$. Finally, suppose i_R is some other derivation of type i_* satisfying $D = i_R \circ d + (-1)^r d \circ i_R$. Then for any $f \in C^\infty(M)$, $i_R df = i_{\overline{D}}df$ so that $R \lrcorner df = \overline{D} \lrcorner df$ and hence for any $a \in E$, $R_a(df_{\pi(a)}) = \overline{D}_a(df_{\pi(a)})$; as a and f are arbitrary, $R = \overline{D}$. ■

Example 3.4.8 Let $E = M$ and $\pi = id_M$. Then if X is a vector field, d_X is just the Lie derivative by X. If I is the identity vector-valued 1-form then $d_I\theta = d\theta$, the exterior derivative of θ. ■

Example 3.4.9 For general bundles (E, π, M), if X is a vector field along π then d_X defines a Lie derivative action of X; for functions, this is just the action described in Proposition 3.2.7. ■

Proposition 3.4.10 *Every derivation along π is the sum of two derivations, one of type i_* and one of type d_*.*

Proof If D is a derivation along π then define a derivation of type d_*, denoted d_D, by $d_D f = Df$ for $f \in C^\infty(M)$. Then $D - d_D$ is a derivation of type i_*. ■

The relationship between vector-valued forms and derivations may be used, in certain circumstances, to define a bracket operation on vector-valued forms. On the bundle (M, id_M, M) this is just a generalisation of the Lie bracket of vector fields; in the more general case it will allow us to define the bracket of vector fields along maps. The background to this construction will, however, be rather more complicated than a single bundle. So suppose there are two bundles, (E_1, π_1, M_1) and (E_2, π_2, M_2) and a bundle morphism (ρ_1, ρ_2) from π_1 to π_2, such that (E_1, ρ_1, E_2) and (M_1, ρ_2, M_2) are themselves both bundles:

$$\begin{array}{ccc} E_1 & \xrightarrow{\rho_1} & E_2 \\ \Big\downarrow \pi_1 & & \Big\downarrow \pi_2 \\ M_1 & \xrightarrow[\rho_2]{} & M_2 \end{array}$$

Definition 3.4.11 If R_1, R_2 are vector-valued r-forms along π_1, π_2 respectively, then R_1, R_2 are said to be *ρ-related* if, for each $a \in E_1$ and every $\xi_1, \ldots, \xi_r \in T_a E_1$,

$$\rho_{2*}((R_1)_a(\xi_1, \ldots, \xi_r)) = (R_2)_{\rho_1(a)}(\rho_{1*}(\xi_1), \ldots, \rho_{1*}(\xi_r)).$$

■

An equivalent statement of this definition would be that R_1, R_2 are ρ-related if, for every $\sigma \in \bigwedge^1_0 \pi_2$, $\rho_1^*(R_2 \lrcorner\, \sigma) = R_1 \lrcorner\, (\rho_1^*(\sigma))$. Note that R_2 (if it exists) is completely determined by R_1.

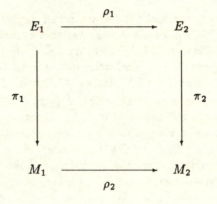

To see what this condition looks like in coordinates, suppose that the following coordinate systems are used:

$$\begin{aligned} x^i &\quad \text{on} \quad M_2 \\ (x^i, u^\alpha) &\quad \text{on} \quad E_2 \\ (x^i, y^a) &\quad \text{on} \quad M_1 \\ (x^i, y^a, u^\alpha) &\quad \text{on} \quad E_1 \end{aligned}$$

and for simplicity suppose that R_1, R_2 are vector-valued 1-forms. Then R_2 has coordinate representation

$$R_2 = (R^i_j dx^j + R^i_\alpha du^\alpha) \otimes \frac{\partial}{\partial x^i}$$

where $R^i_j, R^i_\alpha \in C^\infty(E_2)$. If R_1 is ρ-related to R_2 then we must have

$$\begin{aligned} R_1 &= (R^i_j dx^j + R^i_\alpha du^\alpha) \otimes \frac{\partial}{\partial x^i} \\ &\quad + (R^a_j dx^j + R^a_\alpha du^\alpha + R^a_b dy^b) \otimes \frac{\partial}{\partial y^a} \end{aligned}$$

so that the coefficients of $dy^b \otimes \partial/\partial x^i$ are zero, and the coefficients of $dx^j \otimes \partial/\partial x^i$ and $du^\alpha \otimes \partial/\partial x^i$ are pulled back from E_2 to E_1.

Definition 3.4.12 If R_1, R_2 are ρ-related vector-valued r-forms along π_1, π_2 respectively, and S_1, S_2 are π-related vector-valued s-forms along ρ_1, ρ_2 respectively, then the bracket $[R_1, S_1]$ is the vector-valued $(r+s)$-form along $\pi_2 \circ \rho_1 = \rho_2 \circ \pi_1$ defined by

$$d_{[R_1,S_1]} = d_{R_1} \circ d_{S_2} - (-1)^{rs} d_{S_1} \circ d_{R_2}.$$

∎

It is easy to check that $d_{[R_1,S_1]}$ as specified above is indeed a derivation along $\pi_2 \circ \rho_1$ of type d_* and degree $(r+s)$, so the definition makes sense by Propositions 3.4.4 and 3.4.7. This bracket is sometimes called the Frölicher-Nijenhuis bracket.

Example 3.4.13 Suppose π_i, ρ_i are all identity maps on a single manifold M. If R, S are both vector-valued 0-forms (that is, vector fields) then $[R, S]$ is just the ordinary Lie bracket. More generally, if just R is a vector field then $[R, S]$ is the Lie derivative $\mathcal{L}_R S$. ∎

For the final part of this section we shall restrict attention to vector-valued forms on a single manifold M.

Lemma 3.4.14 *The space $\bigwedge M \otimes \mathcal{X}(M)$ of all vector-valued forms on M is a graded Lie algebra under the Frölicher-Nijenhuis bracket.*

Proof The bracket operation is clearly $\mathbf{R}$-linear; it satisfies

$$[S,R] = (-1)^{rs+1}[R,S]$$

by definition. A simple calculation using this definition verifies the following version of Jacobi's identity with an appropriate combination of minus signs:

$$(-1)^{rt}[R,[S,T]] + (-1)^{rs}[S,[T,R]] + (-1)^{st}[T,[R,S]] = 0.$$

■

A particularly important example of this construction arises when both R and S are vector-valued 1-forms, and then the vector-valued 2-form $[R,S]$ is called the *Nijenhuis tensor* of R and S.

Proposition 3.4.15 *If $X, Y \in \mathcal{X}(M)$ then*

$$\begin{aligned} &(X,Y) \lrcorner [R,S] \\ = \; &[X,Y] \lrcorner R \lrcorner S + [X \lrcorner R, Y \lrcorner S] - [X \lrcorner R, Y] \lrcorner S - [X, Y \lrcorner R] \lrcorner S \\ &+[X,Y] \lrcorner S \lrcorner R + [X \lrcorner S, Y \lrcorner R] - [X \lrcorner S, Y] \lrcorner R - [X, Y \lrcorner S] \lrcorner R. \end{aligned}$$

Proof Each side of the above equation is a vector field. We shall demonstrate equality when each side is contracted with an arbitrary exact 1-form, from which the result will follow: the proof is just a long calculation. Now if $f \in C^\infty(M)$,

$$\begin{aligned} ((X,Y) \lrcorner [R,S]) \lrcorner df &= (X,Y) \lrcorner i_{[R,S]} df \\ &= (X,Y) \lrcorner d_{[R,S]} f \\ &= (X,Y) \lrcorner d_R d_S f + (X,Y) \lrcorner d_S d_R f, \end{aligned}$$

and we may expand the first term in detail. By definition,

$$(X,Y) \lrcorner d_R d_S f = (X,Y) \lrcorner (i_R \circ d - d \circ i_R)(i_S df)$$

and

$$\begin{aligned} (X,Y) \lrcorner i_R d i_S df &= (X \lrcorner R, Y) \lrcorner d(S \lrcorner df) + (X, Y \lrcorner R) \lrcorner d(S \lrcorner df) \\ &= d_{X \lrcorner R}(Y \lrcorner S \lrcorner df) - d_Y(X \lrcorner R \lrcorner S \lrcorner df) \\ &\quad -[X \lrcorner R, Y] \lrcorner S \lrcorner df + d_X(Y \lrcorner R \lrcorner S \lrcorner df) \\ &\quad -d_{Y \lrcorner R}(X \lrcorner S \lrcorner df) - [X, Y \lrcorner R] \lrcorner S \lrcorner df \end{aligned}$$

whereas

$$\begin{aligned} -(X,Y) \lrcorner\, d i_R i_S df &= -d_X(Y \lrcorner R \lrcorner S \lrcorner df) + d_Y(X \lrcorner R \lrcorner S \lrcorner df) \\ &\quad +[X,Y] \lrcorner R \lrcorner S \lrcorner df \end{aligned}$$

so that

$$\begin{aligned} (X,Y) \lrcorner\, d_R d_S f &= [X,Y] \lrcorner R \lrcorner S \lrcorner df \\ &\quad -[X \lrcorner R, Y] \lrcorner S \lrcorner df - [X, Y \lrcorner R] \lrcorner S \lrcorner df \\ &\quad +d_{X \lrcorner R}(Y \lrcorner S \lrcorner df) - d_{Y \lrcorner R}(X \lrcorner S \lrcorner df). \end{aligned}$$

Similarly,

$$\begin{aligned} (X,Y) \lrcorner\, d_S d_R f &= [X,Y] \lrcorner S \lrcorner R \lrcorner df \\ &\quad -[X \lrcorner S, Y] \lrcorner R \lrcorner df - [X, Y \lrcorner S] \lrcorner R \lrcorner df \\ &\quad +d_{X \lrcorner S}(Y \lrcorner R \lrcorner df) - d_{Y \lrcorner S}(X \lrcorner R \lrcorner df) \end{aligned}$$

and noting that (for example)

$$\begin{aligned} d_{X \lrcorner R}(Y \lrcorner S \lrcorner df) - d_{Y \lrcorner S}(X \lrcorner R \lrcorner df) &= d_{X \lrcorner R} d_{Y \lrcorner S} f - d_{Y \lrcorner S} d_{X \lrcorner R} f \\ &= d_{[X \lrcorner R, Y \lrcorner S]} f \\ &= [X \lrcorner R, Y \lrcorner S] \lrcorner df \end{aligned}$$

the required equality is obtained. ∎

The Nijenhuis tensor $[R, R]$ is also denoted N_R, and contains information about the eigenspaces of R. Indeed, at each point $p \in M$ the vector-valued 1-form R gives rise to an endomorphism of the tangent space T_pM which may have eigenvalues and eigenspaces: the "signature" of R at p will be denoted by a multi-index $I_p \in \mathbf{N}^m$ (multi-index notation will be described in more detail in Chapter 6). Here, $I_p(j)$ is the number of distinct eigenspaces of dimension j (so that $0 \leq \sum_{j=1}^m j I_p(j) \leq m$), and we shall require the map $I : M \longrightarrow \mathbf{N}^m$ given by $p \longmapsto I_p$ to be constant. The reason for this condition is that, if it holds, one may define $|I|$ unique eigenfunctions λ which, at each p, yield the $|I_p|$ distinct eigenvalues $\lambda(p)$; the multiplicity of each eigenfunction will be constant. One may correspondingly define $|I|$ unique distributions Δ which, at each p, yield the $|I_p|$ distinct eigenspaces Δ_p.

Proposition 3.4.16 *Suppose the vector-valued 1-form R has constant signature I where $\sum_{j=1}^m j I(j) = m$ (so that R is diagonalisable) and that $N_R = 0$. Then each eigendistribution of R is involutive.*

Proof Let $X, Y \in \mathcal{X}(M)$ belong to the distribution Δ_λ corresponding to the eigenfunction λ. A calculation using the formula from Proposition 3.4.15 shows that

$$(R^2 - 2\lambda R + \lambda^2)[X,Y] = \tfrac{1}{2}N_R(X,Y) = 0.$$

Since R is diagonalisable, so is $R - \lambda I$, and hence

$$\ker(R - \lambda I)^2 = \ker(R - \lambda I) = \Delta_\lambda.$$

Consequently $[X, Y]$ also belongs to Δ_λ. ■

EXERCISES

3.4.1 If R and S are vector-valued 1-forms on M with coordinate representation

$$R = R^i_j dx^j \otimes \frac{\partial}{\partial x^i} \quad \text{and} \quad S = S^k_l dx^l \otimes \frac{\partial}{\partial x^k},$$

show that the Nijenhuis tensor $[R, S]$ has coordinate representation

$$\begin{aligned} [R,S] \;=\; \tfrac{1}{2}\Bigg(& R^l_i \frac{\partial S^k_j}{\partial x^l} - R^l_j \frac{\partial S^k_i}{\partial x^l} + R^k_l \left(\frac{\partial S^l_i}{\partial x^j} - \frac{\partial S^l_j}{\partial x^i} \right) \\ & + S^l_i \frac{\partial R^k_j}{\partial x^l} - S^l_j \frac{\partial R^k_i}{\partial x^l} + S^k_l \left(\frac{\partial R^l_i}{\partial x^j} - \frac{\partial R^l_j}{\partial x^i} \right)\Bigg) (dx^i \wedge dx^j) \otimes \frac{\partial}{\partial x^k}. \end{aligned}$$

3.4.2 Show that the fibre derivative map $\mathcal{F} : C^\infty(TM) \longrightarrow \bigwedge^1_0 \tau_M \subset \bigwedge^1 TM$ described in Exercise 3.3.2 defines a derivation of type d_*, and that the corresponding vector-valued 1-form S has coordinate representation

$$S = dq^i \otimes \frac{\partial}{\partial \dot{q}^i}$$

where $(q^i, \dot{q}^i)$ are coordinates on TM. (The tensor S is called the *almost tangent structure* on TM, and it plays an important part in the geometrical study of the calculus of variations.)

3.4.3 Let G be a Lie group, and let $T \in \mathcal{X}(\tau_G)$ be the "total time derivative" vector field introduced in Exercise 3.2.2, so that d_T is a derivation of type d_* along τ_G. Let $g \in G$ and $\xi \in T_g G$; by associating to every cotangent vector $\eta \in T^*_g G$ the corresponding left-invariant differential form $\overline{\eta} \in \bigwedge^1 G$, use d_T to construct a map $l_\xi : T^*_g G \longrightarrow T^*_\xi TG$. Show that every $\omega \in T^*_\xi TG$ may be written uniquely in the form $l_\xi(\eta_1) + (\tau_G)^*(\eta_2)$ where $\eta_1, \eta_2 \in T^*_g G$. Deduce, using the left translation $L_g : G \longrightarrow G$, that every cotangent space $T^*_\xi TG$ is isomorphic to a direct sum $\mathbf{g}^* \oplus \mathbf{g}^*$ where $\mathbf{g}^*$ is the dual of the Lie algebra $\mathbf{g}$.

3.5 Connections

As we have mentioned on several occasions earlier in this chapter, the vertical bundle $(V\pi, \tau_E|_{V\pi}, E)$ does not in general have a distinguished complement of "horizontal vectors" in the tangent bundle τ_E. In this section we shall see one way of specifying a horizontal bundle, and some of the consequences of making such a specification.

Definition 3.5.1 A *connection on the bundle* π is a vector-valued 1-form $\Gamma \in \bigwedge_0^1 \pi \otimes \mathcal{X}(E)$ which satisfies the condition that $\Gamma \lrcorner \sigma = \sigma$ for every $\sigma \in \bigwedge_0^1 \pi$. ■

It follows immediately from this definition that $\Gamma \lrcorner \Gamma \lrcorner \sigma = \Gamma \lrcorner \sigma$ for any $\sigma \in \bigwedge^1 E$, so that $\Gamma \lrcorner \Gamma = \Gamma$ and hence that each Γ_a may be regarded as a projection operator on $T_a E$. In coordinates, a connection may be written as

$$\Gamma = dx^i \otimes \left(\frac{\partial}{\partial x^i} + \Gamma_i^\alpha \frac{\partial}{\partial u^\alpha} \right).$$

Definition 3.5.2 The *horizontal bundle* defined by the connection Γ is the vector sub-bundle $(H_\Gamma \pi, \tau_E|_{H_\Gamma \pi}, E)$ of τ_E defined by

$$(H_\Gamma \pi)_a = \{\Gamma_a(\xi) : \xi \in T_a E\}$$

where $\Gamma_a \in (\pi^*(T^*M))_a \otimes T_a E \subset T_a^* E \otimes T_a E$ is regarded as a linear map $T_a E \longrightarrow T_a E$. ■

The fact that the horizontal bundle is indeed a vector sub-bundle of τ_E may be seen by letting X_μ be a family of vector fields which span τ_M. Then $X_\mu \lrcorner \Gamma$ is a family of vector fields on E which span $\tau_E|_{H_\Gamma \pi}$ so that, by Proposition 2.1.18, $\tau_E|_{H_\Gamma \pi}$ becomes a vector sub-bundle of τ_E.

Lemma 3.5.3 *Given a connection* Γ *on* π, *the tangent bundle* τ_E *may be written as a direct sum*

$$(V\pi \oplus H_\Gamma \pi, \tau_E, E).$$

Proof If (x^i, u^α) are coordinates around $a \in E$ then the fibre $(H_\Gamma \pi)_a$ has a basis

$$\Gamma_a \left(\left. \frac{\partial}{\partial x^i} \right|_{\pi(a)} \right) = \left. \frac{\partial}{\partial x^i} \right|_a + \Gamma_i^\alpha(a) \left. \frac{\partial}{\partial u^\alpha} \right|_a.$$

If $\eta \in V_a \pi \cap (H_\Gamma \pi)_a$ then

$$\eta = \eta^i \left(\left. \frac{\partial}{\partial x^i} \right|_a + \Gamma_i^\alpha(a) \left. \frac{\partial}{\partial u^\alpha} \right|_a \right)$$

so that $\pi_*(\eta) = 0$ implies

$$\eta^i \left.\frac{\partial}{\partial x^i}\right|_{\pi(a)} = 0 \in T_{\pi(a)}M,$$

demonstrating than $V_a\pi \cap (H_\Gamma\pi)_a = \{0\}$. The result now follows by a dimension argument, because $\dim(H_\Gamma\pi)_a = m$ and $\dim V_a\pi = n$. ∎

It follows from this result that the complementary vector-valued form $I - \Gamma$ is an element of $\bigwedge^1 E \otimes \mathcal{V}(\pi)$, so that if $X \in \mathcal{X}(E)$ then $X \lrcorner (I - \Gamma)$ is vertical. The converse assertion to the lemma, that a complement to the vertical bundle in τ_E determines a connection, is also true.

Lemma 3.5.4 *If the tangent bundle τ_E may be written as a direct sum*

$$(V\pi \oplus H\pi, \tau_E, E)$$

where $(H\pi, \tau_E|_{H\pi}, E)$ is a vector sub-bundle of τ_E, then $H\pi$ determines a unique connection Γ such that $H\pi = H_\Gamma\pi$.

Proof For $a \in E$, let the linear map $\Gamma_a : T_aE \longrightarrow T_aE$ be the projection on $H_a\pi$ along $V_a\pi$. Then Γ_a may be regarded as an element of $T_a^*E \otimes T_aE$, and since $\Gamma_a(V_a\pi) = \{0\}$ it follows from Lemma 3.1.11 that Γ_a may actually be considered to be an element of $(\pi^*(T^*M))_a \otimes T_aE$. We may also define the map $\Gamma : E \longrightarrow \pi^*(T^*M) \otimes TE$ by $a \longmapsto \Gamma_a$.

If coordinates on $V\pi$ are $(x^i, u^\alpha; \dot{u}^\alpha)$ and coordinates on $H\pi$ are $(x^i, u^\alpha; y^j)$, it follows from a dimension argument the range of the index j is from 1 to m; coordinates on $V\pi \oplus H\pi$ are then $(x^i, u^\alpha; \dot{u}^\alpha, y^j)$ and if (e_j, f_α) are the local sections dual to these vector bundle coordinates then

$$\Gamma = dx^i \otimes e_i,$$

showing that Γ is smooth. It is clear that $\Gamma \lrcorner \Gamma = \Gamma$ because each Γ_a is a projection. Finally, it is obvious from the definition that $H\pi = H_\Gamma\pi$. ∎

Example 3.5.5 Let π be the trivial bundle $(M \times F, pr_1, M)$. Then $H\pi$ may be defined by

$$H_a\pi = \{\xi \in T_a(M \times F) : pr_{2*}(\xi) = 0\}$$

and $V\pi \oplus H\pi = T(M \times F)$. The connection defined in this way may be called the *zero connection* determined by the global trivialisation. ∎

Example 3.5.6 If π is the trivial bundle $(M \times \mathbf{R}, pr_1, M)$ with coordinates (x^i, t), where t is the pull-back to $M \times \mathbf{R}$ of the canonical coordinate on $\mathbf{R}$, then the coordinate representation of a connection Γ is

$$\Gamma = dx^i \otimes \left(\frac{\partial}{\partial x^i} + \Gamma_i \frac{\partial}{\partial t} \right).$$

The connection Γ then determines a horizontal 1-form

$$\Gamma \lrcorner\, dt = \Gamma_i dx^i$$

■

Example 3.5.7 If now π is the trivial bundle $(\mathbf{R} \times F, pr_1, \mathbf{R})$ with coordinates (t, q^α) then the coordinate representation of a connection Γ is

$$\Gamma = dt \otimes \left(\frac{\partial}{\partial t} + \Gamma^\alpha \frac{\partial}{\partial q^\alpha} \right).$$

If we consider the vector fields $Y \in \mathcal{X}_{\partial/\partial t}(\pi)$, then any two such vector fields Y differ by a vertical vector field, so that the contraction $Y \lrcorner\, \Gamma$ does not depend on the particular choice of Y. We may therefore write this contraction as $\partial/\partial t \lrcorner\, \Gamma$, and in this way determine a vector field on $\mathbf{R} \times F$ of the particular form

$$\frac{\partial}{\partial t} + \Gamma^\alpha \frac{\partial}{\partial q^\alpha}.$$

■

One of the uses of a connection is to provide a means of lifting entities defined on the base manifold M up to the total space E. This action is called a *horizontal lift*.

Definition 3.5.8 If $a \in E$ and $\xi \in T_{\pi(a)}M$ then the *horizontal lift of* ξ *by* Γ *to* a is the tangent vector

$$\Gamma_a(\xi) \in T_a E$$

where $\Gamma_a \in (\pi^*(T^*M))_a \otimes T_a E$ is regarded as a linear map $T_{\pi(a)}M \longrightarrow T_a E$.

■

The horizontal lift of ξ is then an element of the horizontal bundle $H_\Gamma \pi$, and by definition every element of $H_\Gamma \pi$ is the horizontal lift of a tangent vector on M.

Definition 3.5.9 If $X \in \mathcal{X}(M)$ then the *horizontal lift of X by Γ* is the vector field

$$X \lrcorner \Gamma \in \mathcal{X}(E).$$

■

We may call a vector field which takes its values in the horizontal bundle a *Γ-horizontal vector field*; it is then evident that a horizontal lift is indeed horizontal. Conversely, a *projectable* horizontal vector field is the horizontal lift of its projection. In coordinates, if

$$X = X^i \frac{\partial}{\partial x^i}$$

and

$$\Gamma = dx^i \otimes \left(\frac{\partial}{\partial x^i} + \Gamma^\alpha_i \frac{\partial}{\partial u^\alpha} \right)$$

then the horizontal lift of X by Γ is the projectable vector field on E with coordinate representation

$$X^i \left(\frac{\partial}{\partial x^i} + \Gamma^\alpha_i \frac{\partial}{\partial u^\alpha} \right)$$

where X^i are functions pulled back from M. In contrast, the most general horizontal vector field Y on E has coordinate representation

$$Y^i \left(\frac{\partial}{\partial x^i} + \Gamma^\alpha_i \frac{\partial}{\partial u^\alpha} \right)$$

where Y^i are functions on E which need not have been pulled back from M.

A similar idea to this may sometimes be used to obtain the horizontal lift of a curve in the base manifold M. This may always be done locally; however, our definition of a connection is too general to ensure that a global horizontal lift always exists.

Definition 3.5.10 If $\gamma : (a,b) \longrightarrow M$ is a curve, then the curve $\sigma : (a,b) \longrightarrow E$ is called a *horizontal lift of γ* if $\pi \circ \sigma = \gamma$ and if, for each $s \in (a,b)$, the tangent vector

$$[t \longmapsto \sigma(s+t)] \in T_{\sigma(s)}E$$

is the horizontal lift of $[t \longmapsto \gamma(s+t)] \in T_{\gamma(s)}M$ by Γ. ■

Lemma 3.5.11 *If $\gamma(a,b) \longrightarrow M$ is a curve, if $s \in (a,b)$ and if $p \in E$ satisfies $\pi(p) = \gamma(s)$, then there is an $\varepsilon > 0$ such that $\gamma|_{(s-\varepsilon,s+\varepsilon)}$ has a unique horizontal lift σ satisfying $\sigma(s) = p$.*

Proof In coordinates, σ has to satisfy

$$\frac{d\sigma^\alpha}{dt} = \Gamma_i^\alpha \frac{d\gamma^i}{dt}, \qquad \frac{d\sigma^i}{dt} = \frac{d\gamma^i}{dt}$$

in a neighbourhood of p, and the result follows from the local existence and uniqueness theorem for ordinary differential equations. ∎

Example 3.5.12 Let π be the trivial bundle $(\mathbf{R}\times(0,\infty), pr_1, \mathbf{R})$ with global coordinates (x,u), and let Γ be the connection defined by

$$\Gamma = dx \otimes \left(\frac{\partial}{\partial x} + \frac{\partial}{\partial u}\right).$$

Let $\gamma = id_{\mathbf{R}}$ be the identity curve. Then the curve $\sigma : (-1,\infty) \longrightarrow \mathbf{R} \times (0,\infty)$ given by $\sigma(t) = (t, t+1)$ is the horizontal lift of $\gamma|_{(-1,\infty)}$ through $(0,1)$. However σ cannot be extended to become a horizontal lift of the whole curve γ, and indeed the whole curve γ does not have a horizontal lift through *any* point of $\mathbf{R} \times (0,\infty)$. ∎

The trouble in this last example was that the lifted curve "wanted to leave the total space". This phenomenon is similar to that which arises when an integral curve of a vector field cannot be defined for all real values of its parameter, and such vector fields are termed *incomplete*. We may therefore say that the connection Γ is complete if every curve in M has a horizontal lift to E. Since completeness of a connection is a global property we shall, however, not consider it any further.

One property of a connection Γ which we shall consider is its *curvature*. In Chapter 4 we shall see how certain local sections of the bundle π may be called "integral sections of Γ". These are sections $\phi \in \Gamma_W(\pi)$ with the property that every tangent vector to the image manifold $\phi(W) \subset E$ is horizontal with respect to Γ. In coordinates, the functions ϕ^α must satisfy the partial differential equations

$$\frac{\partial \phi^\alpha}{\partial x^i} = \Gamma_i^\alpha \circ \phi.$$

Solutions to these equations will only exist if the coefficients Γ_i^α satisfy Frobenius' integrability condition; we shall see that this condition is equivalent to the vanishing of the curvature of Γ.

Definition 3.5.13 The *curvature* of the connection Γ is the map $R_\Gamma : \mathcal{X}(E) \times \mathcal{X}(E) \longrightarrow \mathcal{X}(E)$ defined by

$$R_\Gamma(X,Y) = [X \lrcorner \Gamma, Y \lrcorner \Gamma] \lrcorner (I - \Gamma).$$

∎

As may be seen from this definition, the curvature of Γ measures how far the Lie bracket of two Γ-horizontal vector fields deviates from the horizontal. The map R_Γ is evidently skew-symmetric, and we shall see in a moment that it is actually $C^\infty(E)$-linear, so that it defines a vector-valued 2-form on E. In fact, we have the following relationship between the curvature R_Γ and the Nijenhuis tensor N_Γ.

Proposition 3.5.14 *If Γ is a connection on π then $R_\Gamma = \frac{1}{2} N_\Gamma$.*

Proof Let $X, Y \in \mathcal{X}(E)$. If X is vertical then $X \lrcorner \Gamma = 0$, so

$$\begin{aligned}
\tfrac{1}{2}(X,Y) \lrcorner N_\Gamma &= [X,Y] \lrcorner \Gamma \lrcorner \Gamma - [X, Y \lrcorner \Gamma] \lrcorner \Gamma \\
&= [X, Y - Y \lrcorner \Gamma] \lrcorner \Gamma \\
&= 0
\end{aligned}$$

using $\Gamma \lrcorner \Gamma = \Gamma$ and the facts that $Y - Y \lrcorner \Gamma$ is vertical and that the bracket of two vertical vector fields is vertical. If follows from this (and the skew-symmetry of N_Γ) that $(X, Y) \lrcorner N_\Gamma$ depends only on the Γ-horizontal components of X and Y:

$$\begin{aligned}
\tfrac{1}{2}(X,Y) \lrcorner N_\Gamma &= \tfrac{1}{2}(X \lrcorner \Gamma, Y \lrcorner \Gamma) \lrcorner N_\Gamma \\
&= [X \lrcorner \Gamma, Y \lrcorner \Gamma] \lrcorner \Gamma \lrcorner \Gamma + [X \lrcorner \Gamma \lrcorner \Gamma, Y \lrcorner \Gamma \lrcorner \Gamma] \\
&\quad -[X \lrcorner \Gamma \lrcorner \Gamma, Y \lrcorner \Gamma] \lrcorner \Gamma - [X \lrcorner \Gamma, Y \lrcorner \Gamma \lrcorner \Gamma] \lrcorner \Gamma \\
&= [X \lrcorner \Gamma, Y \lrcorner \Gamma] - [X \lrcorner \Gamma, Y \lrcorner \Gamma] \lrcorner \Gamma \\
&= (X, Y) \lrcorner R_\Gamma,
\end{aligned}$$

from which the result follows. ∎

The condition for the existence of integral sections may also be described in more geometric terms using Proposition 3.4.16. Since Γ may be considered as a projection operator, its eigenfunctions are the constant functions zero and one. The distribution corresponding to the eigenfunction zero just contains the vertical vectors, and is always involutive; its integral manifolds are the fibres of π. The distribution corresponding to the eigenfunction one will be involutive when the curvature R_Γ vanishes, and then the image sets of the integral sections will be its integral manifolds.

EXERCISES

3.5.1 If Γ is a connection on π with coordinate representation

$$\Gamma = dx^i \otimes \left(\frac{\partial}{\partial x^i} + \Gamma_i^\alpha \frac{\partial}{\partial u^\alpha} \right),$$

show that its curvature R_Γ, considered as a vector-valued 2-form on E, has coordinate representation

$$R_\Gamma = \tfrac{1}{2}\left(\frac{\partial \Gamma^\alpha_j}{\partial x^i} + \Gamma^\beta_i \frac{\partial \Gamma^\alpha_j}{\partial u^\beta} - \frac{\partial \Gamma^\alpha_i}{\partial x^j} - \Gamma^\beta_j \frac{\partial \Gamma^\alpha_i}{\partial u^\beta}\right)(dx^i \wedge dx^j) \otimes \frac{\partial}{\partial u^\alpha}.$$

3.5.2 Let G be a Lie group, and let $g \in G$ and $\xi \in T_g G$. Use the decomposition

$$T^*_\xi TG = l_\xi(T^*_g G) \oplus (\tau_G)^*(T^*_g G)$$

described in Exercise 3.4.3 to define a connection on the tangent bundle (TG, τ_G, G). Is this the same as the zero connection determined by the trivialisation

$$TG \cong G \times \mathbf{g}$$

constructed in Exercise 1.1.7 using the left translation $L_g : G \longrightarrow G$?

REMARKS

A linear operation on differential forms which satisfies the version of Leibniz' rule with a minus sign is often called an *anti-derivation* rather than a derivation. Our usage follows that of a paper by Frölicher and Nijenhuis [5], where the relationship of derivations to vector-valued forms is studied in detail.

Connections are usually defined on principal fibre bundles, and each connection may be specified by giving an equivariant family of horizontal subspaces. This corresponds directly to Lemma 3.5.4, with the proviso that, in the absence of a particular transformation group, the concept of equivariance is inappropriate.

The connection form of a connection on a principal fibre bundle is a Lie algebra-valued 1-form; since the vertical tangent space at each point of a principal fibre bundle is canonically isomorphic to the Lie algebra, the connection form determines a vertical vector-valued 1-form, and this is just the complement of the vector-valued 1-form used in our definition of a connection in Section 3.5. A useful source of information on connections and their application to physical theories may be found in [2].

Chapter 4

First-order Jet Bundles

In basic differential geometry, a tangent vector to a manifold may be defined as an equivalence class of curves passing through a given point, where two curves are equivalent if they have the same derivative at that point: indeed, this is the definition we have used in earlier chapters. (There are other definitions which may be used instead, but the definition in terms of curves is perhaps the most intuitive.) A first-order jet is a generalisation of this idea to the case of families of higher-dimensional manifolds passing through a point, where the embedding maps have the same first derivatives at that point. We shall, however, choose to consider the graphs of these embeddings rather than the embeddings themselves, in line with our previous policy of considering bundles and sections rather than pairs of manifolds and maps. In the first section of this chapter we shall give a formal definition of a first-order jet, and show that the collection of all such jets is a differentiable manifold. We shall also see that this manifold is the appropriate setting for a general description of a first-order partial differential equation. In subsequent sections, we shall examine some of the properties of this jet manifold which arise when it is regarded as the total space of an affine bundle, and we shall introduce the idea of *prolongation* whereby a bundle morphism may be extended to act upon the jet manifold.

4.1 First-order Jets

Given any bundle (E, π, M), we wish to define the jet of a section ϕ at a point p. Since some bundles do not have any global sections, we shall necessarily have to use local sections, and find a way of dealing with the different domains which these local sections will have. The approach we have chosen is to place an equivalence relation on the set of local sections defined in a neighbourhood of a given point in the base space. The equivalence relation will be specified in terms of local coordinates, so we must first ensure that the particular choice of coordinate system will not matter. In the following lemma we shall write (for example) $u^\alpha \circ \phi$ instead of the more usual ϕ^α to distinguish between the two coordinate systems.

Lemma 4.1.1 *Let (E, π, M) be a bundle, and let $p \in M$. Suppose that $\phi, \psi \in \Gamma_p(\pi)$ satisfy $\phi(p) = \psi(p)$. Let (x^i, u^α) and (y^j, v^β) be two adapted coordinate systems around $\phi(p)$, and suppose also that*

$$\left.\frac{\partial(u^\alpha \circ \phi)}{\partial x^i}\right|_p = \left.\frac{\partial(u^\alpha \circ \psi)}{\partial x^i}\right|_p$$

for $1 \leq i \leq m$ and $1 \leq \alpha \leq n$. Then

$$\left.\frac{\partial(v^\beta \circ \phi)}{\partial y^j}\right|_p = \left.\frac{\partial(v^\beta \circ \psi)}{\partial y^j}\right|_p$$

for $1 \leq j \leq m$ and $1 \leq \beta \leq n$.

Proof From the Chain Rule,

$$\begin{aligned}
\left.\frac{\partial(v^\beta \circ \phi)}{\partial y^j}\right|_p &= \left.\frac{\partial(v^\beta \circ \phi)}{\partial x^i}\right|_p \left.\frac{\partial x^i}{\partial y^j}\right|_p \\
&= \left(\left.\frac{\partial v^\beta}{\partial x^i}\right|_{\phi(p)} + \left.\frac{\partial v^\beta}{\partial u^\alpha}\right|_{\phi(p)} \left.\frac{\partial(u^\alpha \circ \phi)}{\partial x^i}\right|_p \right) \left.\frac{\partial x^i}{\partial y^j}\right|_p
\end{aligned}$$

using the relationship $x^i \circ \phi = x^i$ between similarly-named coordinate functions on E and M. The result follows immediately. ■

Definition 4.1.2 Let (E, π, M) be a bundle and let $p \in M$. Define the local sections $\phi, \psi \in \Gamma_p(\pi)$ to be *1-equivalent at p* if $\phi(p) = \psi(p)$ and if, in some adapted coordinate system (x^i, u^α) around $\phi(p)$,

$$\left.\frac{\partial \phi^\alpha}{\partial x^i}\right|_p = \left.\frac{\partial \psi^\alpha}{\partial x^i}\right|_p$$

for $1 \leq i \leq m$ and $1 \leq \alpha \leq n$. The equivalence class containing ϕ is called the *1-jet of ϕ at p* and is denoted $j^1_p\phi$. ■

Another way of constructing this relation would be in terms of tangent maps.

Lemma 4.1.3 *Let $\phi, \psi \in \Gamma_p(\pi)$ satisfy $\phi(p) = \psi(p)$. Then $j^1_p\phi = j^1_p\psi$ if, and only if, $\phi_*|_{T_pM} = \psi_*|_{T_pM}$.*

Proof Each assertion is just another way of making the statement which, in coordinates around $\phi(p)$, reads

$$\left.\frac{\partial \phi^\alpha}{\partial x^i}\right|_p = \left.\frac{\partial \psi^\alpha}{\partial x^i}\right|_p .$$

For the jets this is just the definition, and for the tangent maps it is obtained from the coefficient of $\partial/\partial u^\alpha$ in the equation $\phi_*(\partial/\partial x^i) = \psi_*(\partial/\partial x^i)$. ■

The set of all the 1-jets of local sections of π has a natural structure as a differentiable manifold. The atlas which describes this structure is constructed from an atlas of adapted coordinate charts on the total space E, in much the same way that the induced atlas on the tangent manifold TM is constructed from an atlas on M.

Definition 4.1.4 The *first jet manifold of* π is the set

$$\{j^1_p\phi : p \in M, \phi \in \Gamma_p(\pi)\}$$

and is denoted $J^1\pi$. The functions π_1 and $\pi_{1,0}$, called the *source* and *target* projections respectively, are defined by

$$\begin{aligned} \pi_1 : J^1\pi &\longrightarrow M \\ j^1_p\phi &\longmapsto p \end{aligned}$$

and

$$\begin{aligned} \pi_{1,0} : J^1\pi &\longrightarrow E \\ j^1_p\phi &\longmapsto \phi(p). \end{aligned}$$

■

Definition 4.1.5 Let (E, π, M) be a bundle, and let (U, u) be an adapted coordinate system on E, where $u = (x^i, u^\alpha)$. The *induced coordinate system* (U^1, u^1) on $J^1\pi$ is defined by

$$\begin{aligned} U^1 &= \{j^1_p\phi : \phi(p) \in U\} \\ u^1 &= (x^i, u^\alpha, u^\alpha_i) \end{aligned}$$

where $x^i(j^1_p\phi) = x^i(p)$, $u^\alpha(j^1_p\phi) = u^\alpha(\phi(p))$ and the mn new functions

$$u^\alpha_i : U^1 \longrightarrow \mathbf{R}$$

are specified by

$$u^\alpha_i(j^1_p\phi) = \left.\frac{\partial \phi^\alpha}{\partial x^i}\right|_p$$

and are known as *derivative coordinates*. ■

Example 4.1.6 Let π be the trivial bundle $(\mathbf{R}^2 \times \mathbf{R}, pr_1, \mathbf{R})$, with global coordinates $(x^1, x^2; u^1)$ on $\mathbf{R}^2 \times \mathbf{R}$, so that global coordinates on $J^1\pi$ are $(x^1, x^2; u^1; u^1_1, u^1_2)$. To each jet $j^1_p\phi \in J^1\pi$, where $p = (p^1, p^2) \in \mathbf{R}^2$, there corresponds an inhomogeneous linear map $\overline{\psi} : \mathbf{R}^2 \longrightarrow \mathbf{R}$, defined as follows:

$$\overline{\psi}(q) = \phi^1(p) + u^1_1(j^1_p\phi)(q^1 - p^1) + u^1_2(j^1_p\phi)(q^2 - p^2)$$

where $q = (q^1, q^2) \in \mathbf{R}^2$ and $\phi^1 = u^1 \circ \phi : \mathbf{R}^2 \longrightarrow \mathbf{R}$. The map $\overline{\psi}$ gives rise to a global section $\psi = (id_{\mathbf{R}^2}, \overline{\psi})$ of π, and it is obvious that $j^1_p\phi = j^1_p\psi$; clearly $\overline{\psi}$ is the unique globally-defined linear inhomogeneous map with this property. The map $\overline{\psi}$ is of course the first-order Taylor polynomial of ϕ, and the jet $j^1_p\phi$ is really no more than a coordinate-free construction which incorporates the same information about derivatives as the polynomial. ■

Proposition 4.1.7 *Given an atlas of adapted charts (U, u) on E, the corresponding collection of charts (U^1, u^1) is a finite-dimensional C^∞ atlas on $J^1\pi$.*

Proof First, note that every 1-jet $j^1_p\phi$ is in the domain of one such chart, namely any chart (U^1, u^1) where $\phi(p) \in U$. We shall show that, if (U, u) and (V, v) are two charts in the atlas on E such that $U^1 \cap V^1$ is non-empty, then the transition function

$$v^1 \circ (u^1)^{-1}\Big|_{u^1(U^1 \cap V^1)}$$

is smooth. For convenience, we shall in future write $v^1 \circ (u^1)^{-1}$ without any indication that $(u^1)^{-1}$ has been restricted to a subset of its domain.

Now the component functions of $v^1 \circ (u^1)^{-1}$ are $y^j \circ (u^1)^{-1}$, $v^\beta \circ (u^1)^{-1}$ and $v^\beta_j \circ (u^1)^{-1}$, and the domain of each of these functions is an open subset of $\mathbf{R}^{m+n} \times \mathbf{R}^{mn}$. From the definition of u^1, we have $pr_1 \circ u^1 = u \circ \pi_{1,0}$, so that

$$\begin{aligned} y^j \circ (u^1)^{-1} &= y^j \circ u^{-1} \circ pr_1 \\ v^\beta \circ (u^1)^{-1} &= v^\beta \circ u^{-1} \circ pr_1 \end{aligned}$$

(on the left-hand side of these equations y^j and v^β are functions defined on $J^1\pi$, whereas on the right-hand side they are functions defined on E). Consequently the first two sets of component functions are smooth. As far as the third set is concerned,

$$\begin{aligned} v^\beta_j(j^1_p\phi) &= \left.\frac{\partial(v^\beta \circ \phi)}{\partial y^j}\right|_p \\ &= \left(\left.\frac{\partial v^\beta}{\partial x^i}\right|_{\phi(p)} + \left.\frac{\partial v^\beta}{\partial u^\alpha}\right|_{\phi(p)} u^\alpha_i(j^1_p\phi) \right) \left.\frac{\partial x^i}{\partial y^j}\right|_p , \end{aligned}$$

and so each $v_j^\beta \circ (u^1)^{-1}$ is also a smooth function, because it is smooth in terms of the first $(m+n)$ coordinates and depends in a linear inhomogeneous manner upon the remaining mn coordinates. ∎

Example 4.1.8 Let $(\mathbf{R}^3-\{0\}, \pi, S^2)$ be the bundle defined in Exercise 1.1.4, and let $(\theta, \phi; \rho)$ be spherical polar coordinates in a neighbourhood U of $(0,1,0) \in \mathbf{R}^3 - \{0\}$. Define new coordinates $(x, z; h)$ in this neighbourhood by

$$\begin{aligned} x &= \sin\theta\cos\phi \\ z &= \cos\theta \\ h &= \rho\sin\theta\sin\phi, \end{aligned}$$

so that if $a \in U$ then x and z are Cartesian coordinates for $\pi(a) \in S^2$. The derivative coordinates on $U^1 \subset J^1\pi$ are then given by the rules

$$\begin{aligned} h_x &= \left(\frac{\partial h}{\partial \theta} + \frac{\partial h}{\partial \rho}\rho_\theta\right)\frac{\partial \theta}{\partial x} + \left(\frac{\partial h}{\partial \phi} + \frac{\partial h}{\partial \rho}\rho_\phi\right)\frac{\partial \phi}{\partial x} \\ &= \rho\cot\phi + \rho_\phi, \\ h_z &= \left(\frac{\partial h}{\partial \theta} + \frac{\partial h}{\partial \rho}\rho_\theta\right)\frac{\partial \theta}{\partial z} + \left(\frac{\partial h}{\partial \phi} + \frac{\partial h}{\partial \rho}\rho_\phi\right)\frac{\partial \phi}{\partial z} \\ &= \sin\phi(\rho\cot\theta + \rho_\theta) + \cot\theta\cos\phi(\rho\cot\phi + \rho_\phi). \end{aligned}$$

∎

To show that $J^1\pi$ satisfies our chosen definition of a manifold, we must now check that the topology induced on $J^1\pi$ by the atlas we have described is Hausdorff, second-countable, and connected. By Proposition 1.1.14, this will follow if we can show that $J^1\pi$ is the total space of a bundle. In fact there are two bundles which may be formed in this way, and they both have interesting properties. The base spaces of these bundles are E and M.

Lemma 4.1.9 *The function $\pi_{1,0} : J^1\pi \longrightarrow E$ is a smooth surjective submersion.*

Proof The function $\pi_{1,0}$ is surjective because, for each $a \in E$, there is always a local section ϕ such that $\phi(\pi(a)) = a$, and then $\pi_{1,0}(j^1_{\pi(a)}\phi) = a$. It is smooth at every $j^1_p\phi \in J^1\pi$ because, using coordinate charts (U, u) around $\pi_{1,0}(j^1_p\phi) = \phi(p)$ and (U^1, u^1) around $j^1_p\phi$, the composite map $u \circ \pi_{1,0} \circ (u^1)^{-1}$ is simply the projection pr_1 from $u^1(U^1) \subset \mathbf{R}^{m+n} \times \mathbf{R}^{mn}$ to $u(U) \subset \mathbf{R}^{m+n}$. It is a submersion for the same reason. ∎

Corollary 4.1.10 *The function $\pi_1 : J^1\pi \longrightarrow M$ is a smooth surjective submersion.* ∎

Given for the moment that the atlas on $J^1\pi$ defines a manifold, we see that the triples $(J^1\pi, \pi_{1,0}, E)$ and $(J^1\pi, \pi_1, M)$ become fibred manifolds. The proof that π_1 is actually a bundle involves the local trivialisations of π, and will be deferred until later. By contrast, the proof that $\pi_{1,0}$ is a bundle does not involve the local trivialisations of π at all (and so would still be valid if we had defined jets of local sections of arbitrary fibred manifolds). Indeed, we can say more: $\pi_{1,0}$ has a natural structure as an *affine bundle*. The reason for this is that the fibre coordinates of $\pi_{1,0}$ are just the derivative coordinates introduced in Definition 4.1.5, and the inhomogeneous linear transformation rule displayed in the proof of Proposition 4.1.7 satisfies the requirements for the local trivialisations of an affine bundle. However, for a precise definition we should give an associated vector bundle, and this will be the bundle over E whose total space is the tensor product $\pi^*(T^*M) \otimes V\pi$: formally, it is the bundle

$$\left(\pi^*(T^*M) \otimes V\pi, (\tau_E^*|_{\pi^*(T^*M)}) \otimes (\tau_E|_{V\pi}), E\right).$$

This rather unusual bundle, and the corresponding affine structure of $J^1\pi$ over E, will turn out to be fundamental to a study of the properties of jet bundles.

Theorem 4.1.11 *The triple $(J^1\pi, \pi_{1,0}, E)$ may be given the structure of an affine bundle modelled on the vector bundle $(\tau_E^*|_{\pi^*(T^*M)}) \otimes (\tau_E|_{V\pi})$ in such a way that, for each adapted chart (U, u) on E, the map*

$$\begin{aligned} t_u : \pi_{1,0}^{-1}(U) &\longrightarrow U \times \mathbf{R}^{mn} \\ j_p^1\phi &\longmapsto (\phi(p), u_i^\alpha(j_p^1\phi)) \end{aligned}$$

is an affine local trivialisation.

Proof We must first define a fibrewise action of the vector bundle on $\pi_{1,0}$, and we shall do this by prescribing the effect of this action upon the derivative coordinates of a given 1-jet. So let $a \in E$ and let (U, u) be an adapted chart around a. A typical element $\xi \in (\pi^*(T^*M) \otimes V\pi)_a$ may be written in coordinates as

$$\xi = \xi_i^\alpha \left(dx^i \otimes \frac{\partial}{\partial u^\alpha} \right)_a .$$

The action of ξ on $j_{\pi(a)}^1\phi$ is then written as $\xi[j_{\pi(a)}^1\phi]$, and is defined by the rule

$$u_i^\alpha(\xi[j_{\pi(a)}^1\phi]) = u_i^\alpha(j_{\pi(a)}^1\phi) + \xi_i^\alpha .$$

We must now check that this definition does not depend upon the choice of chart. So let (y^j, v^β) be another coordinate system around a. Then from the calculation in Proposition 4.1.7,

$$v_j^\beta(j^1_{\pi(a)}\phi) = \left(\left.\frac{\partial v^\beta}{\partial x^i}\right|_a + \left.\frac{\partial v^\beta}{\partial u^\alpha}\right|_a u_i^\alpha(j^1_{\pi(a)}\phi) \right) \left.\frac{\partial x^i}{\partial y^j}\right|_{\pi(a)},$$

whereas, as a tensor,

$$\xi = \xi_j^\beta \left(dy^j \otimes \frac{\partial}{\partial v^\beta} \right)_a = \xi_i^\alpha \left.\frac{\partial v^\beta}{\partial u^\alpha}\right|_a \left.\frac{\partial x^i}{\partial y^j}\right|_a \left(dy^j \otimes \frac{\partial}{\partial v^\beta} \right)_a .$$

It follows immediately that

$$v_j^\beta(\xi[j^1_{\pi(a)}\phi]) = v_j^\beta(j^1_{\pi(a)}\phi) + \xi_j^\beta,$$

as required.

We must also consider the maps t_u. Each such map is a diffeomorphism, for it is just the composite $(u^{-1} \times id_{\mathbf{R}^{mn}}) \circ u^1$, and evidently $pr_1 \circ t_u = \pi_{1,0}|_{U^1}$. Now let $a \in U$; then the map $t_{u;a} : \pi_{1,0}^{-1}(a) \longrightarrow \mathbf{R}^{mn}$ defined by

$$t_{u;a} = pr_2 \circ t_u|_{\pi_{1,0}^{-1}(a)}$$

satisfies $t_{u;a} = (u_i^\alpha)|_{\pi_{1,0}^{-1}(a)}$. Consequently $t_{u;a}$ is an affine morphism, where the fibre $\pi_{1,0}^{-1}(a)$ has the structure of an affine space given by the vector bundle action, and $\mathbf{R}^{mn}$ has its natural affine structure. ■

Corollary 4.1.12 *The total space $J^1\pi$ of $\pi_{1,0}$ is a manifold.* ■

Example 4.1.13 If π is the trivial bundle $(\mathbf{R}^2 \times \mathbf{R}, pr_1, \mathbf{R})$ with global coordinates $(x^1, x^2; u^1)$ on $\mathbf{R}^2 \times \mathbf{R}$, then each 1-jet $j^1_p\phi$ gives rise to the Taylor polynomial

$$(q^1, q^2) \longmapsto \phi^1(q) + u_1^1(j_p^1\phi)(q^1 - p^1) + u_2^1(j_p^1\phi)(q^2 - p^2)$$

as in Example 4.1.6. The affine action of

$$\xi = \left((\xi_1^1 dx^1 + \xi_2^1 dx^2) \otimes \frac{\partial}{\partial u^1} \right)_{\phi(p)}$$

then gives rise to a new 1-jet $\xi[j_p^1\phi]$ with corresponding Taylor polynomial

$$(q^1, q^2) \longmapsto \phi^1(q) + (u_1^1(j_p^1\phi) + \xi_1^1)(q^1 - p^1) + (u_2^1(j_p^1\phi) + \xi_2^1)(q^2 - p^2)$$

■

The following result concerning restricted bundles, although rather obvious, is nevertheless worth recording.

Lemma 4.1.14 *If $W \subset M$ is an open submanifold then*

$$J^1(\pi|_W) \cong \pi_1^{-1}(W).$$

Proof To each $j^1_p\phi \in J^1(\pi|_W)$, where $\phi \in \Gamma_p(\pi|_W)$, there corresponds a unique $j^1_p\overline{\phi} \in \pi_1^{-1}(W)$, where $\overline{\phi} \in \Gamma_p(\pi)$, given by $\overline{\phi} = \phi$. ∎

Example 4.1.15 If π is the trivial bundle $(M \times \mathbf{R}, pr_1, M)$, then there is a canonical diffeomorphism between the first jet manifold $J^1\pi$ and $T^*M \times \mathbf{R}$. To construct this diffeomorphism, for each $\phi \in \Gamma_W(\pi)$ write $\overline{\phi} = pr_2 \circ \phi \in C^\infty(W)$; then whenever $p \in W$,

$$j^1_p\phi = \{\psi : \psi \in \Gamma_p(\pi); \overline{\psi}(p) = \overline{\phi}(p); d\overline{\psi}_p = d\overline{\phi}_p\}.$$

Consequently the mapping

$$\begin{array}{ccc} J^1\pi & \longrightarrow & T^*M \times \mathbf{R} \\ j^1_p\phi & \longmapsto & (d\overline{\phi}_p, \overline{\phi}(p)) \end{array}$$

is well-defined, and is clearly it injective. Writing it out in coordinates shows that it is a diffeomorphism, because if (x^i, u) are coordinates on $M \times \mathbf{R}$ where $u = id_{\mathbf{R}}$ is the identity coordinate, then the derivative coordinates u_i on $J^1\pi$ correspond to the coordinates ∂_i on T^*M. ∎

Example 4.1.16 If π is now the trivial bundle $(\mathbf{R} \times F, pr_1, \mathbf{R})$ then there is a canonical diffeomorphism between $J^1\pi$ and $\mathbf{R} \times TF$. This relationship will be described in more detail in Example 4.1.23. ∎

If we apply Theorem 4.1.11 to (say) Example 4.1.16, we see that $(\mathbf{R} \times TF, id_{\mathbf{R}} \times TF, \mathbf{R} \times F)$ has the structure of an affine bundle. However, this particular example is actually a vector bundle: in general, if the original bundle π is trivial, then $\pi_{1,0}$ may be given the structure of a vector bundle.

Proposition 4.1.17 *If $(M \times F, \pi, M)$ is a trivial bundle, then the trivialisation determines a vector bundle structure on $(J^1\pi, \pi_{1,0}, M \times F)$.*

Proof The vector bundle structure on $\pi_{1,0}$ will be induced from its affine bundle structure by the specification of a zero section. So for each $a \in M \times F$, define the constant section of π through a by

$$\phi_a(p) = (p, pr_2(a)),$$

and then define the zero section of $\pi_{1,0}$ by

$$z(a) = j^1_p(\phi_a).$$

∎

The converse assertion to Proposition 4.1.17 is, however, false: the choice of a distinguished section of $\pi_{1,0}$ does *not* determine a trivialisation of π.

Example 4.1.18 Let (E, π, M) be the Möbius band, regarded as a bundle over the circle. For each $a \in E$, let ϕ_a be a *local* section of π defined to be constant in the local trivialisation around $\pi(a)$ induced from the Cartesian product structure on $[0,1] \times (-1,1)$. Then define a zero section of $\pi_{1,0}$ as before by $z(a) = j^1_p(\phi_a)$. This induces a vector bundle structure on $\pi_{1,0}$; however π is not a trivial bundle. ■

In this example, the distinguished section of $\pi_{1,0}$ was determined essentially by the consistent choice of a "horizontal" direction across the fibre at each point $a \in E$. For a trivial bundle, each Cartesian product structure gives suitable horizontal directions. In general, whenever a connection is given on the bundle π, then the horizontal directions specified by the connection determine a section of $\pi_{1,0}$. This relationship will be examined in detail in Section 4.6.

We shall now return to the fibred manifold $(J^1\pi, \pi_1, M)$ and establish that it, too, has the structure of a bundle, provided that π is locally trivial. To do this, we shall show first that if π is a trivial bundle over $\mathbf{R}^m$, then π_1 is trivial.

Definition 4.1.19 If $p \in M$ then the fibre $\pi_1^{-1}(p)$ is denoted $J^1_p\pi$ rather than $(J^1\pi)_p$. ■

We have already seen that the map π_1 is a submersion, so that $J^1_p\pi$ is a submanifold of $J^1\pi$; if (U, u) is an adapted coordinate system on E, where $p \in \pi(U)$ and $u = (x^i, u^\alpha)$, then (u^α, u^α_i) are coordinates on $U^1 \cap J^1_p\pi$.

Lemma 4.1.20 *Let π be the trivial bundle $(\mathbf{R}^m \times F, pr_1, \mathbf{R}^m)$. Then the first jet bundle $(J^1\pi, \pi_1, \mathbf{R}^m)$ is trivial.*

Proof We shall show that $J^1\pi \cong \mathbf{R}^m \times J^1_0\pi$. So let $j^1_p\phi \in J^1\pi$; define the translation $\tau_p : \mathbf{R}^m \longrightarrow \mathbf{R}^m$ by $\tau_p(q) = p + q$. Since $\phi \in \Gamma_p(\pi)$, we may define $\psi \in \Gamma_0(\pi)$ by $\psi(q) = (q, pr_2(\phi(\tau_p(q))))$, and clearly $j^1_0\psi$ depends only on the value and first derivatives of ϕ at p. Consequently the map

$$\begin{array}{ccc} J^1\pi & \longrightarrow & J^1_0\pi \\ j^1_p\phi & \longmapsto & j^1_0\psi \end{array}$$

is well-defined, and we may construct a map

$$\begin{array}{ccc} J^1\pi & \longrightarrow & \mathbf{R}^m \times J^1_0\pi \\ j^1_p\phi & \longmapsto & (p, j^1_0\psi). \end{array}$$

It is straightforward to check that this map is a diffeomorphism. ■

Proposition 4.1.21 *If (E, π, M) is a bundle then $(J^1\pi, \pi_1, M)$ is a bundle.*

Proof Let $p \in M$ and let (W_p, F, t_p) be a local trivialisation of π around p, where W_p is sufficiently small to be contained in the domain of a single chart on M. Then (t_p, id_{W_p}) is a bundle isomorphism from the restricted bundle $(\pi^{-1}(W_p), \pi|_{W_p}, W_p)$ to the trivial bundle $(W_p \times F, pr_1, W_p)$. Define the map

$$t_p^1 : J^1(\pi|_{W_p}) \longrightarrow J^1(pr_1)$$

by $t_p^1(j_q^1\phi) = j_q^1(t_p \circ \phi)$. Then t_p^1 is well-defined, and it is a diffeomorphism because t_p is a diffeomorphism. Consequently (t_p^1, id_{W_p}) is a bundle isomorphism.

We now observe that, because we have chosen W_p sufficiently small, it is diffeomorphic to an open subset of $\mathbf{R}^m$, so that the bundle $(W_p \times F, pr_1, W_p)$ is isomorphic to the restriction of the bundle $(\mathbf{R}^m \times F, pr_1, \mathbf{R}^m)$ to the image of W_p. The result now follows from Lemma 4.1.20, Lemma 4.1.14 and the fact that triviality is preserved by bundle isomorphisms. ■

The net result of this discussion is that, starting with a bundle (E, π, M), we obtain the following commutative diagram:

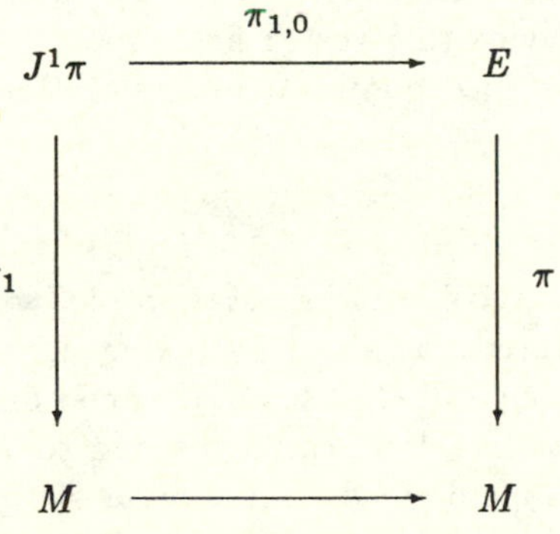

where both the vertical arrows represent bundles, and the horizontal arrow $\pi_{1,0}$ represents, in general, an affine bundle; the points of the jet manifold $J^1\pi$ may be regarded as coordinate-free representations of first-order Taylor polynomials.

Example 4.1.22 If π is the trivial bundle $(\mathbf{R} \times F, pr_1, \mathbf{R})$, then $J_0^1\pi$ is diffeomorphic to TF. To see this, note that if $j_0^1\phi \in J_0^1\pi$, then ϕ is a local section of π defined in a neighbourhood of zero, so that $pr_2 \circ \phi$ is a curve in F which defines the tangent vector $[pr_2 \circ \phi]$. Different representative local sections ϕ give the same tangent vector because the equivalence relations defining both the jet and the tangent involve equality of first derivatives. The correspondence is a diffeomorphism, because the manifold structures on $J^1_0\pi$ and TF are defined in essentially the same way.

In general, when the base manifold of a bundle π is one-dimensional, we shall denote its single coordinate function by t. If q^α are coordinate functions on F, then $(q^\alpha, \dot{q}^\alpha)$ is a coordinate system on TF. On the other hand, (t, q^α) is a coordinate system on $\mathbf{R} \times F$ (where we have, as usual, used the same symbol q^α both for a coordinate function on F and its pull-back to $\mathbf{R} \times F$), and so (q^α, q^α_1) is a coordinate system on $J^1_0\pi$. In these coordinates, the correspondence between $J^1_0\pi$ and TF is just the identity. Furthermore, this diffeomorphism induces a bundle isomorphism between $(J^1_0\pi, \pi_{1,0}|_{\{0\}\times F}, \{0\} \times F)$ and (TF, τ_F, F). ∎

Example 4.1.23 With the same bundle π, the first jet manifold $J^1\pi$ is diffeomorphic to $\mathbf{R} \times TF$. If now $j^1_p\phi \in J^1\pi$ then the corresponding element of $\mathbf{R}\times TF$ is $(p, [pr_2 \circ \phi \circ \tau_p])$, where $\tau_p : \mathbf{R} \longrightarrow \mathbf{R}$ is the translation $q \longmapsto q+p$. Taking coordinates on $\mathbf{R}$ and F as before, the induced coordinate system on $J^1\pi$ is $(t, q^\alpha, q^\alpha_1)$. For this particular bundle π we shall normally identify $J^1\pi$ with $\mathbf{R} \times TF$, and so use coordinates $(t, q^\alpha, \dot{q}^\alpha)$. With this interpretation, a section of $\pi_{1,0}$ corresponds to a vector field along the Cartesian projection $pr_2 : \mathbf{R} \times F \longrightarrow F$, and has coordinate representation

$$X^\alpha \frac{\partial}{\partial q^\alpha}.$$

There is, however, another interpretation of $J^1\pi$, as a submanifold of $T(\mathbf{R} \times F)$. This interpretation arises by taking, for each point $j^1_p\phi \in J^1\pi$, the tangent vector $[\phi \circ \tau_p] \in T_{\phi(p)}(\mathbf{R} \times F)$. The coordinates on $T(\mathbf{R} \times F)$ are $(t, q^\alpha, \dot{t}, \dot{q}^\alpha)$, and the submanifold corresponding to $J^1\pi$ is given by $\dot{t} = 1$. (Note that this submanifold gives a sub-bundle of $\tau_{\mathbf{R}\times F}$ which is an *affine* sub-bundle rather than a vector sub-bundle: although $(J^1\pi, \pi_{1,0}, \mathbf{R} \times F)$ is itself a vector bundle by Proposition 4.1.17, the map $J^1\pi \longrightarrow T(\mathbf{R} \times F)$ is *not* a vector bundle morphism.) Each section X of $\pi_{1,0}$ then gives rise to a section of $\tau_{\mathbf{R}\times F}$, in other words a vector field on $\mathbf{R} \times F$, but in view of the restriction $\dot{t} = 1$ the coordinate representation of the vector field is always of the form

$$\frac{\partial}{\partial t} + X^\alpha \frac{\partial}{\partial q^\alpha}.$$

Such a vector field (in either interpretation) is called a *time-dependent vector field* because the component functions $X^\alpha = \dot{q}^\alpha \circ X$ may depend on the "time" coordinate t as well as the "position" coordinates q^α. If in a particular case the component functions X^α happen to be independent of t, then the vector field is projectable from $\mathbf{R} \times F$ to F, and its projection is just an ordinary vector field on F. ∎

With the machinery of jet bundles at our disposal, we are now in a position to give a coordinate-free definition of a differential equation: it is simply an algebraic equation defined on a jet manifold, where the algebraic equation is expressed as a submanifold.

Definition 4.1.24 Let (E, π, M) be a bundle. A *first-order differential equation* on π is a closed embedded submanifold S of the first jet manifold $J^1\pi$. A *solution* of the differential equation S is a local section $\phi \in \Gamma_W(\pi)$, where W is an open submanifold of M, which satisfies $j^1_p\phi \in S$ for *every* $p \in W$. ∎

Now this definition looks nothing like the usual definition of a differential equation, but we can see the relationship between the two by using coordinates. Choose a point $j^1_p\phi \in S$: note that we are not asserting here that the local section ϕ is a solution of S, because we only know that the jet $j^1_p\phi$ is an element of S for a single $p \in M$. In any event, there is a neighbourhood U^1 of $j^1_p\phi$ and a function $F : U^1 \longrightarrow \mathbf{R}^K$, where $K = \dim J^1\pi - \dim S$, such that $S \cap U^1 = F^{-1}(0)$. We may suppose that U^1 is sufficiently small to be the domain of a jet coordinate system $u^1 : U^1 \longrightarrow \mathbf{R}^N$, where $u^1 = (x^i, u^\alpha, u^\alpha_i)$ and $N = \dim J^1\pi$; the composite map $F \circ (u^1)^{-1}$ then defines a partial differential equation in the traditional sense. The use of a submanifold S is therefore a way of separating the description of the equation from a description of its solutions.

Example 4.1.25 Let π be the trivial bundle $(\mathbf{R}^2 \times \mathbf{R}, pr_1, \mathbf{R})$ with global coordinates $(x^1, x^2; u^1)$. Then the map $F : J^1\pi \longrightarrow \mathbf{R}$ defined by

$$F = u^1_1 u^1_2 - 2x^2 u^1$$

gives rise to the differential equation

$$S = \{j^1_p\phi \in J^1\pi : (u^1_1 u^1_2 - 2x^2 u^1)(j^1_p\phi) = 0\}$$

which in traditional notation would be written

$$\frac{\partial\phi}{\partial x^1}\frac{\partial\phi}{\partial x^2} - 2x^2\phi = 0.$$

The particular section $\phi : \mathbf{R}^2 \longrightarrow \mathbf{R}^2 \times \mathbf{R}$ defined by

$$\phi(p^1, p^2) = (p^1, p^2, p^1(p^2)^2)$$

is a solution of this differential equation, because $j_p^1\phi \in S$ for *every* $p \in \mathbf{R}^2$. ∎

We shall see in later sections how this definition of a differential equation is related to some of the other manifestations of differential equations which appear in differential geometry.

EXERCISES

4.1.1 Let (E, π, M) be a bundle, and let (x^i, u^α) and (y^j, v^β) be two sets of adapted coordinates defined on a neighbourhood U of $a \in E$. Show that, on $\pi_{1,0}^{-1}(U) \subset J^1\pi$, the coordinate differentials dv_j^β transform according to the formula

$$\begin{aligned} dv_j^\beta &= \frac{\partial v^\beta}{\partial u^\alpha}\frac{\partial x^i}{\partial y^j}du_i^\alpha \\ &\quad + \frac{\partial x^i}{\partial y^j}\left(\frac{\partial^2 v^\beta}{\partial u^\alpha\,\partial x^i} + u_i^\gamma \frac{\partial^2 v^\beta}{\partial u^\alpha\,\partial u^\gamma}\right)du^\alpha \\ &\quad + \left(\frac{\partial x^l}{\partial y^j}\left(\frac{\partial^2 v^\beta}{\partial x^i\,\partial x^l} + u_l^\alpha \frac{\partial^2 v^\beta}{\partial u^\alpha\,\partial x^i}\right)\right. \\ &\quad \left. + \frac{\partial y^k}{\partial x^i}\frac{\partial^2 x^l}{\partial y^j\,\partial y^k}\left(\frac{\partial v^\beta}{\partial x^l} + u_l^\alpha \frac{\partial v^\beta}{\partial u^\alpha}\right)\right)dx^i. \end{aligned}$$

Use this formula (and the standard transformation rules for dv^β and dy^j) to determine the corresponding rules for the coordinate vector fields $\partial/\partial y^j$, $\partial/\partial v^\beta$ and $\partial/\partial v_j^\beta$.

4.1.2 Let (E, π, M) be a *vector* bundle. Show directly that $(J^1\pi, \pi_1, M)$ may also be given a natural structure as a vector bundle. (The answer to this exercise will demonstrate why the indirect approach is needed when π is a general bundle.)

4.1.3 Let (E, π, M) and (F, ρ, M) be bundles. Show that there is a canonical diffeomorphism

$$J^1(\pi \times_M \rho) \cong J^1\pi \times_M J^1\rho,$$

where $J^1\pi \times_M J^1\rho$ is the total space of the fibre product bundle

$$(J^1\pi \times_M J^1\rho, \pi_1 \times_M \rho_1, M).$$

4.1.4 Let π be the trivial vector bundle $(M \times \mathbf{R}, pr_1, M)$. Show that π_1 and $\tau_M^* \oplus \pi$ are isomorphic as vector bundles.

4.1.5 Let π be the trivial bundle $(\mathbf{R}^m \times M, pr_1, \mathbf{R}^m)$ and consider the subset of $J^1\pi$ containing those 1-jets $j^1_a\phi$ where the linear map

$$\phi_* : T_p\mathbf{R}^m \longrightarrow T_{\phi(p)}(\mathbf{R}^m \times M)$$

is non-singular. Show that this subset is well-defined and is an open submanifold of $J^1\pi$ which is diffeomorphic to $\mathbf{R}^m \times \mathcal{F}M$, where $\mathcal{F}M$ is the manifold of linear frames on M.

4.1.6 Let ρ be the trivial bundle $(M \times \mathbf{R}^m, pr_1, M)$ and consider the subset of $J^1\rho$ containing those 1-jets $j^1_a\psi$ where the linear map

$$\psi_* : T_aM \longrightarrow T_{\psi(a)}(M \times \mathbf{R}^m)$$

is non-singular. Show that this subset is a well-defined open submanifold of $J^1\rho$ which is diffeomorphic to $\mathcal{F}^*M \times \mathbf{R}^m$, where $\mathcal{F}^*M$ is the manifold of linear coframes on M.

4.1.7 With the same bundles π and ρ as in the previous two exercises, explain how two local sections $\phi \in \Gamma_p(\pi)$, $\psi \in \Gamma_{\phi(p)}(\rho)$ may be considered "mutually inverse" in a neighbourhood of p. Show how this relationship may be used to construct a diffeomorphism between $\mathbf{R}^m \times \mathcal{F}M$ and $\mathcal{F}^*M \times \mathbf{R}^m$ which corresponds to the canonical map from a frame to its dual coframe.

4.1.8 The arguments in Lemma 4.1.20 and Proposition 4.1.21 may be used to show that, if π is the trivial bundle $(M \times F, pr_1, M)$ where M is diffeomorphic to an open subset of $\mathbf{R}^m$, then π_1 is also trivial. Construct an example of a trivial bundle π where π_1 is *not* trivial. (Hint: consider cotangent bundles.)

4.1.9 Let π be the trivial bundle $(M \times \mathbf{R}, pr_1, M)$, so that $J^1\pi$ is diffeomorphic to $T^*M \times \mathbf{R}$. A 1-form ω on M then gives rise to a section $\overline{\omega}$ of the bundle $\pi_{1,0}$, by the rule

$$\overline{\omega}(p, \lambda) = (\omega_p, \lambda).$$

Show that if $\omega = df$, where $f \in C^\infty(M)$, then f is a solution of the differential equation described by the subset $\overline{\omega}(M \times \mathbf{R})$ of $T^*M \times \mathbf{R}$. What happens if ω is not closed?

4.1.10 Let X be a vector field on the manifold F, so that if π is the trivial bundle $(\mathbf{R} \times F, pr_1, \mathbf{R})$ then X defines a section $(id_{\mathbf{R}} \times X)$ of the bundle $\pi_{1,0}$. Show that if $\phi : (a, b) \longrightarrow F$ is an integral curve of X, then the local section $(id_{(a,b)}, \phi)$ of π is a solution of the differential equation $\mathbf{R} \times \mathrm{im}(X) \subset \mathbf{R} \times TF \cong J^1\pi$.

4.2 Prolongations of Morphisms

Corresponding to each local section of the bundle π there is a uniquely determined local section of the bundle π_1. This new section is called the *first prolongation*, and its coordinate representation is obtained by appending to the coordinates of the original section the derivatives of those coordinates. This coordinate representation illustrates that not every section of π_1 is the prolongation of a section of π, and later in this chapter we shall find ways of characterising those sections of π_1 which are prolongations. As a generalisation, we shall also show how to prolong those bundle morphisms which project to diffeomorphisms.

Definition 4.2.1 If (E, π, M) is a bundle, $W \subset M$ is an open submanifold and $\phi \in \Gamma_W(\pi)$ then the *first prolongation of* ϕ is the section $j^1\phi \in \Gamma_W(\pi_1)$ defined by

$$j^1\phi(p) = j^1_p\phi$$

for $p \in W$. ∎

From the definition, $\pi_1 \circ j^1\phi = id_W$, so that $j^1\phi$ is indeed a local section of π_1. Similarly, $\pi_{1,0} \circ j^1\phi = \phi$ so that $j^1(\pi_{1,0} \circ j^1\phi) = j^1\phi$. It is clear that this latter relationship may be used to characterise a prolongation.

Lemma 4.2.2 *If $\psi \in \Gamma_W(\pi_1)$ then there is a local section $\phi \in \Gamma_W(\pi)$ satisfying $\psi = j^1\phi$ if, and only if, $\psi = j^1(\pi_{1,0} \circ \psi)$.* ∎

To find the coordinate representation of $j^1\phi$, we must examine its composition with the fibre coordinate functions u^α and u^α_i. Now

$$\begin{aligned} u^\alpha(j^1\phi(p)) &= u^\alpha(j^1_p\phi) \\ &= u^\alpha(\phi(p)) \\ &= \phi^\alpha(p) \end{aligned}$$

so that $u^\alpha \circ j^1\phi = \phi^\alpha$. Similarly,

$$\begin{aligned} u^\alpha_i(j^1\phi(p)) &= u^\alpha_i(j^1_p\phi) \\ &= \left.\frac{\partial \phi^\alpha}{\partial x^i}\right|_p \end{aligned}$$

so that $u^\alpha_i \circ j^1\phi = \partial\phi^\alpha/\partial x^i$. The coordinate representation of $j^1\phi$ is therefore

$$\left(\phi^\alpha, \frac{\partial \phi^\alpha}{\partial x^i}\right).$$

By contrast, the most general local section $\psi \in \Gamma_W(\pi_1)$ will have coordinates $(\psi^\alpha, \psi^\alpha_i)$ where the functions ψ^α_i need have nothing to do with the functions ψ^α.

Example 4.2.3 Let π be the trivial bundle $(\mathbf{R}^2 \times \mathbf{R}, pr_1, \mathbf{R}^2)$, with global coordinates $(x^1, x^2; u^1)$. If $\phi \in \Gamma(\pi)$ is defined by

$$\phi(p^1, p^2) = (p^1, p^2; p^1 \sin p^2),$$

then in the induced coordinates $(x^1, x^2; u^1; u^1_1, u^1_2)$ on $J^1\pi$ the first prolongation $j^1\phi$ satisfies

$$j^1\phi(p^1, p^2) = (p^1, p^2; p^1 \sin p^2; \sin p^2, p^1 \cos p^2).$$

If, however, $\psi \in \Gamma(\pi_1)$ is defined in these coordinates by

$$\psi(p^1, p^2) = (p^1, p^2; p^1 \sin p^2; p^1 p^2, 0)$$

then, by Lemma 4.2.2, ψ is not the prolongation of a section of π. ∎

Example 4.2.4 More generally, if π is the trivial bundle $(M \times \mathbf{R}, pr_1, M)$, then $J^1\pi \cong T^*M \times \mathbf{R}$, so that a section ψ of π_1 may be written as a pair $(\omega, \overline{\phi})$ where $\omega \in \bigwedge^1 M$ and $\overline{\phi} \in C^\infty(M)$. There may be no relationship between ω and $\overline{\phi}$; if, however, $\psi = j^1\phi$ for some section ϕ of π, then $\overline{\phi} = pr_2 \circ \phi$ and $\omega = d\overline{\phi}$, so that ω is exact. The preceding example may be considered as a special case of this one, where—if $\psi(p^1, p^2) = (p^1, p^2; p^1 \sin p^2; p^1 p^2, 0)$—then $\overline{\phi} = x^1 \sin x^2$ and $\omega = x^1 x^2 dx^1$. ∎

As an application of Definition 4.2.1, we may now restate our definition of the solution of a differential equation. If $S \subset J^1\pi$ is a first-order differential equation, then $\phi \in \Gamma_W(\pi)$ is a solution of S if the first prolongation $j^1\phi$ takes its values in S.

As in example 1.3.10, a section of π may be considered as a special case of a bundle morphism from id_M to π which projects to the identity on M: in other words, the domain of the section consists entirely of "independent variables" with respect to which the differentiation is carried out. A generalisation is to consider the prolongation of a map where only some of the domain variables are considered as independent variables. Such a map would be a bundle morphism projecting to the identity on M. However the generalisation may be extended further, to a bundle morphism between bundles with different base spaces, provided that the projected map is a diffeomorphism.

Definition 4.2.5 Let (E, π, M) and (H, ρ, N) be bundles, and let $(f, \overline{f})$ be a bundle morphism, where $\overline{f}$ is a diffeomorphism. The *first prolongation of* $(f, \overline{f})$ is the map $j^1(f, \overline{f}) : J^1\pi \longrightarrow J^1\rho$ defined by

$$j^1(f, \overline{f})(j^1_p\phi) = j^1_{\overline{f}(p)}(\tilde{f}(\phi))$$

where $\tilde{f}(\phi) = f \circ \phi \circ \overline{f}^{-1}\big|_{\overline{f}^{-1}(\text{domain}\,\phi)}$. If no confusion is possible, the notation $j^1 f$ will be used rather than $j^1(f, \overline{f})$. ∎

For this definition to be valid, we must ensure that choosing a different representative ϕ with the same 1-jet at p gives the same result. As usual, this follows from the Chain Rule, because the right-hand side of the definition just involves the value and first derivatives of ϕ at p. When $M = N$ and $\overline{f} = id_M$, this definition reduces to $j^1 f(j^1_p \phi) = j^1_p(f \circ \phi)$. If $E = M$ (so that $\phi = id_W$ where $W \subset M$, and so that f is a section of π) then the definition collapses completely to $j^1 f(p) = j^1_p f$.

Lemma 4.2.6 *Both* $(j^1 f, f) : \pi_{1,0} \longrightarrow \rho_{1,0}$ *and* $(j^1 f, \overline{f}) : \pi_1 \longrightarrow \rho_1$ *are bundle morphisms.*

Proof If $j^1_p \phi \in J^1\pi$ then

$$\begin{aligned} \rho_{1,0}(j^1 f(j^1_p \phi)) &= \rho_{1,0}(j^1_{\overline{f}(p)}(\tilde{f}(\phi)) \\ &= \tilde{f}(\phi)(\overline{f}(p)) \\ &= (f \circ \phi \circ \overline{f}^{-1})(\overline{f}(p)) \\ &= f(\phi(p)) \\ &= f(\pi_{1,0}(j^1_p \phi)) \end{aligned}$$

so that $\rho_{1,0} \circ j^1 f = f \circ \pi_{1,0}$, as required. If follows that

$$\begin{aligned} \rho_1 \circ j^1 f &= \rho \circ \rho_{1,0} \circ j^1 f \\ &= \rho \circ f \circ \pi_{1,0} \\ &= \overline{f} \circ \pi \circ \pi_{1,0} \\ &= \overline{f} \circ \pi_1. \end{aligned}$$

∎

Lemma 4.2.7 *If* $f : \pi \longrightarrow \rho$ *and* $g : \rho \longrightarrow \sigma$ *are bundle morphisms which project to diffeomorphisms, then* $j^1(g \circ f) = j^1 g \circ j^1 f$ *and* $j^1(id_E) = id_{J^1\pi}$.

Proof Directly from the definitions, using the relationships $\overline{g \circ f} = \overline{g} \circ \overline{f}$, $\widetilde{g \circ f} = \tilde{g} \circ \tilde{f}$ and $\widetilde{id_E} = id_M$. For every $j^1_p \phi \in J^1\pi$,

$$\begin{aligned} j^1(g \circ f)(j^1_p \phi) &= j^1_{\overline{g \circ f}(p)} \widetilde{g \circ f}(\phi) \\ &= j^1_{\overline{g}(\overline{f}(p))} \tilde{g}(\tilde{f}(\phi)) \\ &= j^1 g(j^1_{\overline{f}(p)} \tilde{f}(\phi)) \\ &= j^1 g(j^1 f(j^1_p \phi)) \end{aligned}$$

and

$$\begin{aligned} j^1(id_E)(j^1_p\phi) &= j^1_{\overline{id_E}(p)}\widetilde{id_E}(\phi) \\ &= j^1_p\phi. \end{aligned}$$

■

We may use Lemma 4.2.7 to rewrite the definition of the first prolongation in a very suggestive way. Since $(j^1f, \overline{f})$ is a bundle morphism and $\overline{f}$ is a diffeomorphism, we may write $\widetilde{j^1f}(\psi)$ for $j^1f \circ \psi \circ \overline{f}^{-1}$ whenever ψ is a section of π_1. Using this notation, the definition just becomes

$$\begin{aligned} \widetilde{j^1f}(j^1\phi)(q) &= j^1f(j^1\phi(\overline{f}^{-1}(q))) \\ &= j^1f(j^1_{\overline{f}^{-1}(q)}\phi) \\ &= j^1_q(\tilde{f}(\phi)) \end{aligned}$$

where $q \in N$, so that

$$\widetilde{j^1f}(j^1\phi) = j^1(\tilde{f}(\phi)).$$

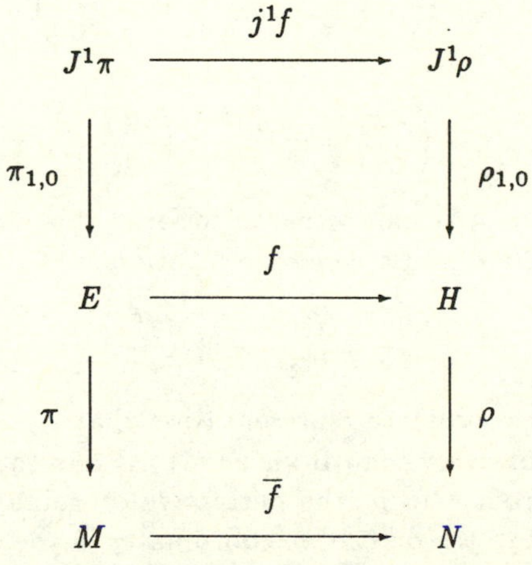

The coordinate representation of j^1f may be obtained by taking its composition with the coordinate functions y^j, v^β and v^β_j on $J^1\rho$ (where $1 \leq j \leq$

$\dim N = \dim M$, and $1 \leq \beta \leq \dim H - \dim N$). With the usual understanding about similarly-named functions related by bundle projections,

$$\begin{aligned} y^j \circ j^1 f &= y^j \circ \rho_1 \circ j^1 f \\ &= y^j \circ \overline{f} \\ &= f^j \end{aligned}$$

and

$$\begin{aligned} v^\beta \circ j^1 f &= v^\beta \circ \rho_{1,0} \circ j^1 f \\ &= v^\beta \circ f \\ &= f^\beta . \end{aligned}$$

Finally, if $j^1_p \phi \in J^1 \pi$ then

$$\begin{aligned} v^\beta_j (j^1 f (j^1_p \phi)) &= v^\beta_j (j^1_{\overline{f}(p)} (\tilde{f}(\phi)) \\ &= \left. \frac{\partial (v^\beta \circ \tilde{f}(\phi))}{\partial y^j} \right|_{\overline{f}(p)} \\ &= \left. \frac{\partial (f^\beta \circ \phi \circ \overline{f}^{-1})}{\partial y^j} \right|_{\overline{f}(p)} \\ &= \left(\left. \frac{\partial f^\beta}{\partial x^i} \right|_p + \left. \frac{\partial \phi^\alpha}{\partial x^i} \right|_p \left. \frac{\partial f^\beta}{\partial u^\alpha} \right|_{\phi(p)} \right) \left. \frac{\partial (\overline{f}^{-1})^i}{\partial y^j} \right|_{\overline{f}(p)} \end{aligned}$$

so that

$$v^\beta_j \circ j^1 f = \left(\frac{\partial f^\beta}{\partial x^i} + u^\alpha_i \frac{\partial f^\beta}{\partial u^\alpha} \right) \left(\frac{\partial (\overline{f}^{-1})^i}{\partial y^j} \circ \overline{f} \right).$$

The expression in the first pair of parentheses in this last equation is often called a *total derivative*, and a common notation is

$$\frac{df^\beta}{dx^i} = \frac{\partial f^\beta}{\partial x^i} + u^\alpha_i \frac{\partial f^\beta}{\partial u^\alpha}.$$

It follows from this coordinate representation that—as with the prolongation of sections—not every map from $J^1\pi$ to $J^1\rho$ is the prolongation of a bundle morphism from π to ρ: the derivative coordinates of the image in $J^1\rho$ must be related to the derivative coordinates in the domain $J^1\pi$ in this inhomogeneous linear way. We shall discuss total derivatives in detail in Section 4.3. For the moment, we simply record that this coordinate representation demonstrates that $(j^1 f, f)$ is always an affine bundle morphism.

Proposition 4.2.8 *The bundle morphism* $(j^1 f, f) : \pi_{1,0} \longrightarrow \rho_{1,0}$ *is an affine bundle morphism.*

Proof The coordinate representation above shows that, on each fibre of $\pi_{1,0}$, $v^\beta_j \circ j^1 f$ is an inhomogeneous linear function of the u^α_i coordinates. ■

Example 4.2.9 Let π be the trivial bundle $(\mathbf{R}^2 \times \mathbf{R}, pr_1, \mathbf{R}^2)$, with global coordinates $(x^1, x^2; u^1)$. Let $(f, id_{\mathbf{R}}) : \pi \longrightarrow \pi$ be defined by

$$f(p^1, p^2; a^1) = (p^1, p^2; p^1 \sin a^1 + 3p^2)$$

so that $f^1 = x^1 \sin u^1 + 3x^2$. Then

$$\begin{aligned} \frac{df^1}{dx^1} &= \sin u^1 + u^1_1 x^1 \cos u^1 \\ \frac{df^1}{dx^2} &= 3 + u^1_2 x^1 \cos u^1 \end{aligned}$$

so that

$$\begin{aligned} & j^1 f(p^1, p^2; a^1; a^1_1, a^1_2) \\ & = (p^1, p^2; p^1 \sin a^1 + 3p^2; \sin a^1 + a^1_1 p^1 \cos a^1, 3 + a^1_2 p^1 \cos a^1). \end{aligned}$$

■

Example 4.2.10 Now let π be the trivial bundle $(\mathbf{R} \times F, pr_1, \mathbf{R})$, and let ρ be the trivial bundle $(\mathbf{R} \times K, pr_1, \mathbf{R})$. Let $(id_{\mathbf{R}} \times f, id_{\mathbf{R}})$ be a bundle morphism from π to ρ. Using the identifications $J^1\pi \cong \mathbf{R} \times TF$ and $J^1\rho \cong \mathbf{R} \times TK$, the prolongation $j^1(id_{\mathbf{R}} \times f)$ may be regarded as a map from $\mathbf{R} \times TF$ to $\mathbf{R} \times TK$. If $(p, \xi) \in \mathbf{R} \times TF$, where $\xi = [\gamma]$ for some curve γ in F, then

$$\begin{aligned} j^1(id_{\mathbf{R}} \times f)(p, \xi) &= j^1(id_{\mathbf{R}} \times f)(j^1_p(id_{\mathbf{R}}, \gamma \circ \tau_{-p})) \\ &= j^1_p(id_{\mathbf{R}}, f \circ \gamma \circ \tau_{-p}) \\ &= (p, [f \circ \gamma]) \\ &= (p, f_*(\xi)), \end{aligned}$$

so that $j^1(id_{\mathbf{R}} \times f) = id_{\mathbf{R}} \times f_*$. Using coordinates (t, r^β) on $\mathbf{R} \times K$,

$$\dot{r}^\beta(j^1(id_{\mathbf{R}} \times f)) = \dot{q}^\alpha \frac{\partial f^\beta}{\partial q^\alpha}$$

since $\partial(id_{\mathbf{R}} \times f)/\partial t = 0$. ■

Example 4.2.11 Now let π be the trivial bundle $(M \times \mathbf{R}, pr_1, M)$, and let ρ be the trivial bundle $(N \times \mathbf{R}, pr_1, N)$. Let $(f \times id_{\mathbf{R}})$ be a bundle morphism from π to ρ, where f is a diffeomorphism. There are now identifications

$J^1\pi \cong T^*M \times \mathbf{R}$ and $J^1\rho \cong T^*N \times \mathbf{R}$, so that the prolongation $j^1(f \times id_{\mathbf{R}})$ may be regarded as a map from $T^*M \times \mathbf{R}$ to $T^*N \times \mathbf{R}$. If $(\eta, q) \in T^*M \times \mathbf{R}$, then there is always a function $\phi \in C^\infty(M)$ such that $\eta = d\phi_{\tau^*_M(\eta)}$ and $q = \phi(\tau^*_M(\eta))$. Then

$$\begin{aligned} j^1(f \times id_{\mathbf{R}})(\eta, q) &= j^1_{f(\tau^*_M(\eta))}(\widetilde{(f \times id_{\mathbf{R}})}(id_M, \phi)) \\ &= j^1_{f(\tau^*_M(\eta))}(id_N, \phi \circ f^{-1}) \\ &= (d(\phi \circ f^{-1})_{f(\tau^*_M(\eta))}, \phi(\tau^*_M(\eta))) \\ &= (f^{-1*}(d\phi_{\tau^*_M(\eta)}), q) \end{aligned}$$

so that $j^1(f \times id_{\mathbf{R}}) = f^{-1*} \times id_{\mathbf{R}}$. ■

Example 4.2.12 If (E, π, M) is a bundle, and if (W_p, F, t_p) is a local trivialisation of π around $p \in M$, then the map t^1_p (used in Proposition 4.1.21 to construct a local trivialisation of π_1) is just the prolongation $j^1(t_p, id_M)$. ■

As an application of this process of prolonging a bundle morphism, we may define a *symmetry* of a differential equation $S \subset J^1\pi$. This is a bundle isomorphism $(f, \bar{f})$ of π with itself, such that $\tilde{f}(\phi)$ is a solution of S exactly when ϕ is a solution. (Strictly speaking, this should be called a *point symmetry*: it is also possible to define *generalised symmetries*, which are not derived from bundle morphisms of π.) Using the definition of a prolongation, we may express the requirement of a symmetry by demanding that $j^1(\tilde{f}(\phi)) = \widetilde{j^1 f}(j^1\phi)$ takes its values in S whenever $j^1\phi$ does, and this will be the case when $j^1 f(S) = S$.

In Lemma 4.2.7 we described the composition of two bundle morphisms, and a particular case of this arises when (E, π, M) and (H, ρ, E) are two bundles, and when $\phi \in \Gamma(\pi)$, $\psi \in \Gamma(\rho)$. For simplicity we shall consider global sections, although the discussion applies equally to local sections where $\operatorname{im}(\phi) \cap \operatorname{domain}(\psi)$ is non-empty, and where one keeps track of all the domains. Now (ϕ, id_M) is a bundle morphism from (M, id_M, M) to (E, π, M), and (ψ, id_M) is a bundle morphism from (E, π, M) to the bundle $(H, \pi \circ \rho, M)$. The composite $\psi \circ \phi$ is a section of $\pi \circ \rho$, so that $(\psi \circ \phi, id_M)$ is a bundle morphism from (M, id_M, M) to $(H, \pi \circ \rho, M)$. By Lemma 4.2.7 we have

$$j^1(\psi \circ \phi, id_M) = j^1(\psi, id_M) \circ j^1(\phi, id_M)$$

where we must use the explicit notation for the prolongations of these bundle morphisms to avoid confusion.

$$
\begin{array}{ccccc}
J^1(id_M) \cong M & \xrightarrow{\quad j^1(\phi, id_M) \quad} & J^1\pi & \xrightarrow{\quad j^1(\psi, id_M) \quad} & J^1(\pi \circ \rho) \\
\Big\downarrow & & \Big\downarrow & & \Big\downarrow \\
M & \xrightarrow{\quad \phi \quad} & E & \xrightarrow{\quad \psi \quad} & H \\
{\scriptstyle id_M}\Big\downarrow & & {\scriptstyle \pi}\Big\downarrow & & \Big\downarrow{\scriptstyle \pi \circ \rho} \\
M & \xrightarrow[\quad id_M \quad]{} & M & \xrightarrow[\quad id_M \quad]{} & M
\end{array}
$$

Now $j^1(\phi, id_M)$ and $j^1(\psi \circ \phi, id_M)$ are just the prolongations of sections $j^1\phi$ and $j^1(\psi \circ \phi)$. However, $j^1(\psi, id_M)$ is *not* the same as $j^1\psi$, for the former is a map $J^1\pi \longrightarrow J^1(\pi \circ \rho)$, whereas the latter is a map $E \longrightarrow J^1\rho$ (and, of course, is just $j^1(\psi, id_E)$). To construct $j^1\psi$, *all* the coordinates in E are regarded as independent variables for the purposes of differentiation; to construct $j^1(\psi, id_M)$, only those coordinates pulled back from M are regarded as independent. To find the relationship between $j^1(\psi \circ \phi)$, $j^1\psi$ and $j^1\phi$, we need to use a canonical map $\kappa_1 : J^1\pi \times_E J^1\rho \longrightarrow J^1(\pi \circ \rho)$ which in effect incorporates the chain rule.

Definition 4.2.13 The map $\kappa_1 : J^1\pi \times_E J^1\rho \longrightarrow J^1(\pi \circ \rho)$ is defined by

$$\kappa_1(j^1_p\phi, j^1_{\phi(p)}\psi) = j^1_p(\psi \circ \phi)$$

where $\phi \in \Gamma_p(\pi)$, $\psi \in \Gamma_{\phi(p)}(\rho)$. ∎

The map is well-defined because it depends only on the value and first derivatives of ϕ and ψ. To see this explicitly, we shall examine its coordinate representation. Let coordinates on M, E and H be x^i, (x^i, u^α) and (x^i, u^α, v^A) respectively. Then coordinates on the jet manifolds are

$$
\begin{aligned}
J^1\pi : &\quad (x^i, u^\alpha, u^\alpha_i) \\
J^1\rho : &\quad (x^i, u^\alpha, v^A, \overline{v}^A_i, \overline{v}^A_\alpha) \\
J^1(\pi \circ \rho) : &\quad (x^i, u^\alpha, v^A, u^\alpha_i, v^A_i)
\end{aligned}
$$

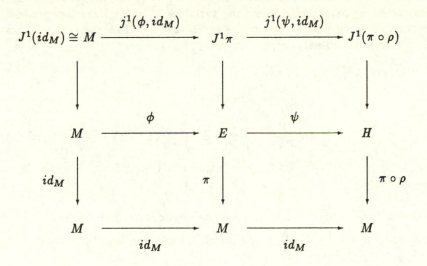

where the bar on the coordinates $\overline{v}^A_i$, $\overline{v}^A_\alpha$ on $J^1\rho$ is to indicate that they are constructed by assuming that *all* the variables x^i and u^α are independent (so that, in particular, $\overline{v}^A_i$ is not the same as v^A_i). Then the map κ_1 does not affect the coordinates x^i, u^α, v^A or u^α_i; furthermore,

$$\begin{aligned} v^A_i(\kappa_1(j^1_p\phi, j^1_{\phi(p)}\psi)) &= v^A_i(j^1_p(\psi \circ \phi)) \\ &= \left.\frac{\partial}{\partial x^i}\right|_p (v^A \circ \psi \circ \phi) \\ &= \left.\frac{\partial}{\partial u^\alpha}\right|_{\phi(p)} (v^A \circ \psi) \left.\frac{\partial}{\partial x^i}\right|_p (u^\alpha \circ \phi) + \left.\frac{\partial}{\partial x^i}\right|_{\phi(p)} (v^A \circ \psi) \\ &= \overline{v}^A_\alpha(j^1_{\phi(p)}\psi)u^\alpha_i(j^1_p\phi) + \overline{v}^A_i(j^1_{\phi(p)}\psi) \\ &= (\overline{v}^A_\alpha u^\alpha_i + \overline{v}^A_i)(j^1_p\phi, j^1_{\phi(p)}\psi), \end{aligned}$$

so that $v^A_i \circ \kappa_1 = \overline{v}^A_\alpha u^\alpha_i + \overline{v}^A_i$, which is the essence of the chain rule.

Proposition 4.2.14 *If (E, π, M) and (H, ρ, E) are bundles, and if $\phi \in \Gamma(\pi)$, $\psi \in \Gamma(\rho)$, then*

$$j^1(\psi \circ \phi) = \kappa_1(j^1\phi, (j^1\psi) \circ \phi).$$

■

EXERCISES

4.2.1 Prove that if (f, id_M) is a bundle morphism then $j^1 f = id_{J^1\pi}$ if, and only if, $f = id_E$.

4.2.2 Suppose that (E, π, M) and (H, ρ, N) are *vector* bundles, and that $(f, \overline{f})$ is a vector bundle morphism (where $\overline{f}$ is a diffeomorphism). Show that $(j^1 f, \overline{f})$ is a vector bundle morphism.

4.2.3 For an arbitrary bundle π, a section X of $\pi_{1,0}$ may be regarded as a bundle morphism (X, id_M) from π to π_1. Suppose that, for each $p \in M$, there is a local section $\phi \in \Gamma_p(\pi)$ satisfying

$$X \circ \phi = j^1\phi.$$

Show that the image of the composite map

$$j^1(X, id_M) \circ X : E \longrightarrow J^1\pi_1$$

must lie in the subset of $J^1\pi_1$ containing points $j^1_p\psi$ ($\psi \in \Gamma_p(\pi_1)$) satisfying

$$j^1(\pi_{1,0}, id_M)(j^1_p\psi) = (\pi_1)_{1,0}(j^1_p\psi)$$

(see Section 5.3).

4.2.4 Let G be a Lie group, let π be the trivial bundle $(\mathbf{R}\times(G\times G), pr_1, \mathbf{R})$, and let ρ be the trivial bundle $(\mathbf{R}\times G, pr_1, \mathbf{R})$. Let $\mu : G\times G \longrightarrow G$ denote group multiplication, and let $f = (id_{\mathbf{R}}\times\mu, id_{\mathbf{R}})$ be the corresponding bundle morphism from π to ρ. Use the identifications $J^1\pi \cong \mathbf{R}\times T(G\times G) \cong \mathbf{R}\times TG\times TG$ and $J^1\rho \cong \mathbf{R}\times TG$ to show that the prolonged map $j^1 f : J^1\pi \longrightarrow J^1\rho$ projects to a map $T\mu : TG\times TG \longrightarrow TG$, and that $T\mu$ defines a group operation on TG. (According to Example 4.2.10, $T\mu$ is just μ_*.) Show that, if $g, h\in G$ and $\xi\in T_gG$, $\eta\in T_hG$ then

$$T\mu(\xi,\eta) = R_{h*}(\xi) + L_{g*}(\eta) \in T_{gh}G,$$

where $L_g, R_h : G \longrightarrow G$ are left and right translations respectively.

4.3 Total Derivatives and Contact Forms

As we hinted at the beginning of this chapter, the bundle $(J^1\pi, \pi_{1,0}, E)$ has a particularly rich structure. We have already seen that it is an affine bundle, and in the present section we shall investigate the pull-back bundles $\pi^*_{1,0}(\tau_E)$ and $\pi^*_{1,0}(\tau^*_E)$. In Chapter 3 we saw that (TE, τ_E, E) and (T^*E, τ^*_E, E) had distinguished sub-bundles, namely the bundles of vertical tangent vectors and horizontal cotangent vectors respectively. We also saw that these sub-bundles did not have distinguished complements in the absence of a connection on π. This fact is clear in coordinates, because tangent and cotangent vectors of the form

$$\xi^i \left.\frac{\partial}{\partial x^i}\right|_a \qquad \text{and} \qquad \eta_\alpha\, du^\alpha|_a$$

need not maintain their form under a general change of adapted coordinates.

Surprisingly, therefore, when these sub-bundles are pulled back to $J^1\pi$ by $\pi_{1,0}$, they *do* have distinguished complements, and these complements are called the bundles of holonomic tangent vectors and contact cotangent vectors. Sections of these latter two bundles are called total derivatives and contact forms. We have already seen the action of a total derivative as a derivation, for if f^α are the coordinates of a bundle morphism (f, id_M) then the derivative coordinates of its prolongation are

$$\left(\frac{\partial}{\partial x^i} + u^\beta_i \frac{\partial}{\partial u^\beta}\right) f^\alpha,$$

where the operator in brackets maps a function on E to a function on $J^1\pi$. It may easily be seen that a change of coordinates maintains the form of these operators, although they are vector fields along $\pi_{1,0}$ (rather than on $J^1\pi$). The dual objects to these have coordinate representation

$$du^\alpha - u^\alpha_j dx^j$$

and they, too, maintain their form under a change of coordinates; these latter objects, however, may legitimately be regarded as differential forms on $J^1\pi$, using the interpretation of the total space $\pi_{1,0}^*(T^*E)$ as a submanifold of $T^*J^1\pi$.

Since all these objects may be constructed without choosing any particular section of $\pi_{1,0}$, they may be regarded as capturing the intrinsic structure of that bundle; in particular, they describe the relationship between the independent coordinate functions $u_i^\alpha \in C^\infty(J^1\pi)$, and those functions pulled back by a prolongation to M, $(j^1\phi)^*(u_i^\alpha) = \partial\phi^\alpha/\partial x^i \in C^\infty(M)$.

We shall start our discussion by considering tangent vectors.

Definition 4.3.1 Let (E, π, M) be a bundle, and let $p \in M$, $\phi \in \Gamma_p(\pi)$ and $\zeta \in T_pM$. The *holonomic lift of ζ by ϕ* is defined to be

$$(\phi_*(\zeta), j_p^1\phi) \in \pi_{1,0}^*(TE).$$

∎

At first sight, there may seem to be no particular reason for considering the pair $(\phi_*(\zeta), j_p^1\phi)$ rather than simply using the tangent vector $\phi_*(\zeta) \in T_{\phi(p)}E$. However, the construction described in the following theorem is not possible on TE because, given $a \in E_p$ and $\zeta \in T_pM$, there are many possible image vectors $\phi_*(\zeta) \in T_aE$ for different sections ϕ satisfying $\phi(p) = a$.

Theorem 4.3.2 *Let (E, π, M) be a bundle, and let $j_p^1\phi \in J^1\pi$. There is then a canonical decomposition of the vector space $\pi_{1,0}^*(TE)_{j_p^1\phi}$ as a direct sum of two subspaces*

$$\pi_{1,0}^*(V\pi)_{j_p^1\phi} \oplus \phi_*(T_pM),$$

where $\phi_*(T_pM)$ denotes the collection of holonomic lifts of tangent vectors in T_pM by ϕ.

Proof Note first that, since ϕ_* depends only upon the value and first derivatives of ϕ at p, the holonomic lift of a tangent vector is completely determined by $j_p^1\phi$, and does not depend on the choice of the section ϕ. In particular, the set $\phi_*(T_pM)$ is well-defined, and is clearly a subspace of $\pi_{1,0}^*(TE)_{j_p^1\phi}$.

Now suppose that $(\xi, j_p^1\phi) \in \pi_{1,0}^*(TE)_{j_p^1\phi}$; then

$$(\phi_*(\pi_*(\xi)), j_p^1\phi) \in \phi_*(T_pM),$$

and from $\pi_*(\xi - \phi_*(\pi_*(\xi))) = 0$ it follows that

$$(\xi - \phi_*(\pi_*(\xi)), j_p^1\phi) \in \pi_{1,0}^*(V\pi)_{j_p^1\phi}.$$

On the other hand, if

$$(\xi, j^1_p\phi) \in \pi^*_{1,0}(V\pi)_{j^1_p\phi} \cap \phi_*(T_pM)$$

then $\pi_*(\xi) = 0$; but $\xi = \phi_*(\zeta)$ for some $\zeta \in T_pM$ so that $\zeta = \pi_*(\phi_*(\zeta)) = 0$, and hence $\xi = 0$. ∎

Corollary 4.3.3 *The vector bundle $(\pi^*_{1,0}(TE), \pi^*_{1,0}(\tau_E), J^1\pi)$ may be written as the direct sum of two sub-bundles*

$$(\pi^*_{1,0}(V\pi) \oplus H\pi_{1,0},\ \pi^*_{1,0}(\tau_E),\ J^1\pi)$$

where $H\pi_{1,0}$ is the union of the fibres $\phi_*(T_pM)$ for $p \in M$. ∎

To obtain the coordinate representation of a holonomic lift, suppose that

$$\zeta = \zeta^i \left.\frac{\partial}{\partial x^i}\right|_p ;$$

then

$$\begin{aligned}
\phi_*(\zeta) &= \zeta^i \phi_*\left(\left.\frac{\partial}{\partial x^i}\right|_p\right) \\
&= \zeta^i \left(\left.\frac{\partial}{\partial x^i}\right|_{\phi(p)} + \left.\frac{\partial \phi^\alpha}{\partial x^i}\right|_p \left.\frac{\partial}{\partial u^\alpha}\right|_{\phi(p)}\right) \\
&= \zeta^i \left(\left.\frac{\partial}{\partial x^i}\right|_{\phi(p)} + u^\alpha_i(j^1_p\phi) \left.\frac{\partial}{\partial u^\alpha}\right|_{\phi(p)}\right).
\end{aligned}$$

The decomposition of $(\xi, j^1_p\phi) \in \pi^*_{1,0}(TE)_{j^1_p\phi}$ may then be found by letting

$$\xi = \xi^i \left.\frac{\partial}{\partial x^i}\right|_{\phi(p)} + \xi^\alpha \left.\frac{\partial}{\partial u^\alpha}\right|_{\phi(p)},$$

so that

$$\begin{aligned}
\xi = {} & \left(\xi^\alpha - \xi^i u^\alpha_i(j^1_p\phi)\right) \left.\frac{\partial}{\partial u^\alpha}\right|_{\phi(p)} \\
& + \xi^i \left(\left.\frac{\partial}{\partial x^i}\right|_{\phi(p)} + u^\alpha_i(j^1_p\phi) \left.\frac{\partial}{\partial u^\alpha}\right|_{\phi(p)}\right).
\end{aligned}$$

One way to describe a holonomic tangent vector is to say that it is in the image of ϕ_* for some local section ϕ. The dual construction is that of the contact cotangent vector, which may be described as being in the kernel of ϕ^*.

Definition 4.3.4 An element $(\eta, j^1_p\phi)$ of $\pi^*_{1,0}(T^*E)$ is called a *contact cotangent vector* if $\phi^*(\eta) = 0$. ∎

It is necessary to check that the vanishing of $\phi^*(\eta)$ does not depend on the particular choice of local section ϕ, but this too is straightforward, because ϕ^* depends only on the value and the first derivatives of ϕ at p, and so is completely determined by $j^1_p\phi$.

The justification for referring to a duality between holonomic tangent vectors and contact cotangent vectors comes from the duality between the pull-back bundles $\pi^*_{1,0}(\tau_E)$ and $\pi^*_{1,0}(\tau^*_E)$; this is a consequence of Lemma 3.1.9. The decomposition of the former bundle in terms of vertical and holonomic tangent vectors is matched by a decomposition of the latter in terms of horizontal and contact cotangent vectors: the contact and holonomic elements annihilate each other, as do the horizontal and vertical elements.

Proposition 4.3.5 *Let (E, π, M) be a bundle and let $j^1_p\phi \in J^1\pi$. Then*

$$\pi^*_{1,0}(\pi^*(T^*M))_{j^1_p\phi} = (\pi^*_{1,0}(V\pi)_{j^1_p\phi})^\circ$$

and

$$\pi^*_{1,0}(\ker\phi^*) = (\phi_*(T_pM))^\circ,$$

where $\pi^*(T^*M)$ is regarded as a submanifold of T^*E, and $\pi^*_{1,0}(\ker\phi^*)$ denotes the set of contact cotangent vectors in $\pi^*_{1,0}(T^*E)_{j^1_p\phi}$.

Proof The first assertion follows from Lemma 3.1.11. To prove the second, suppose that $(\eta, j^1_p\phi) \in \pi^*_{1,0}(\ker\phi^*)$. If $(\xi, j^1_p\phi) \in \phi_*(T_pM)$, then $\xi = \phi_*(\zeta)$ where $\zeta \in T_pM$. It follows that

$$\begin{aligned} \eta(\xi) &= \eta(\phi_*(\zeta)) \\ &= \phi^*(\eta)(\zeta) \\ &= 0, \end{aligned}$$

so that $(\xi, j^1_p\phi) \in \ker(\eta, j^1_p\phi)$, and hence that $\pi^*_{1,0}(\ker\phi^*) \subset \phi_*(T_pM)^\circ$. To prove equality, observe that

$$\dim \pi^*_{1,0}(\ker\phi^*) = \dim\ker\phi^* = n,$$

whereas

$$\dim\phi_*(T_pM) = \dim T_pM = m$$

so that $\dim\phi_*(T_pM)^\circ = (m + n) - m$. ∎

Theorem 4.3.6 *Let (E, π, M) be a bundle, and let $j^1_p\phi \in J^1\pi$. There is then a canonical decomposition of the vector space $\pi^*_{1,0}(T^*E)_{j^1_p\phi}$ as a direct sum*

$$\pi^*_{1,0}(\pi^*(T^*M))_{j^1_p\phi} \oplus \pi^*_{1,0}(\ker\phi^*).$$

Proof This follows directly from Theorem 4.3.2 by duality, using Proposition 4.3.5. ∎

Corollary 4.3.7 *The vector bundle* $(\pi_{1,0}^*(T^*E), \pi_{1,0}^*(\tau_E^*), J^1\pi)$ *may be written as the direct sum of two sub-bundles*

$$(\pi_{1,0}^*(\pi^*(T^*M)) \oplus C^*\pi_{1,0},\ \pi_{1,0}^*(\tau_E^*),\ J^1\pi)$$

where $C^*\pi_{1,0}$ *is the union of the fibres* $\pi_{1,0}^*(\ker\phi^*)$ *for* $p \in M$. ∎

To express a contact cotangent vector in coordinates, suppose that

$$\eta = \eta_\alpha\, du^\alpha|_{j_p^1\phi} + \eta_i\, dx^i\Big|_{j_p^1\phi} \in \pi_{1,0}^*(\ker\phi^*).$$

Then

$$\eta_\alpha\, d\phi^\alpha|_p + \eta_i\, dx^i\Big|_p = 0,$$

so that

$$\eta_\alpha\, \frac{\partial\phi^\alpha}{\partial x^i}\bigg|_p + \eta_i = 0$$

for each index i. Consequently

$$\eta_\alpha u_i^\alpha(j_p^1\phi) + \eta_i = 0,$$

and so

$$\eta = \eta_\alpha(du^\alpha - u_i^\alpha dx^i)_{j_p^1\phi}.$$

We shall need to adopt some notation for the sections of the various bundles we have constructed. We have already denoted the module of vector fields along $\pi_{1,0}$ by $\mathcal{X}(\pi_{1,0})$, where (as in Example 1.4.6) we shall often regard such a vector field as a map $J^1\pi \longrightarrow TE$ rather than $J^1\pi \longrightarrow \pi_{1,0}^*(TE)$. We shall regard the submanifolds $\pi_{1,0}^*(V\pi)$ and $H\pi_{1,0}$ of $\pi_{1,0}^*(TE)$ as containing vertical and horizontal vectors respectively.

Definition 4.3.8 The submodule of $\mathcal{X}(\pi_{1,0})$ corresponding to sections of the bundle $\pi_{1,0}^*(\tau_E)\Big|_{\pi_{1,0}^*(V\pi)}$ will be denoted by $\mathcal{X}^v(\pi_{1,0})$, and the submodule corresponding to sections of the bundle $\pi_{1,0}^*(\tau_E)\Big|_{H\pi_{1,0}}$ will be denoted by $\mathcal{X}^h(\pi_{1,0})$. An element of $\mathcal{X}^h(\pi_{1,0})$ will be called a *total derivative*. ∎

Example 4.3.9 Each vector field $X \in \mathcal{X}(M)$ corresponds to a total derivative $X^0 \in \mathcal{X}^h(\pi_{1,0})$ according to the rule

$$X^0_{j^1_p\phi} = \phi_*(X_p).$$

Not every total derivative is of this form, for the set $\{X^0 : X \in \mathcal{X}(M)\}$ is a module over $\pi_1^*(C^\infty(M))$ rather than over $C^\infty(J^1\pi)$. If, however, (X_i) is a basis for $\mathcal{X}(M)$ over $C^\infty(M)$, then (X_i^0) is a basis for $\mathcal{X}^h(\pi_{1,0})$ over $C^\infty(J^1\pi)$. ■

Theorem 4.3.10 *The module $\mathcal{X}(\pi_{1,0})$ may be written as the direct sum of its two submodules*

$$\mathcal{X}^v(\pi_{1,0}) \oplus \mathcal{X}^h(\pi_{1,0}).$$

Proof This follows from Corollary 4.3.3 by the standard properties of vector bundles. ■

In coordinates, if $X \in \mathcal{X}^h(\pi_{1,0})$ then

$$X = X^i \left(\frac{\partial}{\partial x^i} + u^\alpha_i \frac{\partial}{\partial u^\alpha} \right)$$

for some functions X^i defined locally on $J^1\pi$. The total derivatives

$$\frac{\partial}{\partial x^i} + u^\alpha_i \frac{\partial}{\partial u^\alpha}$$

are called *coordinate total derivatives* and usually written as d/dx^i. If $f \in C^\infty(E)$ then the action of X as a derivation of type d_* yields a function $d_X f \in C^\infty(J^1\pi)$, and in coordinates

$$d_X f = X^i \frac{df}{dx^i}.$$

In particular, the action of d/dx^i on the coordinate functions u^α gives the result one would expect:

$$\frac{du^\alpha}{dx^i} = u^\alpha_i \in C^\infty(J^1\pi).$$

Example 4.3.11 Let π be the trivial bundle $(\mathbf{R} \times F, pr_1, \mathbf{R})$, where q^α are coordinates on F and t is the identity coordinate on $\mathbf{R}$. The canonical vector field d/dt on $\mathbf{R}$ gives rise to the total derivative

$$\frac{d}{dt} = \frac{\partial}{\partial t} + \dot{q}^\alpha \frac{\partial}{\partial q^\alpha},$$

where we have written $(t, q^\alpha, \dot{q}^\alpha)$ for coordinates on $J^1\pi \cong \mathbf{R} \times TF$. This vector field along $\pi_{1,0}$ is called the *total time derivative.*

If $f \in C^\infty(F)$, then f may be pulled back to $\mathbf{R}\times F$; the resulting function

$$\frac{df}{dt} = \dot{q}^\alpha \frac{\partial f}{\partial q^\alpha} \in C^\infty(\mathbf{R} \times TF)$$

is independent of t, and therefore defines a function $df/dt \in C^\infty(TF)$. This is the "total time derivative of f" which is used in the study of autonomous mechanical systems. ∎

As far as the bundles of cotangent vectors are concerned, we have already denoted the module of sections of $\pi_{1,0}^*(T^*E)$ by $\bigwedge_0^1\pi_{1,0}$, which is regarded as a submodule of $\bigwedge^1 J^1\pi$ using the interpretation of $\pi_{1,0}^*(T^*E)$ as a submanifold of $T^*J^1\pi$. We also have a notation for the module of sections of $\pi_{1,0}^*(\tau_E^*|_{\pi^*(T^*M)})$, for this bundle may equally be interpreted as

$$\tau_{J^1\pi}^*|_{\pi_{1,0}^*(\pi^*(T^*M))} = \tau_{J^1\pi}^*|_{\pi_1^*(T^*M)}$$

with a total space containing cotangent vectors pulled back from M to $J^1\pi$ by π_1: the module of its sections is therefore denoted $\bigwedge_0^1\pi_1$.

Definition 4.3.12 The module of sections of $\pi_{1,0}^*(\tau_E^*)\big|_{C^*\pi_{1,0}}$ is denoted by $\bigwedge_C^1\pi_{1,0}$, and an element of this module is called a *contact form.* ∎

Theorem 4.3.13 *The module $\bigwedge_0^1\pi_{1,0}$ may be written as the direct sum of its two submodule*

$$\textstyle\bigwedge_0^1\pi_1 \oplus \bigwedge_C^1\pi_{1,0}.$$

Proof This follows from Corollary 4.3.7 by the standard properties of vector bundles. ∎

The importance of contact forms arises from their relationship with prolongations of sections. It is a straightforward consequence of the definitions that prolongations characterise those 1-forms in $\bigwedge_0^1\pi_{1,0}$ which are contact forms; however, the coordinate representation shows that a similar characterisation holds for arbitrary 1-forms on $J^1\pi$.

Theorem 4.3.14 *If $\sigma \in \bigwedge^1 J^1\pi$ then $\sigma \in \bigwedge_C^1\pi_{1,0}$ if, and only if, for every open submanifold $W \subset M$ and every $\phi \in \Gamma_W(\pi)$,*

$$(j^1\phi)^*(\sigma|_W) = 0.$$

Proof We just have to show that, if $(j^1\phi)^*(\sigma|_W) = 0$ for every $\phi \in \Gamma_W(\pi)$ then $\sigma \in \bigwedge^1_C \pi_{1,0}$. So write σ in coordinates as

$$\sigma = \sigma^i_\alpha du^\alpha_i + \sigma_\alpha du^\alpha + \sigma_i dx^i.$$

If $j^1_p\phi \in J^1\pi$ then $(j^1\phi)^*(\sigma_{j^1_p\phi}) = 0$, so that

$$\begin{aligned} 0 &= \sigma^i_\alpha(j^1_p\phi) d\left(\frac{\partial \phi^\alpha}{\partial x^i}\right)_p + \sigma_\alpha(j^1_p\phi)\, d\phi^\alpha|_p + \sigma_i(j^1_p\phi)\, dx^i\Big|_p \\ &= \left(\sigma^i_\alpha(j^1_p\phi) \left.\frac{\partial^2\phi^\alpha}{\partial x^i\, \partial x^j}\right|_p + \sigma_\alpha(j^1_p\phi) \left.\frac{\partial\phi^\alpha}{\partial x^j}\right|_p + \sigma_j(j^1_p\phi)\right) dx^j\Big|_p . \end{aligned}$$

Now choose particular indices k, l and β with $1 \le k, l \le m$ and $1 \le \beta \le n$, where $k \ne l$. Let $\chi \in \Gamma_W(\pi)$ satisfy $j^1_p\chi = j^1_p\phi$ and, for each $1 \le i, j \le m$ and $1 \le \alpha \le n$,

$$\left.\frac{\partial^2\chi^\alpha}{\partial x^i\, \partial x^j}\right|_p = \left.\frac{\partial^2\phi^\alpha}{\partial x^i\, \partial x^j}\right|_p + \delta^\alpha_\beta(\delta^k_i\delta^l_j + \delta^l_i\delta^k_j),$$

where (for example) δ^α_β equals one if $\alpha = \beta$, and equals zero otherwise. Then, by a similar argument to that above,

$$\begin{aligned} 0 &= \left(\sigma^i_\alpha(j^1_p\chi) \left.\frac{\partial^2\chi^\alpha}{\partial x^i\, \partial x^j}\right|_p + \sigma_\alpha(j^1_p\chi) \left.\frac{\partial\chi^\alpha}{\partial x^j}\right|_p + \sigma_j(j^1_p\chi)\right) dx^j\Big|_p \\ &= \left(\sigma^i_\alpha(j^1_p\phi)\left(\left.\frac{\partial^2\phi^\alpha}{\partial x^i\, \partial x^j}\right|_p + \delta^\alpha_\beta(\delta^k_i\delta^l_j + \delta^l_i\delta^k_j)\right)\right. \\ &\quad \left. + \sigma_\alpha(j^1_p\phi) \left.\frac{\partial\phi^\alpha}{\partial x^j}\right|_p + \sigma_j(j^1_p\phi)\right) dx^j\Big|_p . \end{aligned}$$

Hence, by subtraction,

$$\begin{aligned} 0 &= \sigma^i_\alpha(j^1_p\phi)\delta^\alpha_\beta(\delta^k_i\delta^l_j + \delta^l_i\delta^k_j)\, dx^j\Big|_p \\ &= \sigma^k_\beta(j^1_p\phi)\, dx^l\Big|_p + \sigma^l_\beta(j^1_p\phi)\, dx^k\Big|_p , \end{aligned}$$

demonstrating that each $\sigma^k_\beta(j^1_p\phi) = 0$. Consequently $\sigma_{j^1_p\phi} \in \pi^*_{1,0}(\ker\phi^*)$, so that $\sigma \in \bigwedge^1_C \pi_{1,0}$. ∎

On the other hand, contact forms provide a characterisation of those local sections of π_1 which are the prolongations of sections of π.

Theorem 4.3.15 *Let $\psi \in \Gamma_W(\pi_1)$; then $\psi = j^1\phi$ where $\phi \in \Gamma_W(\pi)$ if, and only if, $\psi^*(\sigma|_W) = 0$ for every $\sigma \in \bigwedge^1_C \pi_{1,0}$.*

Proof First, suppose $\psi = j^1\phi$. Let $p \in W$; then

$$\begin{aligned}(\psi^*(\sigma|_W))_p &= ((j^1\phi)^*(\sigma|_W))_p \\ &= (j^1\phi)^*(\sigma_{j^1_p\phi}) \\ &= \phi^*(\eta)\end{aligned}$$

where $\eta \in \ker\phi^*$, since $\sigma_{j^1_p\phi} \in \pi^*_{1,0}(\ker\phi^*)$. Consequently $\psi^*(\sigma|_W) = 0$.

The converse may be demonstrated using coordinates. We must show that, for $1 \leq i \leq m$ and $1 \leq \alpha \leq n$,

$$\psi^\alpha_i = \frac{\partial\psi^\alpha}{\partial x^i}.$$

But this follows immediately by considering the contact forms σ^α where locally

$$\sigma^\alpha = du^\alpha - u^\alpha_i dx^i,$$

for then

$$\begin{aligned}0 &= \psi^*(du^\alpha - u^\alpha_i dx^i) \\ &= d\psi^\alpha - \psi^\alpha_i dx^i \\ &= \left(\frac{\partial\psi^\alpha}{\partial x^i} - \psi^\alpha_i\right) dx^i.\end{aligned}$$

■

EXERCISES

4.3.1 Let π be the trivial bundle $(\mathbf{R} \times F, pr_1, \mathbf{R})$, where q^α are coordinates on F and t is the identity coordinate on $\mathbf{R}$. Let

$$X = X^\alpha \frac{\partial}{\partial q^\alpha}$$

be a vertical vector field on $\mathbf{R} \times F$, with flow

$$\psi : (\mathbf{R} \times F) \times \mathbf{R} \longrightarrow \mathbf{R} \times F.$$

For $s \in \mathbf{R}$, let $\psi_s : \mathbf{R} \times F \longrightarrow \mathbf{R} \times F$ be defined as usual by $\psi_s(a) = \psi(a, s)$. Show that the vector field on $J^1\pi \cong \mathbf{R} \times TF$ whose flow is $j^1(\psi_s, id_{\mathbf{R}})$ has coordinate representation

$$X^\alpha \frac{\partial}{\partial q^\alpha} + \frac{dX^\alpha}{dt}\frac{\partial}{\partial \dot{q}^\alpha}.$$

4.3.2 Let π be the trivial bundle $(\mathbf{R}^2 \times \mathbf{R}, pr_1, \mathbf{R}^2)$, with global coordinates $(x^1, x^2; u^1)$. Use a suitable contact form on $J^1\pi$ to give an alternative demonstration that the section $\psi \in \Gamma(\pi_1)$ defined in Example 4.2.3 by

$$\psi(p^1, p^2) = (p^1, p^2; p^1 \sin p^2; p^1 p^2, 0)$$

is not the prolongation of a section of π.

4.3.3 If π is an arbitrary bundle, $(f, id_M) : \pi \longrightarrow \pi$ is a bundle morphism, and σ is a contact form on $J^1\pi$, show that $(j^1 f)^*(\sigma)$ is also a contact form. Show in particular that

$$(j^1 f)^*(du^\alpha - u^\alpha_j dx^j) = \frac{\partial f^\alpha}{\partial u^\beta}(du^\beta - u^\beta_j dx^j).$$

4.3.4 Let (E, π, M) be a bundle with $\dim M = m > 1$. Construct (in coordinates) an m-form $\theta \in \bigwedge^m J^1\pi$ which satisfies $(j^1\phi)^*\theta = 0$ for every $\phi \in \Gamma_{loc}(\pi)$, but which does not satisfy $\theta \in \bigwedge^m_0 \pi_{1,0}$. What is the most general coordinate formula for a "contact m-form" on $J^1\pi$?

4.4 Prolongations of Vector Fields

In Section 4.2 we demonstrated how certain bundle morphisms could be prolonged from the total space of a bundle to its first jet manifold, and in the present section we shall consider the "infinitesimal" version of this construction. In other words, we shall start with a vector field X on E, and obtain its prolongation X^1 as a vector field on $J^1\pi$. Such vector fields may appear as infinitesimal symmetries of differential equations, and they also play a part in describing the extremals of variational problems.

If the vector field on E is projectable onto the base manifold M, then prolongation is a straightforward operation, for the flow of X will yield a bundle isomorphism for each value of the time parameter. The prolongations of these isomorphisms will provide a flow on $J^1\pi$, which may then be differentiated with respect to the time parameter to give the required vector field. (This procedure works even when X is not complete, for one then considers bundle isomorphisms between sub-bundles of π.) We intend, however, to define the prolongation of an arbitrary vector field on E, and so the procedure will be rather more complicated. Essentially, one takes the "vertical representative" of X, prolongs this new vector field, and then replaces the "horizontal part" in a suitable way. Now in Chapter 3 we emphasised the point that, in general, an arbitrary vector field on E did not have a vertical representative *as a vector field on* E, and indeed the vertical representative of X is a vector field along $\pi_{1,0}$: its coefficients at a point $\phi(p) \in E$ involve the derivatives of ϕ. Consequently the "prolongation" of the vertical representative will contain second derivatives. However the suitable "horizontal part" to be replaced will contain those same second derivatives with the opposite sign, so the result will be a *bona fide* vector field on $J^1\pi$.

It is possible to describe this process in terms of tangent bundles, and that is the approach we shall adopt in this section; we shall describe the results for vector fields as corollaries. (Details of the approach which deals with vector fields directly will be given in Chapter 6, where we shall also explain how to prolong vector fields along $\pi_{1,0}$.) We start, therefore, with the bundles $(J^1\pi, \pi_1, M)$ and $(V\pi, \nu_\pi, M)$: note that the latter is *not* a vector bundle, because the linear structure on the total space $V\pi$ is defined on its fibres over E, not on its (larger) fibres over M. The vertical bundle of the former (over M) is $(V\pi_1, \nu_{\pi_1}, M)$ and the first jet bundle of the latter is $(J^1\nu_\pi, (\nu_\pi)_1, M)$. These two bundles turn out to be isomorphic.

Theorem 4.4.1 *There is a canonical diffeomorphism $i_1 : J^1\nu_\pi \longrightarrow V\pi_1$ which projects to the identity on M.*

Proof The essence of this proof lies in considering 1-parameter families of local sections of π. Differentiating first with respect to the parameter gives sections of ν_π; taking the prolongation first gives vertical curves in $J^1\pi$. Since prolonging is just a fancy name for differentiating, the two operations commute, and they yield the required bijection between $J^1\nu_\pi$ and $V\pi_1$. Writing this map in coordinates shows that it is a diffeomorphism.

So let W be an open submanifold of M, let $p \in W$, and let the map

$$\gamma : W \times \mathbf{R} \longrightarrow E$$

satisfy $\pi \circ \gamma = pr_1$. If $t \in \mathbf{R}$ and $q \in W$, define the maps $\gamma_t : W \longrightarrow E$, $\gamma_q : \mathbf{R} \longrightarrow E$ by $\gamma_t(q) = \gamma_q(t) = \gamma(q,t)$. (With this definition, the curves γ_q are defined for all t rather than t in some neighbourhood of the origin, but restricting attention to such curves will not affect the possible tangent vectors $[\gamma_q]$.) Then for a given t, $\gamma_t \in \Gamma_W(\pi)$, and so the map

$$\begin{aligned} j^1_p\gamma : \mathbf{R} &\longrightarrow J^1\pi \\ t &\longmapsto j^1_p\gamma_t \end{aligned}$$

is a curve in $J^1\pi$ satisfying $j^1_p\gamma(0) = j^1_p\gamma_0$. Furthermore, $\pi_1(j^1_p\gamma(t)) = p$ for every $t \in \mathbf{R}$, so that this curve lies entirely within the fibre $J^1_p\pi$. Consequently the tangent vector $[j^1_p\gamma]$ is vertical, and so we have

$$[j^1_p\gamma] \in V_{j^1_p\gamma_0}\pi_1 \subset V\pi_1.$$

On the other hand, for a given q, γ_q is a curve in E satisfying $\gamma_q(0) = \gamma_0(q)$ and lying entirely within the fibre E_q, and so defines a tangent vector $[\gamma_q] \in V\pi$. Furthermore, $\nu_\pi[\gamma_q] = q$, so the map

$$\begin{aligned} [\gamma] : W &\longrightarrow V\pi \\ q &\longmapsto [\gamma_q] \end{aligned}$$

is a section of ν_π. We therefore obtain the 1-jet

$$j^1_p[\gamma] \in J^1_p\nu_\pi \subset J^1\nu_\pi.$$

The map $i_1 : J^1\nu_\pi \longrightarrow V\pi_1$ is now given by the correspondence $j^1_p[\gamma] \longmapsto [j^1_p\gamma]$. Technically, of course, each of these objects involves two equivalence relations (as a jet and a tangent vector). However, it should be clear that two maps $\gamma_1 : W_1 \times \mathbf{R} \longrightarrow E$ and $\gamma_2 : W_2 \times \mathbf{R} \longrightarrow E$ will both represent the same $j^1_p[\gamma]$ (where $p \in W_1 \cap W_2$) when

$$\left.\frac{\partial^2 \gamma_1}{\partial t\, \partial x^i}\right|_{t=0;p} = \left.\frac{\partial^2 \gamma_2}{\partial t\, \partial x^i}\right|_{t=0;p}$$

for $1 \leq i \leq m$, and that they will represent the same $[j^1_p\gamma]$ when exactly the same conditions hold. The map i_1 is then a bijection because each element of $J^1\nu_\pi$ may be written in the form $j^1_p[\gamma]$, and each element of $V\pi_1$ may be written in the form $[j^1_p\gamma]$, for a suitable choice of p and γ.

Now let (x^i, u^α) be an adapted coordinate system on E. The induced adapted coordinate system on $V\pi$ is $(x^i, u^\alpha; \dot{u}^\alpha)$, and the corresponding coordinate system on $J^1\nu_\pi$ is

$$(x^i, u^\alpha; \dot{u}^\alpha; u^\alpha_i, \dot{u}^\alpha_i).$$

On the other hand, starting again from the coordinates (x^i, u^α) on E, the induced adapted coordinate system on $J^1\pi$ is $(x^i, u^\alpha, u^\alpha_i)$, and the corresponding coordinate system on $V\pi_1$ is

$$(x^i, u^\alpha, u^\alpha_i; \dot{u}^\alpha, \dot{u}^\alpha_i).$$

In these coordinate systems, i_1 is represented by the linear map which simply transposes the $\dot{u}^\alpha$ and u^α_i sets of coordinates, and so i_1 is a diffeomorphism. The projection of i_1 onto M is clearly just id_M. ■

The coordinate correspondence between $J^1\nu_\pi$ and $V\pi_1$ may also be written in the following way. A typical element $\xi \in V\pi_1$ may (using the linear structure on the fibres over E) be written as

$$\xi = \xi^\alpha \left.\frac{\partial}{\partial u^\alpha}\right|_{j^1_p\phi} + \xi^\alpha_i \left.\frac{\partial}{\partial u^\alpha_i}\right|_{j^1_p\phi}.$$

To find the corresponding element of $J^1\nu_\pi$, let $\phi \in \Gamma(\pi)$ and $X \in \mathcal{V}(\pi)$ satisfy $i_1(j^1_p(X \circ \phi)) = \xi$. Then

$$\begin{aligned} \dot{u}^\alpha(j^1_p(X \circ \phi)) &= \dot{u}^\alpha(X(\phi(p))) \\ &= X^\alpha(\phi(p)) \end{aligned}$$

and

$$\begin{aligned} \dot{u}^\alpha_i(j^1_p(X \circ \phi)) &= \left.\frac{\partial(X^\alpha \circ \phi)}{\partial x^i}\right|_p \\ &= \left.\frac{\partial X^\alpha}{\partial x^i}\right|_{\phi(p)} + \left.\frac{\partial \phi^\beta}{\partial x^i}\right|_p \left.\frac{\partial X^\alpha}{\partial u^\beta}\right|_{\phi(p)} \\ &= \left.\frac{\partial X^\alpha}{\partial x^i}\right|_{\phi(p)} + u^\beta_i(j^1_p\phi) \left.\frac{\partial X^\alpha}{\partial u^\beta}\right|_{\phi(p)} \\ &= \left.\frac{dX^\alpha}{dx^i}\right|_{j^1_p\phi} \end{aligned}$$

so that

$$\xi^\alpha = X^\alpha(\phi(p))$$

and

$$\xi^\alpha_i = \left.\frac{dX^\alpha}{dx^i}\right|_{j^1_p\phi}.$$

The map i_1 may be used directly to prolong vertical vector fields on E, giving vertical vector fields on $J^1\pi$. So suppose that $X \in \mathcal{V}(\pi)$; the pair (X, id_M) may then be regarded as a bundle morphism from (E, π, M) to $(V\pi, \nu_\pi, M)$. The first prolongation of this bundle morphism is the map $j^1X : J^1\pi \longrightarrow J^1\nu_\pi$, and then $i_1 \circ j^1X$ is a map from $J^1\pi$ to $V\pi_1$. Furthermore,

$$\begin{aligned} \tau_{J^1\pi}(i_1(j^1X(j^1_p\phi))) &= \tau_{J^1\pi}(i_1(j^1_p(X \circ \phi))) \\ &= j^1_p\phi, \end{aligned}$$

because if $j^1_p(X \circ \phi) = j^1_p[\gamma]$ then $i_1(j^1_p[\gamma]) = [j^1_p\gamma]$ is a tangent vector to $J^1\pi$ at $j^1_p\gamma_0$; but

$$\left.\frac{\partial \phi^\alpha}{\partial x^i}\right|_p = \left.\frac{\partial \gamma^\alpha}{\partial x^i}\right|_{t=0;p}$$

and $\phi(p) = \gamma_0(p)$, so that $j^1_p\gamma_0 = j^1_p\phi$. Consequently $i_1 \circ j^1X$ is a vertical vector field on $J^1\pi$, and is denoted X^1. It is also clear that X^1 is projectable back to X, for

$$\begin{aligned}\pi_{1,0*}\left(X^1_{j^1_p\phi}\right) &= \pi_{1,0*}[j^1_p\gamma] \\ &= [\pi_{1,0}(j^1_p\gamma)] \\ &= [t \longmapsto \gamma_t(p)] \\ &= X_{\phi(p)}.\end{aligned}$$

We may also see from the coordinate relationship between $J^1\nu_\pi$ and $V\pi_1$ that if

$$X = X^\alpha \frac{\partial}{\partial u^\alpha}$$

then

$$X^1 = X^\alpha \frac{\partial}{\partial u^\alpha} + \frac{dX^\alpha}{dx^i}\frac{\partial}{\partial u^\alpha_i}.$$

If γ is the flow of X, then from the argument in the proof of Theorem 4.4.1 it is clear that $j^1\gamma$ (defined by $j^1\gamma(p,t) = j^1_p\gamma_t$) is the flow of X^1.

There is an important application of this idea to problems in the calculus of variations; these problems may be found in Lagrangian mechanics (where the base manifold M is one-dimensional) and in Lagrangian field theories (where $\dim M > 1$). We shall suppose that M is orientable with volume form Ω, and we shall use the same symbol Ω to denote the pullback $\pi_1^*\Omega$ on $J^1\pi$.

Definition 4.4.2 A *Lagrangian density on* π is a function $L \in C^\infty(J^1\pi)$. The corresponding *Lagrangian* is the m-form $L\Omega \in \bigwedge^m_0 \pi_1$. ∎

In view of our specification of a fixed volume form Ω, we shall usually describe the function L as a Lagrangian, even though this description strictly refers to the m-form $L\Omega$. The function is also sometimes called a *first-order Lagrangian*, to distinguish it from the higher-order Lagrangians which we shall meet in Chapter 6.

Given a fixed Lagrangian L, each $\phi \in \Gamma_W(\pi)$ determines a function $(j^1\phi)^*L : W \longrightarrow \mathbf{R}$, and we wish to study the integrals of such functions.

Definition 4.4.3 If $\phi \in \Gamma_W(\pi)$ and the vector field $X \in \mathcal{V}(\pi)$ has flow ψ_t, then the *variation of* ϕ *induced by* X is the one-parameter family of local sections $\tilde{\psi}_t(\phi) = \psi_t \circ \phi \in \Gamma_W(\pi)$. ∎

For small t, the variation of ϕ is therefore a "nearby" section, a generalisation of the "nearby" curve used in the classical calculus of variations; indeed when $p \in W$ the tangent vector $[t \longmapsto \tilde{\psi}_t(\phi)(p)]$ just equals $(\tilde{X}(\phi))_p$, where $\widetilde{X}(\phi) = X \circ \phi$, just as in Lemma 3.2.18. The vertical vector field X is called a *variation field.*

If C is a compact m-dimensional submanifold of M, and if $\phi \in \Gamma_W(\pi)$ where $C \subset W$, then a function $(-\varepsilon, \varepsilon) \longrightarrow \mathbf{R}$ for some $\varepsilon > 0$ may be defined by

$$t \longmapsto \int_C (j^1(\psi_t \circ \phi))^* L\Omega.$$

The local section ϕ will be called an extremal of L if this function is stationary for every $C \subset W$, and every X which vanishes on $\pi^{-1}(\partial C)$, where ∂C is the boundary of C.

Definition 4.4.4 The local section $\phi \in \Gamma_W(\pi)$ is an *extremal of* L if

$$\left.\frac{d}{dt}\right|_{t=0} \int_C (j^1(\psi_t \circ \phi))^* L\Omega = 0$$

whenever C is a compact m-dimensional submanifold of M with $C \subset W$, and whenever $X \in \mathcal{V}(\pi)$ has flow ψ_t and satisfies $X|_{\pi^{-1}(\partial C)} = 0$. ∎

We should remark at this stage that a detailed study of the calculus of variations would also involve the consideration of extremals which were not necessarily C^∞; we shall, however, not consider those matters here.

Lemma 4.4.5 *The local section* ϕ *is an extremal of* L *if, and only if,*

$$\int_C (j^1\phi)^* d_{X^1} L\Omega = 0.$$

Proof For each $p \in C$,

$$\begin{aligned}
(j^1\phi)^*(d_{X^1}L)(p) &= X^1_{j^1_p\phi} L \\
&= [t \longmapsto \psi^1_t(j^1_p\phi)](L) \\
&= [t \longmapsto j^1(\psi_t \circ \phi)(p)](L) \\
&= \left.\frac{d}{dt}\right|_{t=0} (j^1(\psi_t \circ \phi))^* L(p)
\end{aligned}$$

so that

$$\int_C (j^1\phi)^* d_{X^1} L\Omega = \left.\frac{d}{dt}\right|_{t=0} \int_C (j^1(\psi_t \circ \phi))^* L\Omega.$$

∎

As a consequence of this lemma, we need no longer consider the variation of ϕ (which involves the flow of the variation field X), but may work directly with the first prolongation X^1.

Example 4.4.6 Let π be the trivial bundle $(\mathbf{R} \times \mathbf{R}^n, pr_1, \mathbf{R})$, with global coordinates (t, q^α), so that $J^1\pi \cong \mathbf{R} \times T\mathbf{R}^n \cong \mathbf{R} \times \mathbf{R}^n \times \mathbf{R}^n$ has coordinates $(t, q^\alpha, \dot{q}^\alpha)$. Let $L : J^1\pi \longrightarrow \mathbf{R}$ be defined by

$$L = \tfrac{1}{2}\delta_{\alpha\beta}\dot{q}^\alpha \dot{q}^\beta$$

(where $\delta_{\alpha\beta}$ equals one if $\alpha = \beta$ and is zero otherwise) so that L is just the pull-back of the quadratic function on $T\mathbf{R}^n$ obtained from the standard Euclidean metric on $\mathbf{R}^n$. Let X be a vertical vector field on $\mathbf{R} \times \mathbf{R}^n$, so that

$$X = X^\gamma \frac{\partial}{\partial q^\gamma},$$

and

$$X^1 = X^\gamma \frac{\partial}{\partial q^\gamma} + \frac{dX^\gamma}{dt}\frac{\partial}{\partial \dot{q}^\gamma}.$$

Then

$$\begin{aligned} d_{X^1}L &= \left(X^\gamma \frac{\partial}{\partial q^\gamma} + \frac{dX^\gamma}{dt}\frac{\partial}{\partial \dot{q}^\gamma}\right)\left(\tfrac{1}{2}\delta_{\alpha\beta}\dot{q}^\alpha \dot{q}^\beta\right) \\ &= \delta_{\alpha\beta}\dot{q}^\alpha \frac{dX^\beta}{dt}. \end{aligned}$$

Now suppose that $\phi \in \Gamma(\pi)$ is a parametrised line, so that $q^\alpha(j^1_p\phi) = \phi^\alpha(p) = \lambda^\alpha p + \mu^\alpha$, and that $\dot{q}^\alpha(j^1_p\phi) = \lambda^\alpha$. Then

$$\begin{aligned} (j^1\phi)^*(d_{X^1}L)(p) &= \delta_{\alpha\beta}\dot{q}^\alpha(j^1_p\phi)\left.\frac{dX^\beta}{dt}\right|_{j^1_p\phi} \\ &= \delta_{\alpha\beta}\lambda^\alpha \left.\frac{\partial(X^\beta \circ \phi)}{\partial t}\right|_p . \end{aligned}$$

Integration over the compact interval $[a, b]$ then gives

$$\begin{aligned} \int_a^b (j^1\phi)^*(d_{X^1}L)dt &= \delta_{\alpha\beta}\lambda^\alpha \int_a^b \frac{\partial(X^\beta \circ \phi)}{\partial t}dt \\ &= \delta_{\alpha\beta}\lambda^\alpha(X^\beta(\phi(b)) - X^\beta(\phi(a))), \end{aligned}$$

and this expression vanishes whenever the vertical vector field X vanishes on $\pi^{-1}(\{a, b\})$. It follows that ϕ is an extremal of L. ∎

The general case, where we prolong vector fields on E which need not be vertical, is rather more complicated than our previous considerations might suggest. Although, as we have seen, $J^1\nu_\pi$ is diffeomorphic to $V\pi_1$, it is not true that $J^1(\pi \circ \tau_E)$ is diffeomorphic to $TJ^1\pi$: indeed, the dimension of the former manifold is

$$m + (m + 2n) + m(m + 2n),$$

whereas the dimension of the latter is

$$2(m + n + mn),$$

so that $\dim J^1(\pi \circ \tau_E) - \dim TJ^1\pi = m^2$. (In coordinates, the difference is due to the functions $\dot{x}^i_j$ which do not appear on $TJ^1\pi$.) We shall therefore construct a map $r_1 : J^1(\pi \circ \tau_E) \longrightarrow TJ^1\pi$ which will be surjective rather than bijective.

To define r_1, we shall consider its effect on an arbitrary element of $j^1_p\psi \in J^1(\pi \circ \tau_E)$ by examining in some detail the section $\psi \in \Gamma_W(\pi \circ \tau_E)$. From ψ we may certainly construct a local section of π by composition with τ_E, and we shall write $\phi = \tau_E \circ \psi \in \Gamma_W(\pi)$. We may also, however, use the fact that $\tau_M \circ \pi_* = \pi \circ \tau_E$ to obtain in a similar way a local section of τ_M, and we shall write $X = \pi_* \circ \psi \in \Gamma_W(\tau_M)$. (Of course, X is just a locally-defined vector field on M.) We may then consider the composition $\phi_* \circ X : W \longrightarrow TE$, and since

$$\begin{aligned} \pi \circ \tau_E \circ \phi_* \circ X &= \pi \circ \phi \circ \tau_M \circ X \\ &= id_W, \end{aligned}$$

we have $\phi_* \circ X \in \Gamma_W(\pi \circ \tau_E)$. It would be pleasing to announce that we had thereby recovered our original section ψ, but it turns out that these two sections are not, in general, the same. Nevertheless, for any given $p \in W$,

$$\begin{aligned} \tau_E(\phi_*(X(p))) &= \phi(\tau_M(X(p))) \\ &= \phi(p) \\ &= \tau_E(\psi(p)), \end{aligned}$$

so that $\psi(p)$ and $\phi_*(X(p))$ are in the same fibre of the vector bundle τ_E; it therefore makes sense to consider their difference $\psi(p) - \phi_*(X(p))$. Furthermore,

$$\begin{aligned} \pi_*(\psi(p) - \phi_*(X(p))) &= \pi_*(\psi(p)) - X(p) \\ &= 0 \end{aligned}$$

since $X = \pi_* \circ \psi$, so that the vector $\psi(p) - \phi_*(X(p))$ is vertical over M (and so is an element of $V_{\phi(p)}\pi$). The map $\psi - \phi_* \circ X$ is therefore a local section of the bundle $(V\pi, \nu_\pi, M)$.

Definition 4.4.7 The map $r_1 : J^1(\pi \circ \tau_E) \longrightarrow TJ^1\pi$ is defined by

$$r_1(j^1_p\psi) = i_1(j^1_p(\psi - \phi_* \circ X)) + (j^1\phi)_*(X_p)$$

where $\phi = \tau_E \circ \psi$ and $X = \pi_* \circ \psi$. ∎

As always, we shall need to check that different sections ψ with the same 1-jet at p give the same result, and this will follow from the coordinate representation of the map r_1.

Proposition 4.4.8 *The pair (r_1, id_{TE}) is a bundle morphism from $(J^1(\pi \circ \tau_E), (\pi \circ \tau_E)_{1,0}, TE)$ to $(TJ^1\pi, (\pi_{1,0})_*, TE)$. If $\psi \in \Gamma_W(\pi \circ \tau_E)$ satisfies*

$$\psi(p) = \psi^i(p) \left.\frac{\partial}{\partial x^i}\right|_{\phi(p)} + \psi^\alpha(p) \left.\frac{\partial}{\partial u^\alpha}\right|_{\phi(p)}$$

where $\phi = \tau_E \circ \psi$, $\psi^i = \dot{x}^i \circ \psi$ and $\psi^\alpha = \dot{u}^\alpha \circ \psi$, then

$$r_1(j^1_p\psi) = \psi^i(p) \left.\frac{\partial}{\partial x^i}\right|_{j^1_p\phi} + \psi^\alpha(p) \left.\frac{\partial}{\partial u^\alpha}\right|_{j^1_p\phi} + (\dot{u}^\alpha_i - \dot{x}^j_i u^\alpha_j)(j^1_p\psi) \left.\frac{\partial}{\partial u^\alpha_i}\right|_{j^1_p\phi},$$

so that $r_1(j^1_p\psi)$ does not depend upon the particular choice of ψ to represent the jet $j^1_p\psi$.

Proof To demonstrate the first assertion, notice that

$$\begin{aligned}
\pi_{1,0*}(r_1(j^1_p\psi)) &= \pi_{1,0*}(i_1(j^1_p(\psi - \phi_* \circ X))) + \pi_{1,0*}((j^1\phi)_*(X_p)) \\
&= (\psi - \phi_* \circ X)(p) + \phi_*(X_p) \\
&= \psi(p) \\
&= (\pi \circ \tau_E)_{1,0}(j^1_p\psi)
\end{aligned}$$

so that $\pi_{1,0*} \circ r_1 = (\pi \circ \tau_E)_{1,0}$. This immediately implies that, in the coordinate representation of $r_1(j^1_p\psi)$ as a tangent vector, the first two sets of coefficients remain unchanged as $\psi^i(p)$ and $\psi^\alpha(p)$. To calculate the third set of coefficients, note that

$$\begin{aligned}
\dot{u}^\alpha_i(r_1(j^1_p\psi)) &= \dot{u}^\alpha_i(i_1(j^1_p(\psi - \phi_* \circ X))) + \dot{u}^\alpha_i((j^1\phi)_*(X_p)) \\
&= \left.\frac{\partial}{\partial x^i}\right|_p \left(\dot{u}^\alpha \circ \left(\psi - \phi_* \circ \left(\psi^j \frac{\partial}{\partial x^j}\right)\right)\right) \\
&\quad + \dot{u}^\alpha_i \left((j^1\phi)_* \left(\psi^j(p) \left.\frac{\partial}{\partial x^j}\right|_p\right)\right) \\
&= \left.\frac{\partial \psi^\alpha}{\partial x^i}\right|_p - \left.\frac{\partial}{\partial x^i}\right|_p \left(\psi^j \frac{\partial(u^\alpha \circ \psi)}{\partial x^j}\right) + \left.\frac{\partial}{\partial x^i}\right|_p \left(\frac{\partial(u^\alpha \circ \psi)}{\partial x^j}\right) \psi^j(p)
\end{aligned}$$

$$= \left.\frac{\partial \psi^\alpha}{\partial x^i}\right|_p - \left.\frac{\partial \psi^j}{\partial x^i}\right|_p \left.\frac{\partial (u^\alpha \circ \psi)}{\partial x^j}\right|_p$$
$$= (\dot{u}_i^\alpha - \dot{x}_i^j u_j^\alpha)(j_p^1 \psi).$$

■

The map r_1 may now be used to prolong arbitrary vector fields on E. The method is a direct generalisation of the one used earlier for vertical vector fields, and it reduces to that method when the vector field is vertical.

Definition 4.4.9 For each vector field $X \in \mathcal{X}(E)$, the *prolongation of* X is the vector field $X^1 \in \mathcal{X}(J^1\pi)$ defined by $X^1_{j_p^1\phi} = r_1(j_p^1(X \circ \phi))$. ■

If the coordinate representation of X is

$$X = X^i \frac{\partial}{\partial x^i} + X^\alpha \frac{\partial}{\partial u^\alpha}$$

then we can calculate the coordinate representation of X^1 from Proposition 4.4.8. The coefficients of $\partial/\partial x^i$ and $\partial/\partial u^\alpha$ are just X^i and X^α. The coefficient of $\partial/\partial u_i^\alpha$ is $\dot{u}_i^\alpha \circ X^1$, and

$$\begin{aligned}
\dot{u}_i^\alpha(X^1_{j_p^1\phi}) &= \dot{u}_i^\alpha(r_1(j_p^1(X \circ \phi))) \\
&= (\dot{u}_i^\alpha - \dot{x}_i^j u_j^\alpha)(j_p^1(X \circ \phi)) \\
&= \left.\frac{\partial (X^\alpha \circ \phi)}{\partial x^i}\right|_p - \left.\frac{\partial \phi^\alpha}{\partial x^j}\right|_p \left.\frac{\partial (X^j \circ \phi)}{\partial x^i}\right|_p \\
&= \left.\frac{dX^\alpha}{dx^i}\right|_{j_p^1\phi} - u_j^\alpha(j_p^1\phi) \left.\frac{dX^j}{dx^i}\right|_{j_p^1\phi},
\end{aligned}$$

so that, finally,

$$X^1 = X^i \frac{\partial}{\partial x^i} + X^\alpha \frac{\partial}{\partial u^\alpha} + \left(\frac{dX^\alpha}{dx^i} - u_j^\alpha \frac{dX^j}{dx^i}\right) \frac{\partial}{\partial u_i^\alpha}.$$

Example 4.4.10 Let π be the trivial bundle $(M \times \mathbf{R}, pr_1, M)$, and let X be a projectable vector field whose flow ψ_t is the identity on $\mathbf{R}$ in the given trivialisation, so that $\psi_t = \chi_t \times id_{\mathbf{R}}$. Then X^1 is the vector field on $J^1\pi \cong T^*M \times \mathbf{R}$ given by

$$X^1_{j_p^1\phi} = [t \longmapsto j^1\psi_t(j_p^1\phi)] = [t \longmapsto ((\chi_{-t})^* d\overline{\phi}_p, \overline{\phi}(p))],$$

and this corresponds to the traditional definition of the complete lift of a vector field from a manifold M to its cotangent manifold T^*M. Using (for this example) coordinates q^i on M, t on $\mathbf{R}$ and writing p_i for the derivative coordinates t_i on $J^1\pi$, if $X = X^i\partial/\partial q^i$ then

$$X^1 = X^i\frac{\partial}{\partial q^i} - p_i\frac{\partial X^i}{\partial q^j}\frac{\partial}{\partial p_j}.$$

■

If, as in the last example, the vector field X is projectable onto M, we may also obtain a prolonged vector field by differentiating the prolongation of the flow of X; the result is, of course, equal to X^1.

Theorem 4.4.11 *Let $X \in \mathcal{X}(E)$ be a projectable vector field with projection $\overline{X} \in \mathcal{X}(M)$. Let $a \in E$, let ψ be the flow of X in a neighbourhood of a, and let $\overline{\psi}$ be the flow of $\overline{X}$ in the corresponding neighbourhood of $\pi(a)$. Let the diffeomorphisms ψ_t be defined by $\psi_t(b) = \psi(b,t)$. Then $j^1\psi_t$ is the flow of X^1 in a neighbourhood of all points in $J^1\pi$ which project to a under $\pi_{1,0}$.*

Proof The assertion may be proved using coordinates. For

$$X = \left.\frac{\partial\psi_t^i}{\partial t}\right|_{t=0}\frac{\partial}{\partial x^i} + \left.\frac{\partial\psi_t^\alpha}{\partial t}\right|_{t=0}\frac{\partial}{\partial u^\alpha}$$

and so

$$X^1 = \left.\frac{\partial\psi_t^i}{\partial t}\right|_{t=0}\frac{\partial}{\partial x^i} + \left.\frac{\partial\psi_t^\alpha}{\partial t}\right|_{t=0}\frac{\partial}{\partial u^\alpha} + \left(\frac{d}{dx^i}\left.\frac{\partial\psi_t^\alpha}{\partial t}\right|_{t=0} - u_j^\alpha\frac{\partial}{\partial x^i}\left.\frac{\partial\psi_t^j}{\partial t}\right|_{t=0}\right)\frac{\partial}{\partial u_i^\alpha}$$

where the notation reflects the fact that the functions ψ_t^i may be defined on M. On the other hand,

$$u_i^\alpha \circ j^1\psi_t = \frac{d\psi_t^\alpha}{dx^j}\left(\frac{\partial(\overline{\psi}_t^{-1})^j}{\partial x^i}\circ\overline{\psi}_t\right)$$

and so

$$\left.\frac{\partial(u_i^\alpha \circ j^1\psi_t)}{\partial t}\right|_{t=0} = \left.\frac{\partial}{\partial t}\right|_{t=0}\frac{d\psi_t^\alpha}{dx^i} - u_j^\alpha\left.\frac{\partial}{\partial t}\right|_{t=0}\frac{\partial\psi_t^j}{\partial x^i}$$

since ψ_0 is the identity. ■

An application of this technique of prolonging a general vector field on E arises if we wish to define an infinitesimal symmetry of a differential equation $S \subset J^1\pi$. Such a symmetry will be a vector field $X \in \mathcal{X}(E)$ whose prolongation X^1 is tangent to S: for each $j_p^1\phi \in S$, $X^1_{j_p^1\phi} \in T_{j_p^1\phi}S$. If X

happens to be projectable onto M, then the flow of X^1 is the prolongation of the flow ψ of X, and the tangency condition on X^1 shows that ψ yields a one-parameter family ψ_t of symmetries of S; from any solution ϕ we then obtain a one-parameter family $\widetilde{\psi}_t(\phi)$ of solutions of S for sufficiently small t. However, the definition may also be applied when X is not projectable onto M, and in these circumstances the flow of X^1 has a more complicated relationship to the flow of X which we shall not consider in detail. (It is indeed possible to extend the definition of the prolongation of a bundle morphism to more general diffeomorphisms of the total space E, but the resulting map might not be defined globally on $J^1\pi$.)

Example 4.4.12 Let π be the trivial bundle $(\mathbf{R}\times\mathbf{R}, pr_1, \mathbf{R})$ with coordinates (x,u), and let S be the submanifold of $J^1\pi$ defined by the equation $uu_1 + x = 0$. Solutions of S are the local sections ϕ defined by

$$\phi(p) = \left(p, \sqrt{a^2 - p^2}\right)$$

or

$$\phi(p) = \left(p, -\sqrt{a^2 - p^2}\right)$$

for $p \in (b,c)$ where $a > 0$, $b < c$ and $|b|, |c| \leq a$. The vector field

$$X = u\frac{\partial}{\partial x} - x\frac{\partial}{\partial u}$$

is an infinitesimal symmetry of S, for its prolongation to $J^1\pi$ is

$$X^1 = u\frac{\partial}{\partial x} - x\frac{\partial}{\partial u} - (1 + (u_1)^2)\frac{\partial}{\partial u_1}$$

and

$$\begin{aligned} d_{X^1}(uu_1 + t) &= u - xu_1 - (1 + (u_1)^2)u \\ &= -u_1(uu_1 + x) \end{aligned}$$

which vanishes on S. The flow of X is of course just a family ψ_t of rotations of the total space $\mathbf{R}\times\mathbf{R}$; some of the solutions of S are mapped to other solutions under the flow for sufficiently small values of the time parameter t, whereas there are some solutions (such as the solution $\phi(p) = (p, \sqrt{1-p^2})$ for $p \in (-1,1)$) which become "multi-valued" however small the value of t. ∎

EXERCISES

4.4.1 For an arbitrary bundle (E,π,M) with coordinates (x^i,u^α), show that if $X\in\mathcal{V}(\pi)$ and $f\in C^\infty(E)$ then

$$\frac{d}{dx^i}(d_X f) = d_{X^1}\left(\frac{df}{dx^i}\right)$$

as functions on $J^1\pi$. Find the coordinate representation of the difference between these two functions in the case where X is not necessarily vertical.

4.4.2 For the same bundle π, show that if $X\in\mathcal{X}(E)$ and σ is a contact form on $J^1\pi$, then $d_{X^1}\sigma$ is also a contact form.

4.4.3 Let G be a Lie group, and let $g\in G$ and $\xi\in T_gG$. The vertical lift operation described in Exercise 2.2.2 may be used to map any tangent vector $\zeta\in T_gG$ to $\zeta^{\mathrm{v}}\in T_\xi TG$. On the other hand, ζ determines a unique left-invariant vector field Z on G, and the prolongation Z^1 yields a tangent vector $Z^1_\xi\in T_\xi TG$. Denoting these two maps $T_gG\longrightarrow T_\xi TG$ by v and p respectively, show that every $\eta\in T_\xi TG$ may be written uniquely in the form

$$\eta = \mathrm{v}(\zeta_1)+\mathrm{p}(\zeta_2)$$

where $\zeta_1,\zeta_2\in T_gG$. Is the connection on (TG,τ_G,G) constructed in this way the same as the connection described in Exercise 3.5.2?

4.4.4 Let π be the trivial bundle $(\mathbf{R}\times F, pr_1,\mathbf{R})$, so that there are the standard identifications $V\pi\cong\mathbf{R}\times TF\cong J^1\pi$ and $J^1\nu_\pi\cong\mathbf{R}\times TTF\cong V\pi_1$. Show that, with these identifications, the map $i_1 : J^1\nu_\pi\longrightarrow V\pi_1$ projects to a map $i: TTF\longrightarrow TTF$ which is the bundle isomorphism $\tau_{TF}\longrightarrow\tau_{F*}$ described in Exercise 1.3.2.

4.5 The Contact Structure

In Section 4.3 we saw that the pull-back bundles $(\pi^*_{1,0}(TE),\pi^*_{1,0}(\tau_E),J^1\pi)$ and $(\pi^*_{1,0}(T^*E),\pi^*_{1,0}(\tau^*_E),J^1\pi)$ could be written as direct sums of vector bundles over $J^1\pi$,

$$\begin{aligned}\pi^*_{1,0}(TE) &= \pi^*_{1,0}(V\pi)\oplus H\pi_{1,0}\\ \pi^*_{1,0}(T^*E) &= \pi^*_{1,0}(\pi^*(T^*M))\oplus C^*\pi_{1,0},\end{aligned}$$

where these decompositions were essentially dual to each other. As we saw in Section 2.1, this decomposition determines two complementary vector bundle endomorphisms of each of the bundles $\pi^*_{1,0}(\tau_E)$ and $\pi^*_{1,0}(\tau^*_E)$, and hence two complementary sections of the tensor product bundle $\pi^*_{1,0}(\tau^*_E)\otimes\pi^*_{1,0}(\tau_E)$. We shall use the symbols h and v to denote both the two pairs of endomorphisms and the pair of sections.

Definition 4.5.1 The vector bundle endomorphisms (h, id_E) and (v, id_E) of $\pi^*_{1,0}(\tau_E)$ are defined by

$$\begin{aligned} h(\xi^h + \xi^v) &= \xi^h \\ v(\xi^h + \xi^v) &= \xi^v, \end{aligned}$$

where $\xi^h \in H\pi_{1,0}$ and $\xi^v \in \pi^*_{1,0}(V\pi)$. ■

Definition 4.5.2 The vector bundle endomorphisms (h, id_E) and (v, id_E) of $\pi^*_{1,0}(\tau^*_E)$ are defined by

$$\begin{aligned} h(\eta^h + \eta^v) &= \eta^h \\ v(\eta^h + \eta^v) &= \eta^v, \end{aligned}$$

where $\eta^h \in \pi^*_{1,0}(\pi^*(T^*M))$ and $\eta^v \in C^*\pi_{1,0}$. ■

Definition 4.5.3 The vector-valued 1-forms h, v are the sections of the bundle $\pi^*_{1,0}(\tau^*_E) \otimes \pi^*_{1,0}(\tau_E)$ defined by

$$\begin{aligned} h_{j^1_p\phi}(\xi, \eta) &= \eta(h(\xi)) \\ v_{j^1_p\phi}(\xi, \eta) &= \eta(v(\xi)), \end{aligned}$$

where $\xi \in \pi^*_{1,0}(TE)_{j^1_p\phi}$ and $\eta \in \pi^*_{1,0}(T^*E)_{j^1_p\phi}$. ■

We shall regard $\pi^*_{1,0}(\tau^*_E) \otimes \pi^*_{1,0}(\tau_E)$ as a sub-bundle of $\tau^*_{J^1\pi} \otimes \pi^*_{1,0}(\tau_E)$, and with this identification the sections h and v may be regarded as vector-valued forms along $\pi_{1,0}$ in the sense of Section 3.3. In coordinates, they may be written as

$$\begin{aligned} h &= dx^i \otimes \frac{d}{dx^i} \\ v &= (du^\alpha - u^\alpha_j dx^j) \otimes \frac{\partial}{\partial u^\alpha}. \end{aligned}$$

Since h and v (in their various guises) incorporate the information carried by the contact forms on $J^1\pi$, they are together known as the *contact structure* on π_1, and they may in turn be used to characterise both those sections of π_1 which are prolongations, and (in certain circumstances) those diffeomorphisms and vector fields on $J^1\pi$ which are prolongations. One way of doing this is to consider the vector sub-bundle of $\tau_{J^1\pi}$ containing those tangent vectors to $J^1\pi$ which project to holonomic tangent vectors under $(\pi_{1,0*}, \tau_{J^1\pi})$.

Definition 4.5.4 The *Cartan distribution* is the kernel of the vector bundle morphism over $id_{J^1\pi}$

$$v \circ (\pi_{1,0*}, \tau_{J^1\pi}) : \tau_{J^1\pi} \longrightarrow \pi_{1,0}^*(\tau_E)$$

and is denoted $C\pi_{1,0}$. ■

It is immediate from this definition and from Definition 4.5.1 that $C\pi_{1,0} = (\pi_{1,0*}, \tau_{J^1\pi})^{-1}(H\pi_{1,0})$, and hence that, for each $\phi \in \Gamma_p(\pi)$,

$$C\pi_{1,0}|_{j^1_p\phi} = (j^1\phi)_*(T_pM) \oplus V_{j^1_p\phi}\pi_{1,0}.$$

From the duality relations it also follows that $\tau^*_{J^1\pi}|_{C^*\pi_{1,0}}$ is the annihilator of $\tau_{J^1\pi}|_{C\pi_{1,0}}$, where $C^*\pi_{1,0}$ is regarded here as a submanifold of $T^*J^1\pi$. The fibre dimension of the Cartan distribution at each point of $J^1\pi$ is $m(1+n)$, and a typical element $\xi \in C\pi_{1,0}$ may be written in coordinates as

$$\xi = \xi^i \left(\left.\frac{\partial}{\partial x^i}\right|_{j^1_p\phi} + u^\alpha_i(j^1_p\phi) \left.\frac{\partial}{\partial u^\alpha}\right|_{j^1_p\phi} \right) + \xi^\alpha_i \left.\frac{\partial}{\partial u^\alpha_i}\right|_{j^1_p\phi}.$$

The geometrical significance of the Cartan distribution may be expressed in the following terms. If two local sections ϕ_1 and ϕ_2 *touch* at a given point $\phi_1(p) = \phi_2(p) \in E$, then not only must their prolongations pass through the same point $j^1_p\phi_1 = j^1_p\phi_2 \in J^1\pi$, but whenever $\xi \in T_pM$ then the two images $(j^1\phi_1)_*(\xi)$ and $(j^1\phi_2)_*(\xi)$ must differ by a vector vertical over E. If ψ is an arbitrary local section of π_1 satisfying $\psi(p) = j^1_p\phi_1$, then $\psi_*(\xi) - (j^1\phi_1)_*(\xi)$ need not be vertical over E, in which case ψ will not be a prolongation. The Cartan distribution is just the distribution spanned by tangent vectors to the images of prolongations.

It is now natural to ask whether, as a distribution, $C\pi_{1,0}$ is involutive. Unfortunately, as the following example shows, it is not.

Example 4.5.5 Let X, Y be the local vector fields defined on the domain of the coordinate system (U^1, u^1) by

$$\begin{aligned} X &= \frac{\partial}{\partial x^i} + u^\alpha_i \frac{\partial}{\partial u^\alpha} \\ Y &= \frac{\partial}{\partial u^\beta_i} \end{aligned}$$

for some indices β and i. Then X and Y both belong to $C\pi_{1,0}$, but

$$[X, Y] = -\frac{\partial}{\partial u^\beta}$$

which does not belong to $C\pi_{1,0}$. ■

Example 4.5.6 An alternative characterisation of involutiveness arises when a distribution is specified in terms of an ideal of differential forms. The distribution is then involutive exactly when the ideal is differentially closed. Now since $C^*\pi_{1,0}$ is the annihilator of $C\pi_{1,0}$ and the sections of $\tau^*_{J^1\pi}|_{C^*\pi_{1,0}}$ are the contact forms, the involutiveness of $C\pi$ may be expressed by requiring that $d\sigma \in \bigwedge^1_C\pi_{1,0} \wedge \bigwedge^1 J^1\pi$ whenever $\sigma \in \bigwedge^1_C\pi_{1,0}$. However, if

$$\sigma = du^\alpha - u^\alpha_k dx^k$$

for some index α then

$$d\sigma = -du^\alpha_k \wedge dx^k$$

which is not of the required form. ∎

The reason why $C\pi_{1,0}$ is not involutive is connected with the behaviour of the highest-order derivatives in a differential expression. (This is not terribly obvious here since only first derivatives are involved, but it will become clearer when we consider higher-order jets.) The problem goes away when we discuss infinite jets, and the infinite Cartan distribution is, indeed, involutive. In these circumstances, however, we can no longer apply Frobenius' Theorem.

Returning to the present case, the non-involutiveness of the Cartan distribution implies that there are no $m(1+n)$-dimensional integral manifolds. Nevertheless, there are certainly "integral manifolds" of smaller dimension, and the following proposition describes the most important of these.

Proposition 4.5.7 *For each $a \in E$, the fibre $\pi^{-1}_{1,0}(a)$ is an mn-dimensional integral manifold of $C\pi_{1,0}$. For each $\phi \in \Gamma_W(\pi)$, the image $j^1\phi(W)$ is an m-dimensional integral manifold of $C\pi_{1,0}$; furthermore, if $\psi \in \Gamma_W(\pi_1)$ and $\psi(W)$ is an integral manifold of $C\pi_{1,0}$, then $\psi = j^1\phi$ for some $\phi \in \Gamma_W(\pi)$.*

Proof By Lemma 3.1.2, $T_{j^1_p\phi}(\pi^{-1}_{1,0}(\phi(p))) \cong V_{j^1_p\phi}\pi_{1,0} \subset C\pi_{1,0}|_{j^1_p\phi}$, which establishes the first assertion. The second is just a reformulation of the result in Theorem 4.3.15, that those local sections of π_1 which pull all the contact forms back to zero are just the prolongations of local sections of π. ∎

We shall see shortly that, when the fibre dimension n of the original bundle π is greater than one, then the integral manifolds of $C\pi_{1,0}$ of maximal dimension are just the fibres of $\pi_{1,0}$. When $n = 1$ this is clearly no longer true, and we shall see a curious consequence of this fact when we consider *symmetries* of the Cartan distribution: that is, diffeomorphisms f of $J^1\pi$ which satisfy $f_*(C\pi_{1,0}) \subset C\pi_{1,0}$. If f projects onto a diffeomorphism of M, then it is a symmetry precisely when it is a prolongation. (In particular, therefore, any symmetry of a first-order differential equation on the bundle

π defines a symmetry of the Cartan distribution by prolongation to $J^1\pi$.) The reason for this, of course, is that in these circumstances it maps prolongations of sections to prolongations of sections. If, however, f does not project onto M, then its action on sections need not be defined; nevertheless, it may still be a symmetry of the Cartan distribution. The curiosity is that when $n = 1$ there may be symmetries which do not even project onto E, although when $n > 1$ this is not possible.

Definition 4.5.8 A *symmetry* of the Cartan distribution on $J^1\pi$ is a diffeomorphism f of $J^1\pi$ which satisfies $f_*(C\pi_{1,0}) = C\pi_{1,0}$ ∎

It follows by duality that symmetries of the Cartan distribution are those diffeomorphisms which satisfy $f^*(C^*\pi_{1,0}) = C^*\pi_{1,0}$, and for this reason f is sometimes called a *contact transformation*. Similarly, f may be characterised by the fact that whenever σ is a contact form then so is $f^*(\sigma)$.

Theorem 4.5.9 *Let (E, π, M) and (F, ρ, N) be bundles, and suppose that $(f, \overline{f}) : \pi_1 \longrightarrow \rho_1$ is a bundle morphism, where $\overline{f}$ is a diffeomorphism. Then $f_*(C\pi_{1,0}) \subset C\rho_{1,0}$ if, and only if, f is the prolongation of a bundle morphism $(f_0, \overline{f}) : \pi \longrightarrow \rho$.*

Proof Suppose first that $f = j^1 f_0$, where $(f_0, \overline{f}) : \pi \longrightarrow \rho$ is a bundle morphism. We shall use the decomposition $C\pi_{1,0}|_{j^1_p\phi} = (j^1\phi)_*(T_pM) \oplus V_{j^1_p\phi}\pi_{1,0}$. If $\xi \in (j^1\phi)_*(T_pM)$, we have

$$\begin{aligned} f_*(\xi) &= (j^1 f_0)_*(\xi) \\ &= (j^1 f_0)_*((j^1\phi)_*(\pi_{1*}(\xi))) \\ &= j^1(\tilde{f}_0(\phi))_*(\overline{f}_*(\pi_{1*}(\xi))) \\ &\in j^1(\tilde{f}_0(\phi))_*(T_{\overline{f}(p)}N), \end{aligned}$$

whereas if $\xi \in V_{j^1_p\phi}\pi_{1,0}$ then

$$\begin{aligned} \rho_{1,0*}(f_*(\xi)) &= \rho_{1,0*}((j^1 f_0)_*(\xi)) \\ &= f_{0*}(\pi_{1,0*}(\xi)) \\ &= 0, \end{aligned}$$

so that $f_*(\xi) \in V\rho_{1,0}$. It follows that $f_*(C\pi_{1,0}) \subset C\rho_{1,0}$.

Conversely, suppose that $(f, \overline{f})$ is a bundle morphism with $f_*(C\pi_{1,0}) \subset C\rho_{1,0}$. We shall first establish that f defines a bundle morphism from $\pi_{1,0}$ to $\rho_{1,0}$, and to do this we shall show that $f_*(V\pi_{1,0}) \subset V\rho_{1,0}$ and apply Proposition 3.1.3. So let $\xi \in V_{j^1_p\phi}\pi_{1,0}$ and write $f_*(\xi) = \eta_1 + \eta_2 \in C\rho_{1,0}$, where $\eta_1 \in (j^1\psi)_*(T_{\overline{f}(p)}N)$ for some $\psi \in \Gamma_{loc}(\pi)$, and where $\eta_2 \in V\rho_{1,0}$. Then

$$\begin{aligned} \rho_{1*}(f_*(\xi)) &= \overline{f}_*(\pi_{1*}(\xi)) \\ &= \overline{f}_*(\pi_*(\pi_{1,0*}(\xi))) \\ &= 0, \end{aligned}$$

so that, since $\rho_{1*}(\eta_2) = \rho_*(\rho_{1,0*}(\eta_2)) = 0$,

$$\begin{aligned} \rho_{1*}(\eta_1) &= \rho_{1*}(\eta_1 + \eta_2) \\ &= \rho_{1*}(f_*(\xi)) \\ &= 0. \end{aligned}$$

But $\eta_1 = (j^1\psi)_*(\rho_{1*}(\eta_1))$ so that $\eta_1 = 0$ and hence $f_*(\xi) \in V\rho_{1,0}$. Consequently f defines a bundle morphism $(f, f_0) : \pi_{1,0} \longrightarrow \rho_{1,0}$.

We shall now show that the maps f and $j^1 f_0$ are identical. So let $f(j^1_p\phi) = j^1_q\psi \in J^1\rho$. From $\rho_1 \circ f = \overline{f} \circ \pi_1$ it follows that $q = \overline{f}(p)$, and from $\rho_{1,0} \circ f = f_0 \circ \pi_{1,0}$ it follows that $\psi(\overline{f}(p)) = f_0(\phi(p)) = \tilde{f}_0(\phi)(\overline{f}(p))$. Consequently, both ψ_* and $\tilde{f}_0(\phi)_*$ are maps from $T_{\overline{f}(p)}N$ to $T_{\tilde{f}_0(\phi)(\overline{f}(p))}N$. So let $\xi \in T_{\overline{f}(p)}N$; then

$$\begin{aligned} \psi_*(\xi) &= \rho_{1,0*}((j^1\psi)_*(\xi)) \\ &= \rho_{1,0*}(f_*((j^1\phi)_*(\overline{f}_*^{-1}(\xi)))) \\ &= f_{0*}(\pi_{1,0*}((j^1\phi)_*(\overline{f}_*^{-1}(\xi)))) \\ &= f_{0*}(\phi_*(\overline{f}_*^{-1}(\xi))) \\ &= \tilde{f}_0(\phi)_*(\xi), \end{aligned}$$

so that $\psi_* = \tilde{f}_0(\phi)_*$ on $T_{\overline{f}(p)}N$. We may now use Lemma 4.1.3 to conclude that $f(j^1_p\phi) = j^1_{\overline{f}(p)}\psi = j^1_{\overline{f}(p)}(\tilde{f}_0(\phi)) = j^1 f_0(j^1_p\phi)$. ∎

Corollary 4.5.10 *If $(f, \overline{f})$ is a bundle automorphism of π_1, then f is a symmetry of $C\pi_{1,0}$ if, and only if, $f = j^1 f_0$ for some bundle automorphism $(f_0, \overline{f})$ of π.*

Proof If f is a symmetry of $C\pi_{1,0}$ then $f = j^1 f_0$ for some bundle morphism; equally $f^{-1} = j^1 f_0'$ for some other bundle morphism $(f_0', \overline{f}^{-1})$. But then $j^1(f_0' \circ f_0) = f^{-1} \circ f = id_{J^1\pi}$, and so $f_0' \circ f = id_E$; similarly $f_0 \circ f_0' = id_E$, so that $(f_0, \overline{f})$ is indeed a bundle automorphism. Conversely, if $f = j^1 f_0$ then $f^{-1} = j^1(f_0^{-1})$ for a similar reason, so that f becomes a symmetry of $C\pi_{1,0}$. ∎

The requirement in the theorem that f defines a bundle morphism $\pi_1 \longrightarrow \rho_1$ is essential, as the following example shows.

Example 4.5.11 Let π be the trivial bundle $(\mathbf{R}^m \times \mathbf{R}, pr_1, \mathbf{R}^m)$ with standard coordinates (x^i, u), and let $J^1\pi$ have induced coordinates (x^i, u, u_i). The map $f : J^1\pi \longrightarrow J^1\pi$ defined by

$$\begin{aligned} x^i \circ f &= u_i \\ u \circ f &= x^i u_i - u \\ u_i \circ f &= x^i \end{aligned}$$

is a diffeomorphism, and a calculation using these coordinates demonstrates that $f_*(C\pi_{1,0}) = C\pi_{1,0}$. However, f projects onto neither E nor M. ∎

In that example, the fibre dimension of π was one. When $n > 1$ then every symmetry must project onto E, although it need not project further onto M.

Theorem 4.5.12 *If $n > 1$ and f is a symmetry of $C\pi_{1,0}$, then f defines a bundle automorphism (f, f_0) of $\pi_{1,0}$.*

Proof Let $j^1_p\phi \in J^1\pi$, and let R be an integral manifold of $C\pi_{1,0}$ through $j^1_p\phi$, so that $T_{j^1_p\phi}R \subset C\pi_{1,0}|_{j^1_p\phi}$. Suppose that $\dim \pi_{1,0*}(T_{j^1_p\phi}R) \geq 1$; then there is a non-zero vector $\xi \in T_{j^1_p\phi}R$ where $\xi \in (j^1\phi)_*(T_pM)$.

Now since R is a submanifold of $J^1\pi$, the bracket of two vector fields on R will again be a vector field on R. So let $(\xi_1, \ldots, \xi_r, \eta_1, \ldots, \eta_s)$ be a basis for $T_{j^1_p\phi}R$, where $\xi_\mu \in (j^1\phi)_*(T_pM)$ and $\eta_\nu \in V\pi_{1,0}$; suppose that, in coordinates,

$$\begin{aligned} \xi_\mu &= \xi_\mu{}^i \left(\left.\frac{\partial}{\partial x^i}\right|_{j^1_p\phi} + u^\alpha_i(j^1_p\phi) \left.\frac{\partial}{\partial u^\alpha}\right|_{j^1_p\phi} \right) \\ \eta_\nu &= \eta_\nu{}^\alpha_i \left.\frac{\partial}{\partial u^\alpha_i}\right|_{j^1_p\phi} . \end{aligned}$$

Extend ξ_μ, η_ν to vector fields X_μ, Y_ν on R; then the bracket $[X_\mu, Y_\nu]_{j^1_p\phi}$ must also be an element of $T_{j^1_p\phi}R$. But this bracket will contain a term

$$-\eta_\nu{}^\alpha_i \xi_\mu{}^i \left.\frac{\partial}{\partial u^\alpha}\right|_{j^1_p\phi}$$

which must equal zero to ensure that $[X_\mu, Y_\nu]_{j^1_p\phi} \in C\pi_{1,0}|_{j^1_p\phi}$. Therefore, for each α with $1 \leq \alpha \leq n$, it follows that $\eta_\nu{}^\alpha_i \xi_\mu{}^i = 0$. Since the vectors ξ_μ

are linearly independent, each such vector thereby determines n constraints on the components of each vector η_ν. Since the vectors η_ν are themselves linearly independent, there can therefore be no more than $nm - nr$ of them. The dimension of $T_{j^1_p\phi}R$, and therefore of R, is $r + nm - nr$ which is less than nm since we have supposed $r \geq 1$ and $n > 1$.

Since the fibres of $\pi_{1,0}$ are integral manifolds of $C\pi_{1,0}$, and the dimension of each fibre is mn, it follows that these fibres are integral manifolds of maximal dimension, and that all integral manifolds of this dimension are fibres of $\pi_{1,0}$. Since f is a symmetry of $C\pi_{1,0}$, it maps integral manifolds to integral manifolds, and so maps fibres of $\pi_{1,0}$ to fibres of $\pi_{1,0}$. It therefore defines a bundle morphism from $\pi_{1,0}$ to itself, and a similar argument applied to f^{-1} shows that this must be an automorphism. ■

The reason why this proof does not work when $n = 1$ is that the fibres of $\pi_{1,0}$ then have the same dimension as the integral manifolds of the form $j^1\phi(W)$, namely m.

Definition 4.5.13 An *infinitesimal symmetry* of the Cartan distribution is a vector field X on $J^1\pi$ with the property that, whenever the vector field Y belongs to $C\pi_{1,0}$, then so does the vector field $[X, Y]$. ■

An infinitesimal symmetry is sometimes called an *infinitesimal contact transformation*. By duality, X is such an infinitesimal symmetry precisely when $d_X\sigma$ is a contact form for every contact form σ.

Proposition 4.5.14 *Let X be a complete vector field on $J^1\pi$ with flow ψ_t. Then X is an infinitesimal symmetry of the Cartan distribution if, and only if, for each t the diffeomorphism ψ_t is a symmetry of the Cartan distribution.*

Proof This follows from the characterisation of symmetries and infinitesimal symmetries in terms of contact forms, and the definition of the Lie derivative in terms of pull-backs:

$$d_X\sigma|_{j^1_p\phi} = \left.\frac{d}{dt}\right|_{t=0} \psi_t^*(\sigma)|_{j^1_p\phi},$$

so that, for every $\psi \in \Gamma_W(\pi)$,

$$(j^1\phi)^*(d_X\sigma)(p) = \left.\frac{d}{dt}\right|_{t=0} (j^1\phi)^*(\psi_t^*(\sigma))(p),$$

and hence if each ψ_t is a symmetry then X is an infinitesimal symmetry. Conversely, if X is an infinitesimal symmetry then, by integrating along the flow in the manner described in similar proofs in other chapters, each ψ_t may be seen to be a symmetry. ■

A similar result is true if the vector field E is not complete, as this is essentially a local proposition; the nomenclature is rather more complicated and the (finite) symmetries are only defined on submanifolds of $J^1\pi$.

We shall now obtain a coordinate representation for an infinitesimal symmetry of the Cartan distribution. It follows from Theorem 4.5.12 and Proposition 4.5.14, together with the characterisation in Proposition 3.2.15 of the flow of a projectable vector field as a family of bundle isomorphisms, that when $n > 1$ then every infinitesimal symmetry is projectable onto E; this will in fact become apparent from a closer inspection of the coordinate description. So let $X \in \mathcal{X}(J^1\pi)$ and suppose that if σ is a contact form then so is $d_X\sigma$. By writing X in coordinates as

$$X = X^i \frac{\partial}{\partial x^i} + X^\alpha \frac{\partial}{\partial u^\alpha} + X^\alpha_i \frac{\partial}{\partial u^\alpha_i}$$

and by considering the contact forms $du^\beta - u^\beta_j dx^j$ for $1 \leq \beta \leq n$, we obtain constraint equations

$$\frac{\partial X^\beta}{\partial u^\alpha_i} = u^\beta_j \frac{\partial X^j}{\partial u^\alpha_i}$$

and

$$u^\alpha_i \left(\frac{\partial X^\beta}{\partial u^\alpha} - u^\beta_j \frac{\partial X^j}{\partial u^\alpha} \right) = X^\beta_i - \frac{\partial X^\beta}{\partial x^i} + u^\beta_j \frac{\partial X^j}{\partial x^i}$$

which the components of X must satisfy. When $n > 1$, the first set of constraint equations implies that X must be projectable onto E, as the following argument demonstrates. Differentiate these equations with respect to u^γ_k, giving

$$\frac{\partial^2 X^\beta}{\partial u^\gamma_k \, \partial u^\alpha_i} = u^\beta_j \frac{\partial^2 X^j}{\partial u^\gamma_k \, \partial u^\alpha_i} + \delta^\beta_\gamma \frac{\partial X^k}{\partial u^\alpha_i};$$

by re-labelling we also have

$$\frac{\partial^2 X^\beta}{\partial u^\alpha_i \, \partial u^\gamma_k} = u^\beta_j \frac{\partial^2 X^j}{\partial u^\alpha_i \, \partial u^\gamma_k} + \delta^\beta_\alpha \frac{\partial X^i}{\partial u^\gamma_k}$$

so that

$$\delta^\beta_\gamma \frac{\partial X^k}{\partial u^\alpha_i} = \delta^\beta_\alpha \frac{\partial X^i}{\partial u^\gamma_k}.$$

Choose particular indices α, i and k, and let $\beta = \gamma \neq \alpha$ (which is possible because $n > 1$). The result is

$$\frac{\partial X^k}{\partial u^\alpha_i} = 0$$

which immediately gives

$$\frac{\partial X^\beta}{\partial u_i^\alpha} = 0$$

so that the functions X^k and X^β are all pulled back from E. In these circumstances, the second set of constraint equations may be written as

$$X_i^\beta = \frac{dX^\beta}{dx^i} - u_j^\beta \frac{dX^j}{dx^i},$$

so that X is the prolongation of a vector field on E. If $n = 1$ and X happens to be projectable onto E then a similar result holds. This discussion therefore establishes the following theorem.

Theorem 4.5.15 *If $X \in \mathcal{X}(J^1\pi)$ is projectable onto E, then X is an infinitesimal symmetry of the Cartan distribution if, and only if, X is the prolongation of a vector field on E. If $n > 1$ then every infinitesimal symmetry of the Cartan distribution is necessarily projectable onto E.* ∎

EXERCISES

4.5.1 Verify the coordinate expression given for the vector-valued forms h and v, and for a typical element of the Cartan distribution $C\pi_{1,0}$.

4.5.2 If $\psi \in \Gamma_W(\pi_1)$, write down the coordinate representation of a general element of $T_{\psi(p)}\psi(W)$. Use this to confirm that if $\psi(W)$ is an integral manifold of $C\pi_{1,0}$ then $\psi = j^1\phi$ for some $\phi \in \Gamma_W(\pi)$.

4.5.3 Construct an example of a diffeomorphism f of $J^1\pi$ such that (f, f_0) is a bundle automorphism of $\pi_{1,0}$ and $(f, \bar{f})$ is a bundle automorphism of π_1, but that f is not a symmetry of the Cartan distribution.

4.6 Jet Fields

One of the most important features of the affine bundle $(J^1\pi, \pi_{1,0}, E)$ is that a section of this bundle has many of the features of a vector field on a manifold, with the vital difference that the "flow"—if it exists—is parametrised, not by a one-dimensional time manifold, but by the m-dimensional base manifold M. As some of the examples earlier in this chapter have shown, if π is the trivial bundle $(\mathbf{R} \times M, pr_1, \mathbf{R})$ then $\pi_{1,0}$ is none other than $(\mathbf{R} \times TM, id_{\mathbf{R}} \times \tau_M, \mathbf{R} \times M)$, so it may not be entirely surprising to find that, in general, $\pi_{1,0}$ is like a kind of multi-dimensional tangent bundle, and that a jet (like a tangent vector) may act as a derivation. We shall also see that each section Γ of $\pi_{1,0}$ corresponds to a unique connection $\widetilde{\Gamma}$ on π, and that the correspondence involves the contact structure on π_1. We shall call a section of $\pi_{1,0}$ a jet field.

Definition 4.6.1 Given a 1-jet $j^1_p\phi \in J^1\pi$, the action of the jet on functions is the map $C^\infty(E) \longrightarrow T^*_{\phi(p)}E$ defined by

$$j^1_p\phi[f] = \pi^*(d(\phi^*(f))_p).$$

■

This action is well-defined for different representatives of $j^1_p\phi$, because it depends only on the first derivatives of ϕ. The main difference from the action of a tangent vector on a function is that the resulting entity is a cotangent vector lifted from the base manifold, rather than a number. It is straightforward to check that, in coordinates, one has

$$j^1_p\phi[f] = \left(\left.\frac{\partial f}{\partial x^i}\right|_{\phi(p)} + u^\alpha_i(j^1_p\phi) \left.\frac{\partial f}{\partial u^\alpha}\right|_{\phi(p)} \right) dx^i_{\phi(p)} = \left.\frac{df}{dx^i}\right|_{j^1_p\phi} dx^i_{\phi(p)}.$$

Definition 4.6.2 A *jet field* $\Gamma : E \longrightarrow J^1\pi$ is a section of the bundle $\pi_{1,0}$. The action of Γ on functions is the map $C^\infty(E) \longrightarrow \bigwedge^1_0\pi$ defined by

$$(\Gamma f)_{\phi(p)} = \Gamma(\phi(p))[f].$$

■

If the coordinate representation of Γ is given by $\Gamma^\alpha_i = u^\alpha_i \circ \Gamma$, then the action of Γ on functions may be written in coordinates as

$$\Gamma f = \left(\frac{\partial f}{\partial x^i} + \Gamma^\alpha_i \frac{\partial f}{\partial u^\alpha} \right) dx^i.$$

This action, when extended to differential forms by the rule $\Gamma(d\theta) = -d(\Gamma\theta)$, is a derivation of type d_*, and suggests the following result.

Proposition 4.6.3 *There is a bijective correspondence between the jet fields* $\Gamma : E \longrightarrow J^1\pi$ *and the connections* $R \in \bigwedge^1_0\pi \otimes \mathcal{X}(E)$.

Proof To obtain an explicit proof, suppose the jet field Γ is given. Let $a \in E$, and put $p = \pi(a)$; let ϕ be a local section of π whose domain contains p and which satisfies $a = \phi(p)$ and $j^1_p\phi = \Gamma(\phi(p))$. Define an endomorphism of tangent vectors in T_aE by $\phi_* \circ \pi_*$ (so that the transpose of this map is the endomorphism of cotangent vectors in T^*_aE given by $\pi^* \circ \phi^*$). This endomorphism depends only on the value and first derivatives of ϕ at

p, and is therefore independent of the particular choice of ϕ satisfying the conditions given; in coordinates it may be expressed as

$$\xi^i \left.\frac{\partial}{\partial x^i}\right|_a + \eta^\alpha \left.\frac{\partial}{\partial u^\alpha}\right|_a \longmapsto \xi^i \left(\left.\frac{\partial}{\partial x^i}\right|_a + u_i^\alpha(\Gamma(a)) \left.\frac{\partial}{\partial u^\alpha}\right|_a \right).$$

Taking this endomorphism at each point $a \in E$ yields a vector-valued 1-form $\tilde{\Gamma}$ which is seen to be smooth and to satisfy the conditions of the proposition from its coordinate representation

$$\tilde{\Gamma} = dx^i \otimes \left(\frac{\partial}{\partial x^i} + \Gamma_i^\alpha \frac{\partial}{\partial u^\alpha} \right).$$

Two distinct jet fields will have different coordinate representations at some point in E so that the corresponding vector-valued forms will differ; the correspondence is therefore injective. Furthermore, any vector-valued 1-form R satisfying the conditions of the proposition must have a coordinate representation of the form

$$R = dx^i \otimes \left(\frac{\partial}{\partial x^i} + R_i^\alpha \frac{\partial}{\partial u^\alpha} \right),$$

so locally there is a jet field with coordinates R_i^α which gives rise to R; on overlapping coordinate patches these local jet fields must agree since the correspondence is injective, so that this construction defines a global jet field: the correspondence is therefore also surjective. ∎

We shall call $\tilde{\Gamma}$ the *connection corresponding to the jet field* Γ. It is clear from this result that, for a function f, $\Gamma f = d_{\tilde{\Gamma}} f$. Since the affine bundle $\pi_{1,0}$ only takes the additional structure of a vector bundle in special circumstances, the sum of two jet fields is normally undefined.

Another way of looking at this construction involves the contact structure on π_1. This may be considered as the skeleton upon which all connections on π are built; a jet field then provides the flesh which distinguishes one connection from another. The connection $\tilde{\Gamma}$ is related to the jet field Γ by the formula

$$\tilde{\Gamma}_a(\xi) = pr_1(h(\xi, \Gamma(a)))$$

which describes the action of $\tilde{\Gamma}$ on a tangent vector $\xi \in T_a E$, where (h, id_E) is the horizontal vector bundle endomorphism of $\pi_{1,0}^*(\tau_E)$. Similarly, given a vector field X on M, its holonomic lift is a vector field X^0 along $\pi_{1,0}$; using the jet field Γ we obtain a vector field $X^0 \circ \Gamma$ on the manifold E, and this is just the horizontal lift of X corresponding to the connection $\tilde{\Gamma}$. We may also use a jet field to act on the horizontal and vertical representatives of a vector field Y on E to give horizontal and vertical vector fields $Y^h \circ \Gamma$, $Y^v \circ \Gamma$ defined on E rather than along $\pi_{1,0}$; in fact $Y^h \circ \Gamma = Y \lrcorner \tilde{\Gamma}$, which is the formula we used in Section 3.5 to describe the horizontal component of Y relative to the connection $\tilde{\Gamma}$.

Example 4.6.4 If π is the trivial bundle $(M \times \mathbf{R}, pr_1, M)$ with coordinates (x^i, u), where u is the identity coordinate on $\mathbf{R}$, then the jet field $\Gamma : M \times \mathbf{R} \longrightarrow J^1\pi \cong T^*M \times \mathbf{R}$ may be represented as

$$\tilde{\Gamma} = dx^i \otimes \left(\frac{\partial}{\partial x^i} + \Gamma_i \frac{\partial}{\partial u} \right).$$

If du is used to represent the pull-back $pr_2^*(du)$ of the volume form on $\mathbf{R}$ to $M \times \mathbf{R}$, then Γ may be represented as the horizontal 1-form

$$\tilde{\Gamma} \lrcorner du = \Gamma_i dx^i$$

Conversely, given a 1-form $\sigma \in \bigwedge_0^1 \pi$, σ determines a jet field Γ_σ by $\Gamma_\sigma(p, t) = (\sigma_p, t)$; in coordinates, if $\sigma = \sigma_i dx^i$ then

$$(\Gamma_\sigma)_i = p_i \circ \Gamma_\sigma = \sigma_i$$

(see Example 3.5.6). ■

Example 4.6.5 Now let π be the trivial bundle $(\mathbf{R} \times F, pr_1, \mathbf{R})$ with coordinates (t, q^α). As in Example 4.1.23, a jet field $\Gamma : \mathbf{R} \times F \longrightarrow J^1\pi \cong \mathbf{R} \times TF$ determines a "time-dependent vector field" $X_\Gamma \in \mathcal{X}(\mathbf{R} \times F)$; it also determines a connection $\tilde{\Gamma}$. If $\Gamma^\alpha = \dot{q}^\alpha \circ \Gamma$ then the coordinate representation of X_Γ is

$$X_\Gamma = \frac{\partial}{\partial t} + \Gamma^\alpha \frac{\partial}{\partial q^\alpha},$$

and that of $\tilde{\Gamma}$ is

$$\tilde{\Gamma} = dt \otimes \left(\frac{\partial}{\partial t} + \Gamma^\alpha \frac{\partial}{\partial q^\alpha} \right),$$

so that $\tilde{\Gamma} = dt \otimes X_\Gamma$ (see Example 3.5.7). ■

Example 4.6.6 Retaining π as the bundle $(\mathbf{R} \times F, pr_1, \mathbf{R})$, a jet field on the first jet bundle π_1 is a section of $(\pi_1)_1$ mapping $J^1\pi \cong \mathbf{R} \times TF \longrightarrow J^1\pi_1 \cong \mathbf{R} \times TTF$. If Γ is such a jet field with the additional property that $i_{\tilde{\Gamma}}\theta = 0$ for every $\theta \in \bigwedge_C^1 \pi_{1,0}$ then in coordinates

$$\tilde{\Gamma} = dt \otimes \left(\frac{\partial}{\partial t} + \dot{q}^\alpha \frac{\partial}{\partial q^\alpha} + \Gamma^\alpha \frac{\partial}{\partial \dot{q}^\alpha} \right).$$

The map Γ is then an example of a *second-order jet field*, the general definition of which will be given in Chapter 5. As the base manifold of π_1 is one-dimensional, there is a representation of Γ as a vector field on $J^1\pi$,

$$\frac{\partial}{\partial t} + \dot{q}^\alpha \frac{\partial}{\partial q^\alpha} + \Gamma^\alpha \frac{\partial}{\partial \dot{q}^\alpha};$$

such a vector field is called a *time-dependent second-order vector field* and is used in the study of time-dependent mechanics on $\mathbf{R} \times TF$. ∎

Example 4.6.7 If π happens to be a vector bundle (so that π_1 is also a vector bundle) then we may impose the additional requirement that (Γ, id_M) be a vector bundle morphism from π to π_1. This implies that the coordinate functions Γ_i^α will be linear in the fibre coordinates, so that $\Gamma_i^\alpha = u^\beta \pi^*(\Gamma_{i\beta}^\alpha)$ where $\Gamma_{i\beta}^\alpha$ are functions defined locally on M; in fact $\Gamma_{i\beta}^\alpha = \Gamma_i^\alpha \circ e_\beta$, where e_β are the local sections of π dual to the fibre coordinates u^β. Such a map $E \longrightarrow J^1\pi$ may be called an *affine jet field*. We may then construct the map $K : TE \longrightarrow E$ by taking the composition

$$TE \longrightarrow V\pi \longrightarrow E,$$

where the first map is $I - \widetilde{\Gamma}$ and the second is the map described in Exercise 2.2.1 (rather than $\tau_E|_{V\pi}$). In coordinates, if $\eta \in T_aE$ is given by

$$\eta = \eta^i \left.\frac{\partial}{\partial x^i}\right|_a + \eta^\alpha \left.\frac{\partial}{\partial u^\alpha}\right|_a,$$

then $K(\eta)$ is the element of the fibre through a given by

$$K(\eta) = (\eta^\alpha - \Gamma_{i\beta}^\alpha(\pi(a))u^\beta(a)\eta^i)e_\alpha(\pi(a)).$$

If ϕ is a section of π, then the *covariant differential* of ϕ determined by the affine jet field Γ is the section $\nabla\phi$ of the tensor product bundle $\tau_M^* \otimes \pi$ defined by

$$(\nabla\phi)_p(\xi) = K(\phi_*(\xi))$$

for $\xi \in T_pM$. In coordinates, if ξ is given by $\xi^i\partial/\partial x^i$ then $(\nabla\phi)_p(\xi)$ is the element of the fibre E_p given by

$$(\nabla\phi)_p(\xi) = \left(\left.\frac{\partial \phi^\alpha}{\partial x^i}\right|_p - \Gamma_{i\beta}^\alpha(p)\phi^\beta(p) \right) \xi^i e_\alpha(p).$$

In the particular case when π is actually the tangent bundle (TM, τ_M, M) with coordinates q^α on M and $(q^\alpha, \dot{q}^\alpha)$ on TM, then for a vector field $X \in$

$\Gamma(\tau_M) = \mathcal{X}(M)$ the covariant differential ∇X is the vector-valued 1-form written in coordinates as

$$\nabla X = \left(\frac{\partial X^\alpha}{\partial q^\gamma} - \Gamma^\alpha_{\gamma\beta} X^\beta \right) dq^\gamma \otimes \frac{\partial}{\partial q^\alpha};$$

the functions $\Gamma^\alpha_{\gamma\beta}$ are (apart from sign) the Christoffel symbols of the connection $\tilde{\Gamma}$. ■

If it happens to be the case that π is a trivial bundle $(M \times F, pr_1, M)$ with trivial first jet bundle $(M \times J^1_p\pi, pr_1, M)$ for some $p \in M$, then it makes sense to ask whether a jet field $\Gamma : M \times F \longrightarrow M \times J^1_p\pi$ is a bundle morphism from $(M \times F, pr_2, F)$ to $(M \times J^1_p\pi, pr_2, J^1_p\pi)$. If it is, then one may call Γ a *base-independent jet field*, and obtain an induced map $\overline{\Gamma} : F \longrightarrow J^1_p\pi$. The coordinate representation functions Γ^α_i will then be independent of x^i in any coordinate system (x^i, u^α) which respects the trivialisation $M \times F$.

Example 4.6.8 If π is the trivial bundle $(\mathbf{R} \times F, pr_1, \mathbf{R})$, then a base-independent jet field Γ gives rise to a vector field on $\mathbf{R} \times F$ written in coordinates as

$$\frac{\partial}{\partial t} + \Gamma^\alpha \frac{\partial}{\partial q^\alpha},$$

where the functions Γ^α are independent of t. This vector field is then projectable onto F in the usual sense of a projectable vector field to give

$$\Gamma^\alpha \frac{\partial}{\partial q^\alpha},$$

which of course is just the coordinate representation of the ordinary (time-independent) vector field $\overline{\Gamma} : F \longrightarrow J^1_0\pi \cong TF$. ■

To continue our analogy between jet fields and vector fields, we shall define integral sections of a jet field.

Definition 4.6.9 An *integral section* of the jet field Γ is a local section ϕ of π satisfying $j^1\phi = \Gamma \circ \phi$. ■

This definition clearly mimics the corresponding definition for an integral curve of a vector field. There is, however, an important difference: there is no guarantee that integral sections of a given jet field will exist, even locally. To see this, observe that each jet field Γ defines a first-order differential equation $\operatorname{im}(\Gamma) \subset J^1\pi$, and that an integral section of Γ is a solution of this equation. (However, it should be clear that not every first-order differential equation is the image of a jet field.) In coordinates,

$$\frac{\partial \phi^\alpha}{\partial x^i} = \Gamma^\alpha_i \circ \phi,$$

and this set of partial differential equations must satisfy an integrability condition if solutions ϕ^α are to exist. In fact, the following result is essentially a translation of Frobenius' theorem into the language of jet fields.

Proposition 4.6.10 *The jet field Γ has integral sections if, and only if, the curvature of $\widetilde{\Gamma}$ vanishes; such a jet field is termed integrable.*

Proof From Definition 3.5.13,

$$R_{\widetilde{\Gamma}}(X,Y) = [X \lrcorner \widetilde{\Gamma}, Y \lrcorner \widetilde{\Gamma}] \lrcorner (I - \widetilde{\Gamma}).$$

Consider this expression locally, and let X, Y be coordinate vector fields. Then $R_{\widetilde{\Gamma}}(\partial/\partial u^\alpha, \partial/\partial u^\beta)$ and $R_{\widetilde{\Gamma}}(\partial/\partial x^i, \partial/\partial u^\beta)$ vanish identically; the only non-trivial expression comes from

$$R_{\widetilde{\Gamma}}\left(\frac{\partial}{\partial x^i}, \frac{\partial}{\partial x^j}\right) = \left(\left(\frac{\partial \Gamma^\alpha_j}{\partial x^i} + \Gamma^\beta_i \frac{\partial \Gamma^\alpha_j}{\partial u^\beta}\right) - \left(\frac{\partial \Gamma^\alpha_i}{\partial x^j} + \Gamma^\beta_j \frac{\partial \Gamma^\alpha_i}{\partial u^\beta}\right)\right) \frac{\partial}{\partial u^\alpha}.$$

But the vanishing of this expression is precisely the condition for the equations

$$\frac{\partial \phi^\alpha}{\partial x^i} = \Gamma^\alpha_i \circ \phi$$

to be integrable in the sense of Frobenius. ∎

Finally in this section, we shall consider symmetries of jet fields. In the case of a vector field, a symmetry may be regarded as a diffeomorphism of the manifold which permutes the integral curves without changing their parametrization. It therefore seems natural to consider those bundle isomorphisms (f, id_M) of π which permute the integral sections of Γ. (Such a bundle isomorphism is obviously a symmetry of the differential equation $\mathrm{im}\,(\Gamma) \subset J^1\pi$ as described in Section 4.1; we could, more generally, consider bundle isomorphisms which need not project onto the identity on M.) If (f, id_M) is such a bundle isomorphism, we wish to assert that f is a symmetry of Γ if, whenever ϕ is an integral section, so is $\widetilde{f}(\phi) = f \circ \phi$. Using the characterisation of $\widetilde{\Gamma}_{\phi(p)}$ which was given in Proposition 4.6.3 as an endomorphism of $T_{\phi(p)}E$,

$$\widetilde{\Gamma}_{\phi(p)} = \phi_* \circ \pi_*$$

we obtain

$$\begin{aligned} \widetilde{\Gamma}_{f(\phi(p))} &= (f_* \circ \phi_*) \circ (\pi_* \circ f_*^{-1}) \\ &= f_* \circ \widetilde{\Gamma}_{\phi(p)} \circ f_*^{-1} \end{aligned}$$

or, more generally,

$$\tilde{\Gamma}_{f(a)} = f_* \circ \tilde{\Gamma}_a \circ f_*^{-1}$$

for $a \in E$. This leads to the following definition, which makes sense whether or not Γ is integrable.

Definition 4.6.11 A *symmetry* of the jet field Γ is a bundle isomorphism (f, id_M) of π which satisfies $f_* \circ \tilde{\Gamma} = \tilde{\Gamma} \circ f_*$, where $\tilde{\Gamma}$ is regarded as acting on tangent vectors. ■

Proposition 4.6.12 *If Γ is integrable then (f, id_M) is a symmetry of Γ if, and only if, $\tilde{f}$ permutes the integral sections of Γ.* ■

Corresponding to this idea is the infinitesimal version. We shall consider an infinitesimal symmetry of a jet field to be a vector field whose flow consists of symmetries; as one would expect, this condition may be expressed by the vanishing of the Lie derivative. We shall give an explicit proof of this result in a slightly more general context.

Proposition 4.6.13 *If X is a vector field on E with flow ψ_t, and R is a vector-valued 1-form on E, then $d_X R = 0$ if, and only if, $\psi_{t*} \circ R = R \circ \psi_{t*}$ for each t (where R is regarded as acting on tangent vectors).*

Proof For simplicity we shall assume that X is complete, although this assumption is not necessary for the result to hold.

Suppose first that each ψ_t satisfies $\psi_{t*} \circ R = R \circ \psi_{t*}$. Then for every vector field Y and every point $a \in E$,

$$\begin{aligned}
(Y \lrcorner\, d_X R)_a &= (\mathcal{L}_X(Y \lrcorner\, R))_a - (\mathcal{L}_X Y \lrcorner\, R)_a \\
&= \left.\frac{d}{dt}\right|_{t=0} \psi_{t*}((Y \lrcorner\, R)_{\psi_{-t}(a)}) - R_a \left(\left.\frac{d}{dt}\right|_{t=0} \psi_{t*}(Y_{\psi_{-t}(a)}) \right) \\
&= \left.\frac{d}{dt}\right|_{t=0} \left(\psi_{t*}(R_{\psi_{-t}(a)}(Y_{\psi_{-t}(a)})) - R_a(\psi_{t*}(Y_{\psi_{-t}(a)})) \right)
\end{aligned}$$

using continuity of the endomorphism R_a of $T_a E$. Consequently the right-hand side of this expression vanishes, and so $d_X R = 0$.

The converse involves a proof which effectively integrates along the flow ψ_t. So suppose that $d_X R = 0$. Then for every vector field Y and every point $a \in E$,

$$(\mathcal{L}_X(Y \lrcorner\, R))_a = R_a(\mathcal{L}_X Y)_a$$

which we may write as

$$\left.\frac{d}{dt}\right|_{t=0} \psi_{t*}(R_{\psi_{-t}(a)}(Y_{\psi_{-t}(a)})) = \left.\frac{d}{dt}\right|_{t=0} R_a(\psi_{t*}(Y_{\psi_{-t}(a)})).$$

Fix a and choose an arbitrary real number h, writing $a_{-h} = \psi_{-h}(a)$; then this equation is still true with a replaced by a_{-h}. For each tangent vector $\eta \in T_{a_{-h}}E$ there is certainly a smooth vector field Y satisfying, for sufficiently small t, $Y_{\psi_{-t}(a_{-h})} = \psi_{-t*}(\eta)$; with this choice of Y we obtain

$$\begin{aligned} \left.\frac{d}{dt}\right|_{t=0} \psi_{t*}(R_{\psi_{-t}(a_{-h})}(\psi_{-t*}(\eta))) &= \left.\frac{d}{dt}\right|_{t=0} R_{a_{-h}}(\eta) \\ &= 0 \end{aligned}$$

since $R_{a_{-h}}(\eta)$ is independent of t. Also, $\eta \in T_{a_{-h}}E$ is arbitrary, so

$$\left.\frac{d}{dt}\right|_{t=0} \psi_{t*} \circ R_{\psi_{-t}(a_{-h})} \circ \psi_{-t*} = 0$$

as an endomorphism of $T_{a_{-h}}E$.

Now operate on this equation on the left by ψ_{h*} and on the right by ψ_{-h*}. The result is an equation relating endomorphisms of T_aE, and writing τ for $t+h$ we obtain

$$\left.\frac{d}{d\tau}\right|_{\tau=h} \psi_{\tau*} \circ R_{\psi_{-\tau}(a)} \circ \psi_{-\tau*} = 0.$$

But h, too, is arbitrary and so $\psi_{\tau*} \circ R_{\psi_{-\tau}(a)} \circ \psi_{-\tau*}$ is independent of τ and therefore equals its value when τ is zero:

$$\psi_{\tau*} \circ R_{\psi_{-\tau}(a)} \circ \psi_{-\tau*} = R_a$$

so that $\psi_{\tau*}R = R\psi_{\tau*}$. ∎

Definition 4.6.14 An *infinitesimal symmetry* of the jet field Γ is a vertical vector field X satisfying $d_X\tilde{\Gamma} = 0$. ∎

Proposition 4.6.15 *If Γ is integrable then X is an infinitesimal symmetry of Γ if, and only if, for each t the diffeomorphism ψ_t permutes the integral sections of Γ.* ∎

EXERCISES

4.6.1 If Γ is a jet field and $X \in \mathcal{X}(E)$ satisfies $d_X\tilde{\Gamma} = 0$, show that X is necessarily projectable. (The vector field is then an "infinitesimal symmetry" which need not retain the parametrisation of any integral sections of Γ.)

4.6.2 If X is an infinitesimal symmetry of Γ, show that

$$\frac{\partial X^\beta}{\partial x^i} = X^\alpha \frac{\partial \Gamma^\beta_i}{\partial u^\alpha} - \Gamma^\alpha_i \frac{\partial X^\beta}{\partial u^\alpha}$$

where $X = X^\alpha \partial / \partial u^\alpha$ and $\Gamma^\alpha_i = u^\alpha_i \circ \Gamma$.

4.6.3 Let Γ be an affine jet field on the vector bundle (T^*M, τ^*_M, M). Let q^i be coordinates on M, and let (q^i, p_i) be the corresponding coordinates on T^*M; let the coordinate representation of Γ be $\Gamma_{ij} = (p_i)_j \circ \Gamma$, and let $\Gamma^k_{ij} = \Gamma_{ij} \circ dq^k$. If ∇ is the covariant differential defined by Γ, show that

$$\nabla dq^k = -\Gamma^k_{ji} dq^i \otimes dq^j.$$

4.7 Vertical Lifts

There is one further property of the bundle $(J^1\pi, \pi_{1,0}, E)$ which is important in the study of the calculus of variations, and which may be viewed as a generalisation of the "almost tangent structure" on a tangent manifold TM introduced in Exercise 3.4.2. This latter object is a vector-valued 1-form S on TM having the properties that rank $S = \dim M$, that $S \lrcorner S = 0$ and that the Nijenhuis tensor $N_S = 0$. In coordinates, S takes the form

$$S = dx^i \otimes \frac{\partial}{\partial \dot{x}^i}.$$

The pointwise action of S on tangent vectors may be defined by the rule

$$S_a(\xi) = (\tau_{M*}(\xi))^{\mathrm{v}}$$

where $\xi \in T_a TM$. The symbol v denotes the vertical lift of a tangent vector mentioned in Exercise 2.2.2, and arises as a consequence of the vector bundle isomorphism between $(V\tau_M, \tau_M|_{V\tau_M}, TM)$ and $(\tau^*_M(TM), \tau^*_M(\tau_M), TM)$ (where τ^*_M here denotes the pull-back of an object by the tangent bundle projection τ_M, and not the cotangent bundle projection).

The generalisation of this construction to the bundle $\pi_{1,0}$ takes advantage of the vector bundle isomorphism between the vertical bundle to an affine bundle, and the pull-back of the vector bundle on which the affine bundle is modelled. As explained in Theorem 4.1.11, the bundle $\pi_{1,0}$ is an affine bundle modelled on the vector bundle

$$\left(\pi^*(T^*M) \otimes V\pi, (\tau^*_E|_{\pi^*(T^*M)}) \otimes (\tau_E|_{V\pi}), E\right),$$

so for each point $j^1_p\phi \in J^1\pi$ there is a vertical lift operation from a *pair* of elements (η, ζ) where $\eta \in T^*_p M$, $\zeta \in V_{\phi(p)}\pi$, to give a tangent vector in $V_{j^1_p\phi}\pi_{1,0}$.

Each 1-form ω on the base manifold M therefore yields a vector-valued 1-form S_ω on $J^1\pi$ by a somewhat more complicated version of composing the projection $\pi_{1,0}$ with the vertical lift.

There is, however, a rather different way of carrying out the vertical lift operation on $\pi_{1,0}$ which generalises more easily to higher-order jet bundles. We shall therefore adopt this alternative approach, and subsequently demonstrate the equivalence of the two operations using coordinates.

Theorem 4.7.1 *Suppose given a point $j^1_p\phi \in J^1\pi$, a cotangent vector $\eta \in T^*_pM$ and a tangent vector $\zeta \in V_{\phi(p)}\pi$. Let W be a neighbourhood of $p \in M$ and let $\gamma : W \times \mathbf{R} \longrightarrow E$ satisfy $[t \longmapsto \gamma(p,t)] = \zeta$, $j^1_p(q \longmapsto \gamma(q,0)) = j^1_p\phi$. Let $f \in C^\infty(M)$ satisfy $f(p) = 0$, $df_p = \eta$. Then the new tangent vector*

$$[t \longmapsto j^1_p(q \longmapsto \gamma(q, tf(q)))],$$

denoted by the symbol $\eta \circledS_{j^1_p\phi} \zeta$, is an element of $V_{j^1_p\phi}\pi_{1,0}$ which is independent of the choices of γ and f.

Proof We note first that the existence of maps γ satisfying the required conditions may be seen easily in coordinates. The new tangent vector $\eta \circledS_{j^1_p\phi} \zeta$ is an element of $T_{j^1_p\phi}J^1\pi$ because $j^1_p(q \longmapsto \gamma(q,0)) = j^1_p\phi$. It is vertical over $\pi_{1,0}$ because

$$\begin{aligned}
\pi_{1,0*}[t \longmapsto j^1_p(q \longmapsto \gamma(q, tf(q)))] &= [t \longmapsto \pi_{1,0}(j^1_p(q \longmapsto \gamma(q, tf(q))))] \\
&= [t \longmapsto \gamma(p, tf(p))] \\
&= 0
\end{aligned}$$

since $f(p) = 0$. Finally, if in coordinates

$$\eta = \eta_i \, dx^i \Big|_p , \qquad \zeta = \zeta^\alpha \frac{\partial}{\partial u^\alpha}\bigg|_{\phi(p)}$$

then

$$\begin{aligned}
\eta \circledS_{j^1_p\phi} \zeta &= \frac{\partial}{\partial t}\bigg|_{t=0} \frac{\partial}{\partial x^i}\bigg|_{q=p} \gamma^\alpha(q, tf(q)) \frac{\partial}{\partial u^\alpha_i}\bigg|_{j^1_p\phi} \\
&= \frac{\partial}{\partial x^i}\bigg|_p \left(f \frac{\partial \gamma^\alpha}{\partial t}\bigg|_{t=0} \right) \frac{\partial}{\partial u^\alpha_i}\bigg|_{j^1_p\phi} \\
&= \frac{\partial f}{\partial x^i}\bigg|_p \frac{\partial \gamma^\alpha}{\partial t}\bigg|_{t=0} \frac{\partial}{\partial u^\alpha_i}\bigg|_{j^1_p\phi} \\
&= \eta_i \zeta^i \frac{\partial}{\partial u^\alpha_i}\bigg|_{j^1_p\phi} ,
\end{aligned}$$

demonstrating that the new tangent vector depends on ζ and η rather than γ and f. ■

Corollary 4.7.2 *The tangent vector $\eta \circledvee_{j^1_p\phi} \zeta$ is the image of $(\pi^*(\eta) \otimes \zeta, j^1_p\phi)$ under the canonical vector bundle isomorphism*

$$\pi^*_{1,0}(\pi^*(T^*M) \otimes V\pi) \longrightarrow V\pi_{1,0}.$$

Proof The vector bundle isomorphism is given explicitly by

$$(\lambda, j^1_p\phi) \longmapsto [t \longmapsto (t\lambda)[j^1_p\phi]]$$

where $\lambda \in (\pi^*(T^*M) \otimes V\pi)_{\phi(p)}$ and $(t\lambda)[j^1_p\phi]$ denotes the affine action of $t\lambda$ on $j^1_p\phi$ described in Theorem 4.1.11. In coordinates, if

$$\lambda = \pi^*(\eta) \otimes \zeta = \eta_i \zeta^\alpha \left(dx^i \otimes \frac{\partial}{\partial u^\alpha} \right)_{\phi(p)}$$

then

$$u^\alpha_i((t\lambda)[j^1_p\phi]) = u^\alpha_i(j^1_p\phi) + t\eta_i\zeta^\alpha,$$

so that

$$[t \longmapsto (t\lambda)[j^1_p\phi]] = \eta_i\zeta^\alpha \left. \frac{\partial}{\partial u^\alpha_i} \right|_{j^1_p\phi}.$$

■

To obtain the vector-valued 1-form from this vertical lift operation, we shall define a pointwise action upon tangent vectors to $J^1\pi$. Since, however, the vertical lift is only defined for vectors vertical over M, we must project each tangent vector to E, and then take its "vertical representative over M" using the decomposition of $\pi^*_{1,0}(TE)_{j^1_p\phi}$ described in Theorem 4.3.2. To do this, we shall use the representation of the contact structure on π_1 as a vector bundle endomorphism (v, id_E) of $\pi \circ \tau_E$.

Definition 4.7.3 If $\omega \in \bigwedge^1 M$, then the vector-valued 1-form $S_\omega \in \bigwedge^1_C \pi_{1,0} \otimes \mathcal{X}^v(\pi_{1,0})$ is defined by

$$(S_\omega)_{j^1_p\phi}(\xi) = \omega_p \circledvee_{j^1_p\phi} pr_1(v(\pi_{1,0*}(\xi), j^1_p\phi))$$

where $\xi \in T_{j^1_p\phi} J^1\pi$. ■

In this definition, $v(\pi_{1,0*}(\xi), j^1_p\phi) \in \pi^*_{1,0}(V\pi) = V\pi \times_E J^1\pi$, so that projection on the first factor gives an element of $V\pi$ as required. In coordinates,

$$S_\omega = \omega_i(du^\alpha - u^\alpha_j dx^j) \otimes \frac{\partial}{\partial u^\alpha_i}.$$

Example 4.7.4 If π is the trivial bundle $(\mathbf{R} \times F, pr_1, \mathbf{R})$, then it is natural to use the volume form dt on the base manifold $\mathbf{R}$; the vector-valued 1-form S_{dt} then has the coordinate representation

$$S_{dt} = (dq^\alpha - \dot{q}^\alpha dt) \otimes \frac{\partial}{\partial \dot{q}^\alpha}.$$

This is the operator used in the classical Hamilton-Cartan formalism for problems in the calculus of variations (in one independent variable) which involve time explicitly. ∎

In general, the vector-valued 1-form S_ω retains some (but not all) of the properties of the almost tangent structure on a tangent manifold. The rank of S_ω is constant (and equal to the fibre dimension n of the bundle π) at all points of $J^1\pi$ where $\pi_1^*(\omega)$ does not vanish; at the remaining points its rank is obviously zero. Equally obviously, if $\omega_1, \omega_2 \in \bigwedge^1 M$ then $S_{\omega_1} \lrcorner S_{\omega_2} = 0$. However the Nijenhuis tensor of S_ω does *not* vanish (unless ω is identically zero), and a quick calculation shows that, for example,

$$\left(\frac{\partial}{\partial x^i}, \frac{\partial}{\partial u^\alpha}\right) \lrcorner N_{S_\omega} = \omega_i \omega_j \frac{\partial}{\partial u_j^\alpha}.$$

A disadvantage of this construction (at least when the base manifold has dimension greater than one) is its dependence upon a 1-form $\omega \in \bigwedge^1 M$. Since the dependence is linear, it is possible to find a single object S which is a section of the bundle $T^*J^1\pi \otimes TJ^1\pi \otimes \pi_1^*(TM)$ over $J^1\pi$, and may be called a "type (2,1) tensor field along π_1". The tensor S is defined by the rule

$$C(S \otimes \omega) = S_\omega,$$

where C denotes contraction of the second contravariant index of S with the 1-form ω. In coordinates,

$$S = (du^\alpha - u_j^\alpha dx^j) \otimes \frac{\partial}{\partial u_i^\alpha} \otimes \frac{\partial}{\partial x^i}.$$

However, when M is orientable with a given volume form Ω, then a more convenient version of this entity may be obtained by contracting S with Ω to give a vector-valued m-form S_Ω on $J^1\pi$. In fact we can avoid the use of the tensor S altogether by defining, for each $\sigma \in \bigwedge^1 J^1\pi$, the vector-valued 1-form $S\sigma$ along π_1 according to the rule

$$(S\sigma) \lrcorner \omega = S_\omega \lrcorner \sigma,$$

and then setting

$$S_\Omega \,\lrcorner\, \sigma = i_{S\sigma}\Omega$$

where $i_{S\sigma}$ is the derivation of type i_* corresponding to $S\sigma$. In coordinates,

$$S_\Omega = (du^\alpha - u^\alpha_j dx^j) \wedge \left(\frac{\partial}{\partial x^i} \,\lrcorner\, \Omega\right) \otimes \frac{\partial}{\partial u^\alpha_i}.$$

This vector-valued m-form, together with a generalisation of the contact structure introduced in Section 4.5, contains all the information necessary for a study of the first-order calculus of variations.

Example 4.7.5 For the trivial bundle $(\mathbf{R} \times F, pr_1, \mathbf{R})$, where the volume form on the base manifold is $\Omega = dt$, the vector-valued m-form S_Ω is identical to the vector-valued 1-form S_{dt}. This identity is one of the reasons for the relative simplicity of the calculus of variations in a single independent variable. It also provides, as we shall see later, a reason why the natural generalisation to the "higher-order" calculus of variations which may be used for a single independent variable is inappropriate for multiple independent variables. ■

EXERCISES

4.7.1 Show that the image distribution of S_ω is involutive, and that its integral manifolds are affine subspaces of the fibres of $\pi_{1,0}$. Is the kernel distribution of S_ω ever involutive?

4.7.2 Let π be the bundle $(\mathbf{R} \times F, pr_1, \mathbf{R})$, so that π_1 is the bundle $(\mathbf{R} \times TF, pr_1, \mathbf{R})$. Show that a section of $(\pi_1)_{1,0}$ may be represented as a vector field on $J^1\pi = \mathbf{R} \times TF$, and that if such a vector field X has the additional property that $X \,\lrcorner\, S_{dt} = 0$ then X has the coordinate representation

$$X = \frac{\partial}{\partial t} + \dot{q}^\alpha \frac{\partial}{\partial q^\alpha} + X^\alpha \frac{\partial}{\partial \dot{q}^\alpha}.$$

and so is a time-dependent second-order vector field as described in Example 4.6.6.

4.7.3 Let π again be the bundle $(\mathbf{R} \times F, pr_1, \mathbf{R})$, and let $L \in C^\infty(J^1\pi)$ be a Lagrangian; define $\omega_L \in \bigwedge^2 J^1\pi$ by

$$\omega_L = d(S \,\lrcorner\, dt + L\, dt).$$

If ω_L has constant rank $2n$ (where $n = \dim F$), show that there is a unique time-dependent second-order vector field X_L satisfying

$$X_L \lrcorner\, \omega_L = 0.$$

(The vector field X_L is called the *Euler-Lagrange field* of the Lagrangian L.) Show that the coefficient functions X_L^α of X_L satisfy

$$\frac{\partial^2 L}{\partial \dot{q}^\alpha \, \partial \dot{q}^\beta} X_L^\alpha = \frac{\partial L}{\partial q^\beta} - \left(\frac{\partial}{\partial t} + \dot{q}^\alpha \frac{\partial}{\partial q^\alpha} \right) \frac{\partial L}{\partial \dot{q}^\beta},$$

where the expression in brackets is regarded as an operator on $C^\infty(J^1\pi)$.

REMARKS

Although we have defined jets of local sections of a bundle, it is clearly possible to define jets of functions $f : M \longrightarrow F$, where M and F are manifolds; the jet of f then just corresponds to the jet of the section gr_f of the trivial bundle $(M \times F, pr_1, F)$. This restriction does not simplify the theory because, according to Exercise 4.1.8, the global triviality of π does not imply the global triviality of π_1. If, instead, we consider functions $f : \mathbf{R}^m \longrightarrow F$, then this *does* introduce a simplification, because $J^1\pi$ is then diffeomorphic to $\mathbf{R}^m \times J^1_0\pi$ by Lemma 4.1.20. This is the approach adopted in, for example, [4].

It is also possible to generalise the idea of a jet to include jets of "multi-valued sections" or "sections with infinite derivatives" by considering arbitrary local embeddings of M in E. These "extended jets" can be useful in the study of differential equations: for instance, in the context of Example 4.4.12, the map $\phi : \mathbf{R} \longrightarrow \mathbf{R} \times \mathbf{R}$ given by $\phi(p) = (\cos p, \sin p)$ is a "solution" of the equation which cannot be represented by a local section of the bundle π. A discussion of extended jets may be found in [14]; this reference also contains a great deal of information about symmetries of differential equations.

Chapter 5

Second-order Jet Bundles

This chapter is something of a half-way house, where we start the process of generalising the idea of a jet to take account of derivatives higher than the first. There are several advantages to be gained by restricting our attention at this stage to second-order jets. We may continue with the coordinate notation used in the previous chapter, and so see quite clearly the difference between "holonomic" and "non-holonomic" jets; the integrability condition for a jet field may be expressed in terms of second-order jets; the Euler-Lagrange equations for a variational problem may be expressed in the form of a "second-order jet field". It should be clear, however, that many of the constructions involving second-order jets may be generalised further to jets of arbitrary order.

5.1 Second-order Jets

If (E, π, M) is a bundle, we may define the *second jet manifold* $J^2\pi$ using a similar method to the one we employed when defining $J^1\pi$. The elements of $J^2\pi$ will be 2-jets $j^2_p\phi$ of local sections $\phi \in \Gamma_p(\pi)$, where a 2-jet is an equivalence class containing those local sections with the same value and first *two* derivatives at p. As with 1-jets, we shall specify the equivalence relation using coordinates, and so we must ensure that the particular choice of coordinate system will not matter.

Lemma 5.1.1 *Let (E, π, M) be a bundle and let $p \in M$. Suppose that $\phi, \psi \in \Gamma_p(\pi)$ satisfy $\phi(p) = \psi(p)$. Let (x^i, u^α) and (y^j, v^β) be two adapted coordinate systems around $\phi(p)$, and suppose also that*

$$\left.\frac{\partial(u^\alpha \circ \phi)}{\partial x^i}\right|_p = \left.\frac{\partial(u^\alpha \circ \psi)}{\partial x^i}\right|_p$$

and

$$\left.\frac{\partial^2(u^\alpha\circ\phi)}{\partial x^i\,\partial x^j}\right|_p = \left.\frac{\partial^2(u^\alpha\circ\psi)}{\partial x^i\,\partial x^j}\right|_p$$

for $1\le i,j\le m$ *and* $1\le\alpha\le n$. *Then*

$$\left.\frac{\partial(v^\beta\circ\phi)}{\partial y^k}\right|_p = \left.\frac{\partial(v^\beta\circ\psi)}{\partial y^k}\right|_p$$

and

$$\left.\frac{\partial^2(v^\beta\circ\phi)}{\partial y^k\,\partial y^l}\right|_p = \left.\frac{\partial^2(v^\beta\circ\psi)}{\partial y^k\,\partial y^l}\right|_p$$

for $1\le k,l\le m$ *and* $1\le\beta\le n$.

Proof The first assertion is just Lemma 4.1.1. To prove the second, note that

$$\begin{aligned}\frac{\partial(v^\beta\circ\phi)}{\partial y^k} &= \left(\frac{\partial v^\beta}{\partial x^i}\circ\phi+\left(\frac{\partial v^\beta}{\partial u^\alpha}\circ\phi\right)\frac{\partial(u^\alpha\circ\phi)}{\partial x^i}\right)\frac{\partial x^i}{\partial y^k}\\ &= F^\beta_k\circ\left(x^i,u^\alpha\circ\phi,\frac{\partial(u^\alpha\circ\phi)}{\partial x^i}\right),\end{aligned}$$

where the functions F^β_k do not depend on the local section ϕ. A second application of the Chain Rule then gives

$$\frac{\partial^2(v^\beta\circ\phi)}{\partial y^k\,\partial y^l} = F^\beta_{kl}\circ\left(x^i,u^\alpha\circ\phi,\frac{\partial(u^\alpha\circ\phi)}{\partial x^i},\frac{\partial^2(u^\alpha\circ\phi)}{\partial x^i\,\partial x^j}\right),$$

where again the functions F^β_{kl} do not depend on ϕ. The result then follows by evaluating this new equation at p, and using the equality of ϕ and ψ (and their respective first- and second-order derivatives) at p. ∎

Although the exact form of the functions F^β_{kl} is not important at this stage, it is worth noting that we may write

$$\frac{\partial^2(v^\beta\circ\phi)}{\partial y^k\,\partial y^l} = \left(\frac{\partial v^\beta}{\partial u^\alpha}\circ\phi\right)\frac{\partial x^i}{\partial y^k}\frac{\partial x^j}{\partial y^l}\frac{\partial^2(u^\alpha\circ\phi)}{\partial x^i\,\partial x^j}+G^\beta_{kl}\circ\left(x^i,u^\alpha\circ\phi,\frac{\partial(u^\alpha\circ\phi)}{\partial x^i}\right),$$

so that F^β_{kl} is an inhomogeneous linear function of its final argument.

Definition 5.1.2 Let (E,π,M) be a bundle, and let $p\in M$. Define the local sections $\phi,\psi\in\Gamma_p(\pi)$ to be *2-equivalent at p* if $\phi(p)=\psi(p)$ and if, in some adapted coordinate system (x^i,u^α) around $\phi(p)$,

$$\left.\frac{\partial\phi^\alpha}{\partial x^i}\right|_p = \left.\frac{\partial\psi^\alpha}{\partial x^i}\right|_p \quad\text{and}\quad \left.\frac{\partial^2\phi^\alpha}{\partial x^i\,\partial x^j}\right|_p = \left.\frac{\partial^2\psi^\alpha}{\partial x^i\,\partial x^j}\right|_p$$

for $1 \le i, j \le m$ and $1 \le \alpha \le n$. The equivalence class containing ϕ is called the *2-jet of ϕ at p* and is denoted $j^2_p\phi$. ∎

Definition 5.1.3 The *second jet manifold of π* is the set

$$\{j^2_p\phi : p \in M, \phi \in \Gamma_p(\pi)\}$$

and is denoted $J^2\pi$. The functions π_2, $\pi_{2,0}$ and $\pi_{2,1}$, called the *source*, *target* and *1-jet* projections respectively, are defined by

$$\begin{aligned} \pi_2 : J^2\pi &\longrightarrow M \\ j^2_p\phi &\longmapsto p; \end{aligned}$$

$$\begin{aligned} \pi_{2,0} : J^2\pi &\longrightarrow E \\ j^2_p\phi &\longmapsto \phi(p) \end{aligned}$$

and

$$\begin{aligned} \pi_{2,1} : J^2\pi &\longrightarrow J^1\pi \\ j^2_p\phi &\longmapsto j^1_p\phi. \end{aligned}$$

∎

So far, our construction of 2-jets has been similar to that of 1-jets, although of course it has been slightly more complicated. The first major difference arises in the next definition, where the commutativity of repeated partial differentiation plays an important rôle.

Definition 5.1.4 Let (E, π, M) be a bundle, and let (U, u) be an adapted coordinate system on E, where $u = (x^i, u^\alpha)$. The *induced coordinate system* (U^2, u^2) on $J^2\pi$ is defined by

$$\begin{aligned} U^2 &= \{j^2_p\phi : \phi(p) \in U\} \\ u^2 &= (x^i, u^\alpha, u^\alpha_i, u^\alpha_{ij}), \end{aligned}$$

where $x^i(j^2_p\phi) = x^i(p)$; $u^\alpha(j^2_p\phi) = u^\alpha(\phi(p))$; $u^\alpha_i(j^2_p\phi) = u^\alpha_i(j^1_p\phi)$; and the $\frac{1}{2}mn(m+1)$ new functions

$$u^\alpha_{ij} : U^2 \longrightarrow \mathbf{R}$$

are specified by

$$u^\alpha_{ij}(j^2_p\phi) = \left.\frac{\partial^2\phi^\alpha}{\partial x^j\,\partial x^i}\right|_p .$$

The functions u^α_i and u^α_{ij} are known as *derivative coordinates*. ∎

The reason why there are only $\frac{1}{2}mn(m+1)$ different functions of the form u^α_{ij}, rather than m^2n as the notation might suggest, is of course that $u^\alpha_{ij} = u^\alpha_{ji}$ (we have deliberately interchanged the indices in the partial derivative used in this definition for, later convenience). This symmetry in the derivative indices gives rise to complications in coordinate formulæ, and in Chapter 6, when we study higher-order jets, we shall introduce an alternative notation which is rather easier to handle. For the moment, however, we shall continue with our existing notation.

Proposition 5.1.5 *Given an atlas of adapted charts (U, u) on E, the corresponding collection of charts (U^2, u^2) is a finite-dimensional C^∞ atlas on $J^2\pi$.*

Proof This is essentially the same as the proof of Proposition 4.1.7. The only additional step is to show that, if the two charts on E are (U, u) and (V, v), then each function $v^\beta_{kl} \circ (u^2)^{-1}$ is smooth. But

$$v^\beta_{kl}(j^2_p\phi) = F^\beta_{kl}(u^2(j^2_p\phi)),$$

where F^β_{kl} are the smooth functions obtained in Lemma 5.1.1 by applying the Chain Rule twice to the change-of-coordinates formula; it follows that $v^\beta_{kl} \circ (u^2)^{-1} = F^\beta_{kl}$ is smooth. ■

Lemma 5.1.6 *The functions $\pi_2 : J^2\pi \longrightarrow M$, $\pi_{2,0} : J^2\pi \longrightarrow E$ and $\pi_{2,1} : J^2\pi \longrightarrow J^1\pi$ are smooth surjective submersions.*

Proof The proof for $\pi_{2,1}$ is similar to the proof of the part of Lemma 4.1.9 referring to $\pi_{1,0}$. The results for $\pi_{2,0} = \pi_{1,0} \circ \pi_{2,1}$ and $\pi_2 = \pi_1 \circ \pi_{2,1}$ then follow immediately. ■

When studying first-order jets, we saw that $(J^1\pi, \pi_{1,0}, E)$ had the structure of an affine bundle, modelled on the vector bundle

$$\left(\pi^*(T^*M) \otimes V\pi, (\tau^*_E|_{\pi^*(T^*M)}) \otimes (\tau_E|_{V\pi}), E\right).$$

For second-order jets, it is the bundle $(J^2\pi, \pi_{2,1}, J^1\pi)$ which is an affine bundle. This is the significance of our earlier remark that each function $F^\beta_{kl} = v^\beta_{kl} \circ (u^2)^{-1}$ is an inhomogeneous linear function of its final argument. The associated vector bundle in this case has total space $\pi^*_1(S^2T^*M) \otimes \pi^*_{1,0}(V\pi)$, where S^2T^*M is the total space of *symmetric* 2-covectors: formally, it is the bundle

$$\left(\pi^*_1(S^2T^*M) \otimes \pi^*_{1,0}(V\pi), \left(S^2\tau^*_{J^1\pi}\Big|_{\pi^*_1(S^2T^*M)}\right) \otimes \pi^*_{1,0}(\tau_E|_{V\pi}), J^1\pi\right).$$

Theorem 5.1.7 *The triple $(J^2\pi, \pi_{2,1}, J^1\pi)$ may be given the structure of an affine bundle modelled on the vector bundle*

$$\left(S^2\tau^*_{J^1\pi}\Big|_{\pi_1^*(S^2T^*M)} \right) \otimes \pi^*_{1,0}(\tau_E|_{V\pi})$$

in such a way that, for each adapted coordinate system (U, u) on E, the map

$$\begin{aligned} t_u : \pi_{2,1}^{-1}(U^1) &\longrightarrow U^1 \times \mathbf{R}^N \\ j^2_p\phi &\longmapsto (j^1_p\phi, u^\alpha_{ij}(j^2_p\phi)), \end{aligned}$$

where $N = \frac{1}{2}mn(m+1)$, is an affine local trivialisation.

Proof As in Theorem 4.1.11, we shall define a fibrewise action of the vector bundle on $\pi_{2,1}$. So let $a \in J^1\pi$, and let (U, u) be an adapted coordinate system around $\pi_{1,0}(a)$. A typical element $\xi \in (\pi_1^*(S^2T^*M) \otimes \pi^*_{1,0}(V\pi))_a$ may be written in coordinates as

$$\xi = \xi^\alpha_{ij} \left((dx^i \odot dx^j) \otimes \frac{\partial}{\partial u^\alpha} \right)_a ,$$

where $dx^i \odot dx^j$ denotes the symmetric product $dx^i \otimes dx^j + dx^j \otimes dx^i$, and where $\xi^\alpha_{ij} = \xi^\alpha_{ji}$. If $\phi \in \Gamma_{\pi_1(a)}(\pi)$ satisfies $j^1_{\pi_1(a)}\phi = a$, then the action of ξ on $j^2_{\pi_1(a)}\phi$ is written as $\xi[j^2_{\pi_1(a)}\phi]$, and is defined by the rule

$$u^\alpha_{ij}(\xi[j^2_{\pi_1(a)}\phi]) = u^\alpha_{ij}(j^2_{\pi_1(a)}\phi) + \xi^\alpha_{ij},$$

where the symmetry of the coordinates ξ^α_{ij} in the subscript indices is obviously necessary. As before, however, we must check that this definition does not depend upon the choice of coordinate system. So let (y^j, v^β) be another family of coordinates around $\pi_{1,0}(a)$. Then

$$\begin{aligned} v^\beta_{kl}(j^2_{\pi_1(a)}\phi) &= F^\beta_{kl}(u^2(j^2_{\pi_1(a)}\phi)) \\ &= \left.\frac{\partial v^\beta}{\partial u^\alpha}\right|_{\pi_{1,0}(a)} \left.\frac{\partial x^i}{\partial y^k}\right|_{\pi_1(a)} \left.\frac{\partial x^j}{\partial y^l}\right|_{\pi_1(a)} u^\alpha_{ij}(j^2_{\pi_1(a)}\phi) + G^\beta_{kl}(u^1(a)), \end{aligned}$$

using the inhomogeneous linearity of the functions F^β_{kl} described earlier. On the other hand, as a tensor,

$$\begin{aligned} \xi &= \xi^\beta_{kl} \left((dy^k \odot dy^l) \otimes \frac{\partial}{\partial v^\beta} \right)_a \\ &= \xi^\alpha_{ij} \left.\frac{\partial v^\beta}{\partial u^\alpha}\right|_{\pi_{1,0}(a)} \left.\frac{\partial x^i}{\partial y^k}\right|_{\pi_1(a)} \left.\frac{\partial x^j}{\partial y^l}\right|_{\pi_1(a)} \left((dy^k \odot dy^l) \otimes \frac{\partial}{\partial v^\beta} \right)_a , \end{aligned}$$

from which it follows immediately that

$$v^\beta_{kl}(\xi[j^2_{\pi_1(a)}\phi]) = v^\beta_{kl}(j^2_{\pi_1(a)}\phi) + \xi^\beta_{kl},$$

as required.

We must also consider the maps t_u. Each such map is a diffeomorphism, for it is just the composite $((u^1)^{-1} \times id_{\mathbf{R}^N}) \circ u^2$, and evidently $pr_1 \circ t_u = \pi_{2,1}|_{U^2}$. Now let $a \in U^1$; then the map $t_{u;a} : \pi^{-1}_{2,1}(a) \longrightarrow \mathbf{R}^N$ defined by

$$t_{u;a} = pr_2 \circ t_u|_{\pi^{-1}_{2,1}(a)}$$

satisfies $t_{u;a} = (u^\alpha_{ij})\Big|_{\pi^{-1}_{2,1}(a)}$. Consequently $t_{u;a}$ is an affine morphism, where the fibre $\pi^{-1}_{2,1}(a)$ has the structure of an affine space given by the vector bundle action, and $\mathbf{R}^N$ has its natural affine structure. ∎

Corollary 5.1.8 *The total space $J^2\pi$ of $\pi_{2,1}$ is a manifold.*

Proof From Proposition 1.1.14. ∎

It is also the case that the fibred manifolds $(J^2\pi, \pi_{2,0}, E)$ and $(J^2\pi, \pi_2, M)$ are bundles, although we shall leave this to be deduced from more general results in Chapter 6.

Example 5.1.9 Let π be the trivial bundle $(\mathbf{R}^2 \times \mathbf{R}, pr_1, \mathbf{R})$ with global coordinates $(x^1, x^2; u^1)$ on $\mathbf{R}^2 \times \mathbf{R}$. Then global coordinates on $J^2\pi$ are $(x^1, x^2; u^1; u^1_1, u^1_2, u^1_{11}, u^1_{12}, u^1_{22})$. To each jet $j^2_p\phi \in J^2\pi$, where $p = (p^1, p^2) \in \mathbf{R}^2$, there corresponds an inhomogeneous quadratic map $\overline{\psi} : \mathbf{R}^2 \longrightarrow \mathbf{R}$, defined as follows:

$$\begin{aligned} \overline{\psi}(q) \quad = \quad & \phi^1(p) + u^1_1(j^2_p\phi)(q^1 - p^1) + u^1_2(j^2_p\phi)(q^2 - p^2) \\ & + \tfrac{1}{2}\left(u^1_{11}(j^2_p\phi)(q^1 - p^1)^2 + 2u^1_{12}(j^2_p\phi)(q^1 - p^1)(q^2 - p^2)\right. \\ & \left. + u^1_{22}(j^2_p\phi)(q^2 - p^2)^2\right), \end{aligned}$$

where $q = (q^1, q^2) \in \mathbf{R}^2$ and $\phi^1 = u^1 \circ \phi : \mathbf{R}^2 \longrightarrow \mathbf{R}$. The map $\overline{\psi}$ gives rise to a global section $\psi = (id_{\mathbf{R}^2}, \overline{\psi})$ of π, and it is obvious that $j^2_p\phi = j^2_p\psi$; clearly $\overline{\psi}$ is the unique globally-defined inhomogeneous quadratic map with this property. The map $\overline{\psi}$ is of course the second-order Taylor polynomial of ϕ. ∎

EXERCISES

5.1.1 Let (E, π, M) be a bundle, and let (x^i, u^α) and (y^k, v^β) be two sets of adapted coordinates defined in a neighbourhood W of $a \in E$. Show that the explicit formula for the coordinate transformation functions $F^\beta_{kl} = v^\beta_{kl} \circ (u^2)^{-1}$ is given by

$$\begin{aligned} v^\beta_{kl} &= \frac{\partial^2 x^i}{\partial y^k \partial y^l}\left(\frac{\partial v^\beta}{\partial x^i} + u^\alpha_i \frac{\partial v^\beta}{\partial u^\alpha}\right) \\ &\quad + \frac{\partial x^i}{\partial y^k}\frac{\partial x^j}{\partial y^l}\left(\frac{\partial^2 v^\beta}{\partial x^i \partial x^j} + u^\alpha_i \frac{\partial^2 v^\beta}{\partial x^j \partial u^\alpha} + u^\alpha_j \frac{\partial^2 v^\beta}{\partial x^i \partial u^\alpha}\right. \\ &\quad \left. + u^\alpha_i u^\gamma_j \frac{\partial^2 v^\beta}{\partial u^\alpha \partial u^\gamma} + u^\alpha_{ij}\frac{\partial v^\beta}{\partial u^\alpha}\right). \end{aligned}$$

5.1.2 Let $L \in C^\infty(J^2\pi)$. Why is the coordinate representation of the 1-form dL *not* always given by

$$dL = \frac{\partial L}{\partial x^i}dx^i + \frac{\partial L}{\partial u^\alpha}du^\alpha + \frac{\partial L}{\partial u^\alpha_i}du^\alpha_i + \frac{\partial L}{\partial u^\alpha_{ij}}du^\alpha_{ij}?$$

(Hint: just one term is incorrect, and this involves the omission of a numerical factor.)

5.1.3 Let π be the trivial bundle $(\mathbf{R} \times F, pr_1, \mathbf{R})$. Show that there is a canonical diffeomorphism $J^2\pi \cong \mathbf{R} \times T^2F$, where T^2F is the second-order tangent manifold introduced in Exercise 1.4.3.

5.1.4 Let (E, π, M) be a bundle. Show that the correspondence

$$\xi \longmapsto ((j^1\phi)_*(\xi), j^2_p\phi),$$

where $\xi \in T_pM$, gives a well-defined map $TM \longrightarrow \pi^*_{2,1}(TJ^1\pi)$, and consequently a well-defined map $\mathcal{X}(M) \longrightarrow \mathcal{X}(\pi_{2,1})$. (The images of these maps are called holonomic lifts and total derivatives, and provide a direct generalisation of the corresponding maps from M to $J^1\pi$ introduced in Section 4.3.) Show that the image of the coordinate vector field $\partial/\partial x^i$ on M under the second of these maps is the vector field along $\pi_{2,1}$ with coordinate representation

$$\frac{d}{dx^i} = \frac{\partial}{\partial x^i} + u^\alpha_i \frac{\partial}{\partial u^\alpha} + u^\alpha_{ij}\frac{\partial}{\partial u^\alpha_j}.$$

5.2 Repeated Jets

If (E, π, M) is an arbitrary bundle, then its first jet bundle $(J^1\pi, \pi_1, M)$ is a bundle in its own right, and so we may consider *its* first jet bundle $(J^1\pi_1, (\pi_1)_1, M)$. An element of the total space $J^1\pi_1$ is then a 1-jet $j^1_p\psi$, where $\psi \in \Gamma_p(\pi_1)$. If local coordinates on E are (x^i, u^α) and on $J^1\pi$ are $(x^i, u^\alpha; u^\alpha_i)$, then coordinates on $J^1\pi_1$ will be

$$(x^i, u^\alpha; u^\alpha_i; u^\alpha_{;j}, u^\alpha_{i;j}),$$

where the additional mn coordinate functions $u^\alpha_{;j}$ and m^2n coordinate functions $u^\alpha_{i;j}$ are defined by

$$u^\alpha_{;j}(j^1_p\psi) = \left.\frac{\partial\psi^\alpha}{\partial x^j}\right|_p$$

and

$$u^\alpha_{i;j}(j^1_p\psi) = \left.\frac{\partial\psi^\alpha_i}{\partial x^j}\right|_p,$$

using the standard coordinate representation $\psi^\alpha = u^\alpha \circ \psi$, $\psi^\alpha_i = u^\alpha_i \circ \psi$. Notice that the functions u^α_i and $u^\alpha_{;i}$ are distinct, as are the functions $u^\alpha_{i;j}$ and $u^\alpha_{j;i}$ for $i \neq j$. The dimension of the manifold $J^1\pi_1$ is therefore $m + n(1+m)^2$.

There is, however, a distinguished subset of $J^1\pi_1$ containing those elements $j^1_p\psi$ where the local section ψ is itself the prolongation $j^1\phi$ of a local section $\phi \in \Gamma_p(\pi)$. It is not immediate from this description that the subset is well-defined, but we shall see that it is by considering it as the image of the second jet manifold $J^2\pi$ under a canonical map.

Definition 5.2.1 The map $\iota_{1,1} : J^2\pi \longrightarrow J^1\pi_1$ is defined by

$$\iota_{1,1}(j^2_p\phi) = j^1_p(j^1\phi).$$

Elements of $\iota_{1,1}(J^2\pi)$ are called *holonomic jets.* ∎

We may show that the map $\iota_{1,1}$ is well-defined by considering it in coordinates. It is clear that $x^i \circ \iota_{1,1} = x^i$, $u^\alpha \circ \iota_{1,1} = u^\alpha$ and $u^\alpha_i \circ \iota_{1,1} = u^\alpha_i$. To calculate the remaining coordinates, we observe that

$$\begin{aligned} u^\alpha_{;j}(\iota_{1,1}(j^2_p\phi)) &= u^\alpha_{;j}(j^1_p(j^1\phi)) \\ &= \left.\frac{\partial(u^\alpha \circ j^1\phi)}{\partial x^j}\right|_p \\ &= \left.\frac{\partial\phi^\alpha}{\partial x^j}\right|_p \\ &= u^\alpha_j(j^2_p\phi), \end{aligned}$$

and that

$$\begin{aligned} u^\alpha_{i;j}(\iota_{1,1}(j^2_p\phi)) &= u^\alpha_{i;j}(j^1_p(j^1\phi)) \\ &= \left.\frac{\partial(u^\alpha_i \circ j^1\phi)}{\partial x^j}\right|_p \\ &= \left.\frac{\partial}{\partial x^j}\right|_p \frac{\partial \phi^\alpha}{\partial x^i} \\ &= \left.\frac{\partial^2\phi^\alpha}{\partial x^j\,\partial x^i}\right|_p \\ &= u^\alpha_{ij}(j^2_p\phi). \end{aligned}$$

In these coordinates, the map $\iota_{1,1}$ therefore corresponds to a map from an open subset of $\mathbf{R}^M$ to an open subset of $\mathbf{R}^N$ (where $M = m + \frac{1}{2}(m+1)(m+2)n$ and $N = m + n(m+1)^2$) which is a linear injection; it follows that $\iota_{1,1}$ is an embedding. It also follows from these calculations that, when restricted to the submanifold $\iota_{1,1}(J^2\pi)$, we *do* have equalities $u^\alpha_i = u^\alpha_{;i}$ and $u^\alpha_{i;j} = u^\alpha_{j;i}$. Indeed, this local coordinate description characterises $\iota_{1,1}(J^2\pi)$, for if there is a point $j^1_p\psi \in J^1\pi_1$ where all these equalities hold then we may construct a local section $\phi \in \Gamma_p(\pi)$ which satisfies $j^1_p\psi = j^1_p(j^1\phi)$: we simply use the common values of the coordinates to form a quadratic Taylor polynomial along the lines of Example 5.1.9.

Example 5.2.2 Let π be the trivial bundle $(\mathbf{R}^2 \times \mathbf{R}, pr_1, \mathbf{R}^2)$, with global coordinates $(x^1, x^2; u^1)$ on $\mathbf{R}^2 \times \mathbf{R}$. Let $j^1_p\psi \in J^1\pi_1$ have coordinates

$$x^1(j^1_p\psi) = p^1, \quad x^2(j^1_p\psi) = p^2;$$

$$u^1(j^1_p\psi) = a^1;$$

$$u^1_1(j^1_p\psi) = u^1_{;1}(j^1_p\psi) = a^1_1, \quad u^1_2(j^1_p\psi) = u^1_{;2}(j^1_p\psi) = a^1_2;$$

$$u^1_{1;1}(j^1_p\psi) = a^1_{11}, \quad u^1_{1;2}(j^1_p\psi) = u^1_{2;1}(j^1_p\psi) = a^1_{12}, \quad u^1_{2;2}(j^1_p\psi) = a^1_{22}.$$

Then the local section $\phi \in \Gamma_p(\pi)$ defined by

$$\begin{aligned} \phi^1(q) &= a^1 + a^1_1(q^1 - p^1) + a^1_2(q^2 - p^2) \\ &\quad + \tfrac{1}{2}\left(a^1_{11}(q^1 - p^1)^2 + 2a^1_{12}(q^1 - p^1)(q^2 - p^2) + a^1_{22}(q^2 - p^2)^2\right) \end{aligned}$$

satisfies $j^1_p(j^1\phi) = j^2_p\psi$. ■

Example 5.2.3 With the same bundle π, the section $\psi \in \Gamma(\pi_1)$ defined in Example 4.2.3 by

$$\psi(p^1, p^2) = (p^1, p^2; p^1 \sin p^2; p^1p^2, 0)$$

is not the prolongation of a section of π, and if we consider *its* prolongation $j^1\psi$ we find that

$$u_1^1(j_p^1\psi) = p^1p^2, \quad u_2^1(j_p^1\psi) = 0,$$

whereas

$$u_{;1}^1(j_p^1\psi) = \sin p^2, \quad u_{;2}^1(j_p^1\psi) = p^1 \cos p^2.$$

We also obtain

$$u_{1;1}^1(j_p^1\psi) = p^2, \quad u_{1;2}^1(j_p^1\psi) = p^1$$

$$u_{2;1}^1(j_p^1\psi) = 0, \quad u_{2;2}^1(j_p^1\psi) = 0.$$

It follows from these calculations that, except at discrete points of the form $(0, n\pi)$, $j_p^1\psi$ does not take its values in $\iota_{1,1}(J^2\pi)$. ■

There are several uses of the map $\iota_{1,1}$. For instance, we may use it to define the prolongation of a bundle morphism $(f, \overline{f})$ from $(J^1\pi, \pi_1, M)$ to (H, ρ, N) as a map $J^2\pi \longrightarrow J^1\rho$ rather than as a map $J^1\pi_1 \longrightarrow J^1\rho$: we simply consider $j^1f \circ \iota_{1,1}$ instead of j^1f. In this slightly tautological sense, $\iota_{1,1}$ may be considered as the prolongation of the identity $id_{J^1\pi}$. We shall see other uses of this map (and of its generalisation to higher-order jet bundles) in Chapter 6, although it will sometimes instead be convenient to identify $J^2\pi$ with its image in $J^1\pi_1$.

In general, there is no canonical projection of the repeated jet manifold $J^1\pi_1$ onto its submanifold $\iota_{1,1}(J^2\pi)$: this is one reason why the construction of a higher-order Cartan form in the Calculus of Variations is rather more complicated than in the first-order theory. There is, however, a map $J^1\pi_1 \longrightarrow J^1\pi$ which is distinct from the target projection $(\pi_1)_{1,0}$, and this is is obtained by taking the map $\pi_{1,0} : J^1\pi \longrightarrow E$ (regarded as a bundle morphism $(\pi_{1,0}, id_M) : \pi_1 \longrightarrow \pi$) and prolonging it to a map $j^1(\pi_{1,0}) : J^1\pi_1 \longrightarrow J^1\pi$. The distinction between $(\pi_1)_{1,0}$ and $j^1(\pi_{1,0})$ may be seen quite easily in coordinates: the x^i and u^α coordinates of the two maps are equal, but

$$\begin{aligned} u_i^\alpha((\pi_1)_{1,0}(j_p^1\psi)) &= u_i^\alpha(\psi(p)) \\ &= \psi_i^\alpha(p), \end{aligned}$$

so that $u_i^\alpha \circ (\pi_1)_{1,0} = u_i^\alpha$, whereas

$$\begin{aligned} u_i^\alpha(j^1(\pi_{1,0})(j_p^1\psi)) &= u_i^\alpha(j_p^1(\pi_{1,0} \circ \psi)) \\ &= \left.\frac{\partial \psi^\alpha}{\partial x^i}\right|_p \end{aligned}$$

so that $u^\alpha_i \circ j^1(\pi_{1,0}) = u^\alpha_{;i}$. It follows that the two maps are equal when restricted to $\iota_{1,1}(J^2\pi)$ (although that submanifold is not characterised by this equality). The existence of these two maps $J^1\pi_1 \longrightarrow J^1\pi$ is a direct generalisation of the existence of two distinct maps τ_{TF}, τ_{F*} from the repeated tangent manifold TTF to TF.

We may summarise how all these maps fit together by the following diagram:

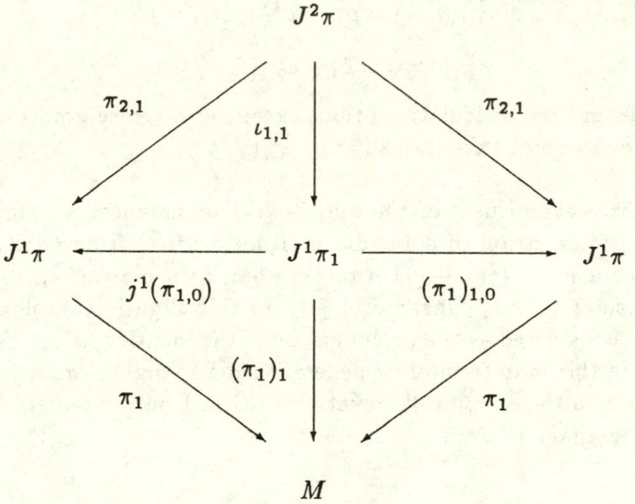

EXERCISES

5.2.1 Let (E, π, M) be a bundle, and let (x^i, u^α) and (y^k, v^β) be two sets of adapted coordinates defined in a neighbourhood W of $a \in E$. Show that the coordinate transformation rule for $v^\beta_{k;l}$ on $J^1\pi_1$ is given by

$$\begin{aligned} v^\beta_{k;l} &= \frac{\partial^2 x^i}{\partial y^k\,\partial y^l}\left(\frac{\partial v^\beta}{\partial x^i} + u^\alpha_i \frac{\partial v^\beta}{\partial u^\alpha}\right) \\ &\quad + \frac{\partial x^i}{\partial y^k}\frac{\partial x^j}{\partial y^l}\left(\frac{\partial^2 v^\beta}{\partial x^i\,\partial x^j} + u^\alpha_i \frac{\partial^2 v^\beta}{\partial x^j\,\partial u^\alpha} + u^\alpha_{;j}\frac{\partial^2 v^\beta}{\partial x^i\,\partial u^\alpha}\right. \\ &\quad \left. + u^\alpha_i u^\gamma_{;j}\frac{\partial^2 v^\beta}{\partial u^\alpha\,\partial u^\gamma} + u^\alpha_{i;j}\frac{\partial v^\beta}{\partial u^\alpha}\right). \end{aligned}$$

5.2.2 Show that $(J^2\pi, \pi_{2,1}, J^1\pi)$ may be identified with an affine subbundle of $J^1\pi_1, (\pi_1)_{1,0}, J^1\pi)$ using the map $\iota_{1,1}$.

5.2.3 Let π be the trivial bundle $(\mathbf{R} \times F, pr_1, \mathbf{R})$. Construct a canonical diffeomorphism $J^1\pi_1 \cong \mathbf{R} \times TTF$, and show how this is related to the diffeomorphism $J^2\pi \cong \mathbf{R} \times T^2F$ of Exercise 5.1.3. If coordinates on $\mathbf{R} \times F$ are (t, q^α), explain how the relationship between the coordinates on $J^1\pi_1$ and $J^2\pi$ is simpler than in the general case where the base manifold has dimension greater than one.

5.2.4 Give an example to show that, in general, the subset of $J^1\pi_1$ where $(\pi_1)_{1,0} = j^1(\pi_{1,0})$ strictly contains $\iota_{1,1}(J^2\pi)$.

5.2.5 Let (E, π, M) be a bundle, and let $(f, \overline{f}) : \pi \longrightarrow \pi$ be a bundle morphism, where $\overline{f}$ is a diffeomorphism. Let $j^1 f : J^1\pi \longrightarrow J^1\pi$ be the prolongation of $(f, \overline{f})$, and let $j^1(j^1 f) : J^1\pi_1 \longrightarrow J^1\pi_1$ be the prolongation of $(j^1 f, \overline{f})$. Show that the composite map $j^1(j^1 f) \circ \iota_{1,1} : J^2\pi \longrightarrow J^1\pi_1$ takes its values in $J^2\pi$. (This map is the *second prolongation* of f, and a direct definition of higher-order prolongations will be given in Chapter 6.)

5.2.6 Let X be a vertical vector field on E with flow ψ_t. The second prolongation of ψ_t defines a flow on $J^2\pi$, and the corresponding vector field may be called the *second prolongation* of X and denoted X^2. (A direct definition of higher-order prolongations of vertical vector fields will be given in Chapter 6.) If the coordinate representation of X is $X^\alpha \partial / \partial u^\alpha$, show that the coordinate representation of X^2 is

$$X^2 = X^\alpha \frac{\partial}{\partial u^\alpha} + \frac{dX^\alpha}{dx^i} \frac{\partial}{\partial u^\alpha_i} + \frac{d}{dx^j} \left(\frac{dX^\alpha}{dx^i} \right) \frac{\partial}{\partial u^\alpha_{ij}},$$

where dX^α / dx^i is a function on $J^1\pi$, and where the operator d/dx^j is the second-order total derivative described in Exercise 5.1.4, which (as a derivation) maps functions on $J^1\pi$ to functions on $J^2\pi$.

5.3 Integrability and Semi-holonomic Jets

When we introduced jet fields in Section 4.6, we described an integral section of a jet field $\Gamma : E \longrightarrow J^1\pi$ as a local section of π satisfying $\Gamma \circ \phi = j^1\phi$, and we said that a jet field which admitted integral sections was integrable. It was obviously desirable to have a characterisation of integrability in terms of the jet field itself, and we saw that Γ was integrable exactly when the associated connection $\widetilde{\Gamma}$ had zero curvature.

Since the coordinate expression for the curvature of $\widetilde{\Gamma}$ involves the derivatives of the coefficients of Γ, it seems reasonable to try to express this condition directly in terms of the jet field Γ. Our aim will be to reproduce, in the language of jets, the traditional statement of Frobenius' Theorem in terms of the commutativity of partial differentiation operators. We start, therefore, by considering the bundle morphism (Γ, id_M) from (E, π, M) to $(J^1\pi, \pi_1, M)$. The prolongation of this bundle morphism is the map $j^1(\Gamma, id_M) : J^1\pi \longrightarrow J^1\pi_1$. In this section, we shall identify $J^2\pi$ with its image $\iota_{1,1}(J^2\pi)$ in $J^1\pi_1$.

Proposition 5.3.1 *The jet field* Γ *is integrable if, and only if, the composite map* $j^1(\Gamma, id_M) \circ \Gamma$ *takes its values in the submanifold* $J^2\pi \subset J^1\pi_1$.

Proof If we regard the formula

$$\Gamma \circ \phi = j^1\phi$$

as describing a composition of bundle morphisms (Γ, id_W) and (ϕ, id_W), where W is the domain of ϕ, then we may use the composition formula for prolongations to obtain

$$j^1(\Gamma, id_W) \circ j^1\phi = j^1(j^1\phi) = j^2\phi,$$

so that

$$j^1(\Gamma, id_W) \circ \Gamma \circ \phi = j^2\phi \in J^2\pi \subset J^1\pi_1.$$

If there is an integral section of Γ through every point of the total space E, it follows that the composite map $j^1(\Gamma, id_M) \circ \Gamma$ must take its values in $J^2\pi$.

We shall demonstrate the converse using the coordinate formula for the prolongation of a bundle morphism. If $a \in E$, we find that

$$\begin{aligned} u^\alpha_{i;j}(j^1(\Gamma, id_M)(\Gamma(a))) &= \left.\frac{d\Gamma^\alpha_i}{dx^j}\right|_{\Gamma(a)} \\ &= \left.\frac{\partial\Gamma^\alpha_i}{\partial x^j}\right|_a + \Gamma^\beta_j(a) \left.\frac{\partial\Gamma^\alpha_i}{\partial u^\beta}\right|_a, \end{aligned}$$

so that if $\operatorname{im}(j^1(\Gamma, id_M) \circ \Gamma) \subset J^2\pi$ then

$$\frac{\partial\Gamma^\alpha_i}{\partial x^j} + \Gamma^\beta_j \frac{\partial\Gamma^\alpha_i}{\partial u^\beta} = \frac{\partial\Gamma^\alpha_j}{\partial x^i} + \Gamma^\beta_i \frac{\partial\Gamma^\alpha_j}{\partial u^\beta};$$

it follows that the curvature of the connection $\widetilde{\Gamma}$ must vanish. ■

If Γ is a non-integrable jet field, the composite map $j^1(\Gamma, id_M) \circ \Gamma$ will not take its values in $J^2\pi$: nevertheless,

$$\begin{aligned} u^\alpha_{;j}(j^1(\Gamma, id_M)(\Gamma(a))) &= \left.\frac{du^\alpha}{dx^j}\right|_{\Gamma(a)} \\ &= \Gamma^\alpha_j(a) \\ &= u^\alpha_j(j^1(\Gamma, id_M)(\Gamma(a))). \end{aligned}$$

We may see from this calculation that there is a restricted subset of $J^1\pi_1$ in which the image of the composite map must always lie, namely the subset where the coordinate functions u^α_j and $u^\alpha_{;j}$ are equal, but where the coordinate functions $u^\alpha_{i;j}$ and $u^\alpha_{j;i}$ are allowed to differ for $i \neq j$. This subset may be defined independently of coordinates, and is called the *semi-holonomic 2-jet manifold* and denoted $\widehat{J}^2\pi$.

Lemma 5.3.2 *There is a unique map $D_1 : J^1\pi_1 \longrightarrow \pi^*(T^*M) \otimes V\pi$ which satisfies*

$$D_1(j^1_p\psi)[(\pi_1)_{1,0}(j^1_p\psi)] = j^1(\pi_{1,0})(j^1_p\psi),$$

where the square brackets denote the affine action of an element of $\pi^*(T^*M) \otimes V\pi$ on $J^1\pi$. This map is called the Spencer operator.

Proof We just have to confirm that $(\pi_1)_{1,0}(j^1_p\psi)$ and $j^1(\pi_{1,0})(j^1_p\psi)$ are in the same fibre of $J^1\pi$ over E, for then we simply let $D_1(j^1_p\psi)$ be the unique element of $\pi^*(T^*M) \otimes V\pi$ which maps one to the other. But

$$\pi_{1,0}((\pi_1)_{1,0}(j^1_p\psi)) = \pi_{1,0}(\psi(p)),$$

whereas

$$\begin{aligned} \pi_{1,0}(j^1(\pi_{1,0})(j^1_p\psi)) &= \pi_{1,0}(j^1_p(\pi_{1,0} \circ \psi)) \\ &= \pi_{1,0}(\psi(p)). \end{aligned}$$

■

Definition 5.3.3 The *semi-holonomic 2-jet manifold* $\widehat{J}^2\pi$ is the submanifold $D_1^{-1}(0)$ of $J^1\pi_1$. ■

It is easy to see that $\widehat{J}^2\pi$ is indeed the submanifold of $J^1\pi_1$ where $u^\alpha_i = u^\alpha_{;i}$; we just examine the coordinate representation of the Spencer operator D_1. Since $u^\alpha_i \circ (\pi_1)_{1,0} = u^\alpha_i$, whereas $u^\alpha_i \circ j^1(\pi_{1,0}) = u^\alpha_{;i}$, it follows that

$$\begin{aligned} D_1(j^1_p\psi) &= (u^\alpha_{;i}(j^1_p\psi) - u^\alpha_i(j^1_p\psi)) \left(dx^i \otimes \frac{\partial}{\partial u^\alpha} \right)_{\pi_{1,0}(\psi(p))} \\ &= \left(\left.\frac{\partial \psi^\alpha}{\partial x^i}\right|_p - \psi^\alpha_i(p) \right) \left(dx^i \otimes \frac{\partial}{\partial u^\alpha} \right)_{\pi_{1,0}(\psi(p))}, \end{aligned}$$

so that $D_1(j^1_p\psi) = 0$ (and thus $j^1_p\psi \in \widehat{J}^2\pi$) if, and only if,

$$\left.\frac{\partial \psi^\alpha}{\partial x^i}\right|_p = \psi^\alpha_i(p).$$

If, in addition, we have

$$\left.\frac{\partial \psi_i^\alpha}{\partial x^j}\right|_p = \left.\frac{\partial \psi_j^\alpha}{\partial x^i}\right|_p$$

(which would be the case if, for example, $\partial\psi^\alpha/\partial x^i = \psi_i^\alpha$ in a neighbourhood of p rather than at a single point) then we would also have $j_p^1\psi \in J^2\pi$. We therefore have the inclusions

$$J^2\pi \subset \widehat{J}^2\pi \subset J^1\pi_1,$$

and we shall normally use the functions $(x^i, u^\alpha, u_i^\alpha, u_{i;j}^\alpha)$ as coordinates on $\widehat{J}^2\pi$. We shall also define the map $\widehat{\pi}_{2,1} : \widehat{J}^2\pi \longrightarrow J^1\pi$ by $\widehat{\pi}_{2,1} = (\pi_1)_{1,0}|_{\widehat{J}^2\pi}$.

Theorem 5.3.4 *The triple $(\widehat{J}^2\pi, \widehat{\pi}_{2,1}, J^1\pi)$ is a bundle, and is isomorphic to the fibred product of the affine bundle*

$$(J^2\pi, \pi_{2,1}, J^1\pi)$$

and the vector bundle

$$\left(\pi_1^*(\textstyle\bigwedge^2 T^*M) \otimes \pi_{1,0}^*(V\pi), \pi_1^*(\textstyle\bigwedge^2 \tau_M^*) \otimes \pi_{1,0}^*(\tau_E|_{V\pi}), J^1\pi\right).$$

Proof We shall specify a bundle isomorphism

$$(\Psi, id_{J^1\pi}) : J^2\pi \times_{J^1\pi} (\pi_1^*(\textstyle\bigwedge^2 T^*M) \otimes \pi_{1,0}^*(V\pi)) \longrightarrow \widehat{J}^2\pi$$

by describing, in coordinates, its action on the fibres over $J^1\pi$. So let $j_p^2\phi \in J^2\pi$, and let $\theta \in (\pi_1^*(\bigwedge^2 T^*M) \otimes \pi_{1,0}^*(V\pi))_{j_p^1\phi}$. Let (x^i, u^α) be a coordinate system around $\phi(p)$, and suppose that

$$\theta = \theta_{ij}^\alpha \left((dx^i \wedge dx^j) \otimes \frac{\partial}{\partial u^\alpha} \right)_{j_p^1\phi},$$

where $\theta_{ij}^\alpha + \theta_{ji}^\alpha = 0$. Set

$$u_{i;j}^\alpha(\Psi(j_p^2\phi, \theta)) = u_{ij}^\alpha(j_p^2\phi) + \theta_{ij}^\alpha,$$

so that $u_{ij}^\alpha(j_p^2\phi)$ and θ_{ij}^α are respectively the symmetric and antisymmetric parts of $u_{i;j}^\alpha(\Psi(j_p^2\phi, \theta))$, and put

$$(\pi_1)_{1,0}(\Psi(j_p^2\phi, \theta)) = j_p^1\phi,$$

so that Ψ projects to the identity on $J^1\pi$. We must check that this definition is independent of the choice of coordinates; but if (y^k, v^β) is another coordinate system around $\phi(p)$, then the coordinate transformation rule for v^β_{kl} is

$$\begin{aligned} v^\beta_{kl} &= \frac{\partial^2 x^i}{\partial y^k \, \partial y^l}\left(\frac{\partial v^\beta}{\partial x^i} + u^\alpha_i \frac{\partial v^\beta}{\partial u^\alpha}\right) \\ &+ \frac{\partial x^i}{\partial y^k}\frac{\partial x^j}{\partial y^l}\left(\frac{\partial^2 v^\beta}{\partial x^i \, \partial x^j} + u^\alpha_i \frac{\partial^2 v^\beta}{\partial x^j \, \partial u^\alpha} + u^\alpha_j \frac{\partial^2 v^\beta}{\partial x^i \, \partial u^\alpha}\right. \\ &\left. + u^\alpha_i u^\gamma_j \frac{\partial^2 v^\beta}{\partial u^\alpha \, \partial u^\gamma} + u^\alpha_{ij} \frac{\partial v^\beta}{\partial u^\alpha}\right) \end{aligned}$$

from Exercise 5.1.1, whereas the rule for $v^\beta_{k;l}$ (when restricted to $\widehat{J}^2\pi$) is

$$\begin{aligned} v^\beta_{k;l} &= \frac{\partial^2 x^i}{\partial y^k \, \partial y^l}\left(\frac{\partial v^\beta}{\partial x^i} + u^\alpha_i \frac{\partial v^\beta}{\partial u^\alpha}\right) \\ &+ \frac{\partial x^i}{\partial y^k}\frac{\partial x^j}{\partial y^l}\left(\frac{\partial^2 v^\beta}{\partial x^i \, \partial x^j} + u^\alpha_i \frac{\partial^2 v^\beta}{\partial x^j \, \partial u^\alpha} + u^\alpha_j \frac{\partial^2 v^\beta}{\partial x^i \, \partial u^\alpha}\right. \\ &\left. + u^\alpha_i u^\gamma_j \frac{\partial^2 v^\beta}{\partial u^\alpha \, \partial u^\gamma} + u^\alpha_{i;j} \frac{\partial v^\beta}{\partial u^\alpha}\right) \end{aligned}$$

from Exercise 5.2.1. The transformation rule for the difference $v^\beta_{k;l} - v^\beta_{kl}$ is then just the standard tensor transformation rule. Since θ transforms as a tensor, the result follows. ∎

One consequence of this theorem is that, although there is no canonical projection $J^1\pi_1 \longrightarrow J^2\pi$, there *is* a projection $\widehat{J}^2\pi \longrightarrow J^2\pi$ which picks out the symmetric part of each fibre coordinate $u^\alpha_{i;j}$. For our present purposes, however, we shall be more concerned with the projection on the second factor

$$pr_2 \circ \Psi^{-1} : \widehat{J}^2\pi \longrightarrow \pi_1^*(\textstyle\bigwedge^2 T^*M) \otimes \pi_{1,0}^*(V\pi) \cong \pi_{1,0}^*(\pi^*(\textstyle\bigwedge^2 T^*M) \otimes V\pi)$$

which describes the antisymmetric part of each $u^\alpha_{i;j}$. If Γ is a jet field, the sequence of maps

$$E \overset{j^1(\Gamma, id_M)\circ\Gamma}{\longrightarrow} \widehat{J}^2\pi \overset{pr_2\circ\Psi^{-1}}{\longrightarrow} \pi_{1,0}^*(\pi^*(\textstyle\bigwedge^2 T^*M) \otimes V\pi) \longrightarrow \pi^*(\textstyle\bigwedge^2 T^*M) \otimes V\pi$$

defines a map $A_\Gamma : E \longrightarrow \pi^*(\bigwedge^2 T^*M) \otimes V\pi$ which is a vector-valued 2-form on E, and which measures the deviation of $j^1(\Gamma, id_M) \circ \Gamma$ from symmetry. As we might expect, this vector-valued 2-form is very closely related to the curvature of the associated connection $\widetilde{\Gamma}$, and provides a geometric explanation for the relationship between the vanishing of the curvature of the connection, and the integrability of the jet field.

Theorem 5.3.5 *The map $A_\Gamma : E \longrightarrow \pi^*(\bigwedge^2 T^*M) \otimes V\pi$ satisfies $A_\Gamma = -R_{\widetilde{\Gamma}}$, where $R_{\widetilde{\Gamma}}$ is the curvature of the connection $\widetilde{\Gamma}$.*

Proof In coordinates, if

$$A_\Gamma = A^\alpha_{ij}(dx^i \wedge dx^j) \otimes \frac{\partial}{\partial u^\alpha}$$

where $A^\alpha_{ij} + A^\alpha_{ji} = 0$, then

$$\begin{aligned} A^\alpha_{ij} &= \tfrac{1}{2}(u^\alpha_{i;j} - u^\alpha_{j;i}) \circ j^1(\Gamma, id_M) \circ \Gamma \\ &= \tfrac{1}{2}\left(\frac{\partial \Gamma^\alpha_i}{\partial x^j} + \Gamma^\beta_j \frac{\partial \Gamma^\alpha_i}{\partial u^\beta} - \frac{\partial \Gamma^\alpha_j}{\partial x^i} - \Gamma^\beta_i \frac{\partial \Gamma^\alpha_j}{\partial u^\beta}\right) \\ &= -R^\alpha_{ij}. \end{aligned}$$

∎

EXERCISES

5.3.1 If $\Gamma : E \longrightarrow J^1\pi$ is a jet field, show (without using coordinates) that

$$(\pi_1)_{1,0}(j^1(\Gamma, id_M)(j^1_p\phi)) = \Gamma(\phi(p))$$

and

$$j^1(\pi_{1,0})(j^1(\Gamma, id_M)(j^1_p\phi)) = j^1_p\phi$$

for each point $j^1_p\phi \in J^1\pi$. Deduce that if $j^1_p\phi \in \operatorname{im}(\Gamma)$ then

$$(\pi_1)_{1,0}(j^1(\Gamma, id_M)(j^1_p\phi)) = j^1(\pi_{1,0})(j^1(\Gamma, id_M)(j^1_p\phi)),$$

so that $D_1 \circ j^1(\Gamma, id_M) \circ \Gamma = 0$; conclude that $j^1(\Gamma, id_M) \circ \Gamma$ takes its values in $\widehat{J}^2\pi$.

5.3.2 Let π be the trivial bundle $(M \times \mathbf{R}, pr_1, M)$, and let $\Gamma : M \times \mathbf{R} \longrightarrow J^1\pi \cong T^*M \times \mathbf{R}$ be a jet field which is projectable to a map $\overline{\Gamma} : M \longrightarrow T^*M$ (so that $\overline{\Gamma}$ is just a differential form on M). Show that the curvature $R_{\widetilde{\Gamma}}$ of the associated connection $\widetilde{\Gamma}$ satisfies

$$R_{\widetilde{\Gamma}} = d\overline{\Gamma} \otimes \frac{\partial}{\partial t},$$

where t is the identity coordinate on $\mathbf{R}$. Deduce that the jet field Γ is integrable precisely when the 1-form $\overline{\Gamma}$ is closed.

5.3.3 Let π be the trivial bundle $(M \times \mathbf{R}, pr_1, M)$. Show that the repeated jet manifold $J^1\pi_1$ is diffeomorphic to $J^1\tau^*_M \times_M (T^*M \times \mathbf{R})$. Hence show that the semi-holonomic jet manifold $\widehat{J}^2\pi$ is diffeomorphic to $J^1\tau^*_M \times \mathbf{R}$, and that this diffeomorphism defines a bundle isomorphism

$$(\widehat{J}^2\pi, \widehat{\pi}_{2,1}, J^1\pi) \cong (J^1\tau^*_M \times \mathbf{R}, ((\tau^*_M)_{1,0}, id_{\mathbf{R}}), T^*M \times \mathbf{R}).$$

5.4 Second-order Jet Fields

If (E, π, M) is a bundle, we may consider jet fields defined on the first jet bundle $(J^1\pi, \pi_1, M)$. Such a jet field is then a section Γ of the bundle $(J^1\pi_1, (\pi_1)_{1,0}, J^1\pi)$, and the associated connection is the vector-valued 1-form on $J^1\pi$ given in coordinates by

$$\tilde{\Gamma} = dx^j \otimes \left(\frac{\partial}{\partial x^j} + \Gamma^\alpha_j \frac{\partial}{\partial u^\alpha} + \Gamma^\alpha_{ij} \frac{\partial}{\partial u^\alpha_i} \right),$$

where $\Gamma^\alpha_j = u^\alpha_{;j} \circ \Gamma$ and $\Gamma^\alpha_{ij} = u^\alpha_{i;j} \circ \Gamma$. If the jet field is integrable, then its integral sections ψ will satisfy

$$\Gamma \circ \psi = j^1\psi,$$

where each ψ is, of course, a local section of π_1. We may, however, take advantage of the fact that π_1 is a jet bundle, and consider those jet fields which take their values in the submanifolds $\widehat{J}^2\pi$ or $J^2\pi$ of $J^1\pi_1$.

Definition 5.4.1 A *semi-holonomic second-order jet field on* π is a section of the bundle $(\widehat{J}^2\pi, \widehat{\pi}_{2,1}, J^1\pi)$. ∎

If Γ is a semi-holonomic second-order jet field then, from the definition of the semi-holonomic jet manifold $\widehat{J}^2\pi$, it follows that $j^1(\pi_{1,0}) \circ \Gamma = id_{J^1\pi}$. We may express this in coordinates as

$$\Gamma^\alpha_j = u^\alpha_{;j} \circ \Gamma = u^\alpha_j \circ \Gamma = u^\alpha_j,$$

so that the coordinate representation of the connection $\tilde{\Gamma}$ will be

$$\tilde{\Gamma} = dx^j \otimes \left(\frac{\partial}{\partial x^j} + u^\alpha_j \frac{\partial}{\partial u^\alpha} + \Gamma^\alpha_{ij} \frac{\partial}{\partial u^\alpha_i} \right).$$

We may use this coordinate representation to find a characterisation of semi-holonomic jet fields in terms of contact forms.

Proposition 5.4.2 *The jet field* $\Gamma : J^1\pi \longrightarrow J^1\pi_1$ *is a semi-holonomic second-order jet field if, and only if,* $\tilde{\Gamma} \lrcorner \sigma = 0$ *for every contact form* σ *on* $J^1\pi$.

Proof Suppose first that $\tilde{\Gamma} \lrcorner \sigma = 0$ for each contact form σ. Choose an adapted coordinate system on E, and take σ to be given locally by $du^\beta - u^\beta_k dx^k$; it follows that $(\Gamma^\alpha_j - u^\alpha_j)dx^j = 0$, so that $\Gamma^\alpha_j = u^\alpha_j$, and therefore that Γ takes its values in $\widehat{J}^2\pi$. The argument may clearly be reversed to show that, if Γ satisfies this condition, then $\tilde{\Gamma} \lrcorner \sigma$ vanishes for every contact form σ. ∎

If Γ is a semi-holonomic jet field, we may use the decomposition of $\widehat{J}^2\pi$ as a fibred product $J^2\pi \times_{J^1\pi} \pi_{1,0}^*(\pi^*(\bigwedge^2 T^*M) \otimes V\pi)$ to define the *torsion* of Γ as its antisymmetric part.

Definition 5.4.3 The *torsion* of the semi-holonomic second-order jet field $\Gamma : J^1\pi \longrightarrow \widehat{J}^2\pi$ is the composite map

$$pr_2 \circ \Psi^{-1} \circ \Gamma : J^1\pi \longrightarrow \pi_{1,0}^*(\pi^*(\bigwedge^2 T^*M) \otimes V\pi).$$

■

In coordinates, the torsion of Γ is the map

$$j_p^1\phi \longmapsto \left(\tfrac{1}{2}(\Gamma_{ij}^\alpha(j_p^1\phi) - \Gamma_{ji}^\alpha(j_p^1\phi)) \left((dx^i \wedge dx^j) \otimes \frac{\partial}{\partial u^\alpha} \right)_{j_p^1\phi}, j_p^1\phi \right),$$

which we may write more simply as

$$\tfrac{1}{2}(\Gamma_{ij}^\alpha - \Gamma_{ji}^\alpha)(dx^i \wedge dx^j) \otimes \frac{\partial}{\partial u^\alpha}.$$

The reason for calling this map the torsion of Γ may be seen from the following example.

Example 5.4.4 Let π be the trivial bundle $(M \times \mathbf{R}, pr_1, M)$, so that we have the identifications $J^1\pi \cong T^*M \times \mathbf{R}$ and $\widehat{J}^2\pi \cong J^1\tau_M^* \times \mathbf{R}$. Let $\Gamma : J^1\pi \longrightarrow \widehat{J}^2\pi$ be a semi-holonomic jet field which is projectable to a map $\overline{\Gamma} : T^*M \longrightarrow J^1\tau_M^*$ (so that, as may readily be checked, $\overline{\Gamma}$ is itself a jet field on the vector bundle (T^*M, τ_M^*, M)). For this example, we shall let t be the identity coordinate on $\mathbf{R}$, and choose coordinates q^j on M and (q^j, p_i) on T^*M; the coordinates on $J^1\tau_M^*$ will then be $(q^j, p_i, p_{i;j})$, where

$$p_{i;j}(j_a^1\omega) = \left.\frac{\partial \omega_i}{\partial q^j}\right|_a = \left.\frac{\partial(p_i \circ \omega)}{\partial q^j}\right|_a.$$

The coordinate representation of $\overline{\Gamma}$ will then be

$$\Gamma_{ij} = p_{i;j} \circ \overline{\Gamma}.$$

If $\overline{\Gamma}$ is an *affine* jet field as described in Example 4.6.7, we may write $\Gamma_{ij}^k = \Gamma_{ij} \circ dq^k$, where dq^k are of course the local sections of τ_M^* dual to the fibre coordinates p_k. As in Exercise 4.6.3, the covariant differential of dq^k then satisfies

$$\nabla dq^k = -\Gamma_{ji}^k dq^i \otimes dq^j.$$

This covariant differential may also be defined to act on vector fields by duality, and the resulting coordinate formula is then

$$\nabla \frac{\partial}{\partial q^j} = \Gamma^k_{ji} dq^i \otimes \frac{\partial}{\partial q^k}.$$

The standard definition of the torsion T is

$$T(X, Y) = \nabla_X Y - \nabla_Y X - [X, Y],$$

where X, Y are vector fields on M, and the covariant derivative $\nabla_X Y$ is defined to equal the contraction $X \lrcorner \nabla Y$. In coordinates, this is just

$$T = (\Gamma^k_{ji} - \Gamma^k_{ij}) dq^i \otimes dq^j \otimes \frac{\partial}{\partial q^k} = \tfrac{1}{2}(\Gamma^k_{ji} - \Gamma^k_{ij})(dq^i \wedge dq^j) \otimes \frac{\partial}{\partial q^k},$$

and a connection whose covariant differential has vanishing torsion is called (for obvious reasons) a symmetric connection.

Now let us return to the original semi-holonomic jet field Γ. According to our definition, the torsion of Γ is the map

$$pr_2 \circ \Psi^{-1} \circ \Gamma : T^*M \times \mathbf{R} \longrightarrow \pi^*_{1,0}(\pi^*(\textstyle\bigwedge^2 T^*M) \otimes V\pi),$$

and from our hypothesis about the existence of $\overline{\Gamma}$, this projects to a map

$$T^*M \longrightarrow \pi^*_{1,0}(\pi^*(\textstyle\bigwedge^2 T^*M) \otimes V\pi).$$

We may now form the composite map

$$T^*M \longrightarrow \pi^*_{1,0}(\pi^*(\textstyle\bigwedge^2 T^*M) \otimes V\pi) \longrightarrow \pi^*(\textstyle\bigwedge^2 T^*M) \otimes V\pi$$
$$\longrightarrow \pi^*(\textstyle\bigwedge^2 T^*M) \longrightarrow \textstyle\bigwedge^2 T^*M,$$

where the map $\pi^*(\bigwedge^2 T^*M) \otimes V\pi \longrightarrow \pi^*(\bigwedge^2 T^*M)$ is given by contraction with the canonically-defined cotangent vector dt_p in the appropriate fibre of $T^*(M \times \mathbf{R})$. This composite map $T^*M \longrightarrow \bigwedge^2 T^*M$ is linear on the fibres of T^*M because $\overline{\Gamma}$ is an affine jet field, and so it defines a vector-valued 2-form which in coordinates may be written

$$\tfrac{1}{2}(\Gamma^k_{ji} - \Gamma^k_{ij})(dq^i \wedge dq^j) \otimes \frac{\partial}{\partial q^k}.$$

■

If the torsion of a semi-holonomic second-order jet field Γ is zero, then Γ must take its values in $J^2\pi$ rather than $\widehat{J}^2\pi$. We may call such a map a holonomic second-order jet field on π, or, more simply, just a second-order jet field.

Definition 5.4.5 A *second-order jet field on π* is a section of the bundle $(J^2\pi, \pi_{2,1}, J^1\pi)$. ∎

In coordinates, a second-order jet field Γ gives rise to a connection

$$\tilde{\Gamma} = dx^j \otimes \left(\frac{\partial}{\partial x^j} + u^\alpha_j \frac{\partial}{\partial u^\alpha} + \Gamma^\alpha_{ij} \frac{\partial}{\partial u^\alpha_i} \right),$$

where now $\Gamma^\alpha_{ij} = \Gamma^\alpha_{ji}$ because the coordinate functions u^α_{ij} are symmetric in the derivative indices. As with semi-holonomic jet fields, it is possible to characterise second-order jet fields using differential forms on $J^1\pi$; now, however, we need to use m-forms θ which are $(m-1)$-horizontal over M and which have the property that

$$(j^1\phi)^*(\theta) = 0$$

for every local section of π.

Proposition 5.4.6 *The jet field $\Gamma : J^1\pi \longrightarrow J^1\pi_1$ is a second-order jet field on π if, and only if, $\tilde{\Gamma} \lrcorner\, \theta = (m-1)\theta$ for every $\theta \in \bigwedge^m_1 \pi_1$ having the property that $(j^1\phi)^*\theta = 0$ for every local section ϕ of π.*

Proof Again we shall give a proof in coordinates, and so we note first that if θ satisfies the conditions of the proposition then it must be represented locally in coordinates as

$$\begin{aligned}\theta &= \theta^i_\alpha (du^\alpha - u^\alpha_k dx^k) \wedge \left(\frac{\partial}{\partial x^i} \lrcorner\, \Omega \right) \\ &\quad + \theta^{ij}_\alpha \left(du^\alpha_i \wedge \left(\frac{\partial}{\partial x^j} \lrcorner\, \Omega \right) - du^\alpha_j \wedge \left(\frac{\partial}{\partial x^i} \lrcorner\, \Omega \right) \right),\end{aligned}$$

where $\theta^{ij}_\alpha + \theta^{ji}_\alpha = 0$, and where $\Omega = dx^1 \wedge \ldots \wedge dx^m$ (the orientability of M is not assumed because this is only a local description). So let $\tilde{\Gamma} \lrcorner\, \theta = (m-1)\theta$ for every such θ. By taking θ to be given locally by

$$(du^\alpha - u^\alpha_k dx^k) \wedge \left(\frac{\partial}{\partial x^i} \lrcorner\, \Omega \right),$$

we see that $(\Gamma^\alpha_i - u^\alpha_i)\Omega = 0$, so that $\operatorname{im}(\Gamma) \subset \hat{J}^2\pi$; by taking θ to be given instead by

$$\left(du^\alpha_i \wedge \left(\frac{\partial}{\partial x^j} \lrcorner\, \Omega \right) - du^\alpha_j \wedge \left(\frac{\partial}{\partial x^i} \lrcorner\, \Omega \right) \right),$$

we obtain $(\Gamma^\alpha_{ij} - \Gamma^\alpha_{ji})\Omega = 0$, so that $\operatorname{im}(\Gamma) \subset J^2\pi$ as required. If, conversely, we are given that Γ is a second-order jet field, then the coordinate representation shows that $\tilde{\Gamma} \lrcorner\, \theta = (m-1)\theta$ for every m-form θ satisfying the conditions of the proposition. ∎

As a jet field on π_1, a second-order jet field Γ may have integral sections ψ satisfying $\Gamma \circ \psi = j^1\psi$; we may see, however, that any such integral section is itself always a prolongation. Since Γ is automatically a semi-holonomic second-order jet field, each contact form σ on $J^1\pi$ satisfies $\widetilde{\Gamma} \lrcorner \sigma = 0$, so that $\pi_1^*(\psi^*(\sigma)) = 0$ using the relationship between connections and integral sections described in Section 4.6; since π_1^* is injective, it follows that $\psi^*(\sigma) = 0$. Since σ is arbitrary, we then have $\psi = j^1\phi$ where $\phi = \pi_{1,0} \circ \psi$, and so

$$\Gamma \circ j^1\phi = j^1(j^1\phi) = j^2\phi.$$

In these circumstances, we shall normally regard ϕ (rather than $j^1\phi$) as the integral section of Γ, so that Γ may be considered as defining the family of second-order partial differential equations

$$\frac{\partial^2\phi^\alpha}{\partial x^i\,\partial x^j} = \Gamma^\alpha_{ij}\left(x^k, \phi^\beta, \frac{\partial\phi^\beta}{\partial x^k}\right).$$

The submanifold $\operatorname{im}(\Gamma) \subset J^2\pi$ may also be considered as an example of a coordinate-free "second-order differential equation" in the same way as we regard a submanifold of $J^1\pi$ as a first-order differential equation. We shall see the importance of second-order jet fields in the next section, when we apply some of these ideas to the calculus of variations.

To finish the present section, we shall show how a second-order jet field defines a complement to $V\pi_{1,0}$ in $V\pi_1 \subset TJ^1\pi$, so that every tangent vector to $J^1\pi$ which is vertical over M may be assigned a unique component vertical over E. Since a second-order jet field is automatically an (ordinary) jet field on π_1, so that any tangent vector on $J^1\pi$ is assigned a unique tangent vector vertical over M, it follows that $TJ^1\pi$ may be written as a direct sum of three components $V\pi_{1,0} \oplus H_{1\Gamma} \oplus H_\Gamma$.

Theorem 5.4.7 *Let (E, π, M) be a bundle, where the base manifold M is orientable with volume form Ω. Each second-order jet field Γ on $J^1\pi$ then determines a decomposition of the bundle $(V\pi_1, \tau_{J^1\pi}|_{V\pi_1}, J^1\pi)$ as a direct sum*

$$(V\pi_{1,0} \oplus H_{1\Gamma}, \tau_{J^1\pi}|_{V\pi_1}, J^1\pi).$$

Proof Let S_Ω be the vector-valued m-form defined on $J^1\pi$ in Section 4.7. We shall consider the Frölicher-Nijenhuis bracket $[S_\Omega, \widetilde{\Gamma}]$, which is a vector-valued $(m{+}1)$-form on $J^1\pi$. For every 1-form σ on $J^1\pi$, we have $[S_\Omega, \widetilde{\Gamma}] \lrcorner \sigma \in \bigwedge_1^{m+1}\pi_1$; if, in addition, $\sigma \in \bigwedge_0^1\pi_1$, then $[S_\Omega, \widetilde{\Gamma}] \lrcorner \sigma = 0$. We may see this from the coordinate representation of $[S_\Omega, \tilde{\Gamma}]$:

$$[S_\Omega, \tilde{\Gamma}] = (du_i^\alpha \wedge \Omega) \otimes \frac{\partial}{\partial u_i^\alpha} - (du^\alpha \wedge \Omega) \otimes \left(\frac{\partial}{\partial u^\alpha} + \frac{\partial \Gamma_{ij}^\beta}{\partial u_i^\alpha} \frac{\partial}{\partial u_j^\beta} \right).$$

Writing $I \wedge \Omega$ for the vector-valued $(m+1)$-form defined by

$$(I \wedge \Omega) \lrcorner \sigma = \sigma \wedge \Omega,$$

and putting $Q = \frac{1}{2}(I \wedge \Omega - [S_\Omega, \tilde{\Gamma}])$, then again $Q \lrcorner \sigma \in \bigwedge_1^{m+1} \pi_1$; if, in addition, $\sigma \in \bigwedge_0^1 \pi_1$, then $Q \lrcorner \sigma = 0$. In coordinates,

$$Q = (du^\alpha \wedge \Omega) \otimes \left(\frac{\partial}{\partial u^\alpha} + \frac{1}{2} \frac{\partial \Gamma_{ij}^\beta}{\partial u_i^\alpha} \frac{\partial}{\partial u_j^\beta} \right).$$

We shall now use the canonical isomorphism between $V^*\pi_1$ and $T^*J^1\pi \wedge \bigwedge^m \pi_1^*(T^*M)$. The vector-valued $(m+1)$-form Q defines a mapping (which we shall also call Q) from $T^*J^1\pi$ to $T^*J^1\pi \wedge \bigwedge^m \pi_1^*(T^*M)$ by the rule $Q(\sigma_{j_p^1\phi}) = (Q \lrcorner \sigma)_{j_p^1\phi}$, where $\sigma_{j_p^1\phi} \in T^*_{j_p^1\phi} J^1\pi$; if it so happens that $\sigma_{j_p^1\phi} \in \pi_1^*(T_p^*M)$ then $Q(\sigma_{j_p^1\phi}) = 0$. But each $\theta_{j_p^1\phi} \in T^*J^1\pi \wedge \bigwedge^m \pi_1^*(T^*M)$ has a representative $\sigma_{j_p^1\phi}$ satisfying $\theta_{j_p^1\phi} = \sigma_{j_p^1\phi} \wedge \Omega_{j_p^1\phi}$, and any two such representatives differ by an element of $\pi_1^*(T^*M)$. We may therefore define $Q(\theta_{j_p^1\phi})$ to equal $Q(\sigma_{j_p^1\phi})$, where $\sigma_{j_p^1\phi}$ is a representative of $\theta_{j_p^1\phi}$. The resulting endomorphism of $T^*J^1\pi \wedge \bigwedge^m \pi_1^*(T^*M)$ (and hence of $V^*\pi_1$) yields the dual endomorphism of $V\pi_1$ which is a projection operator expressed in coordinates as

$$\xi^\alpha \left.\frac{\partial}{\partial u^\alpha}\right|_{j_p^1\phi} + \xi_i^\alpha \left.\frac{\partial}{\partial u_i^\alpha}\right|_{j_p^1\phi} \longmapsto \xi^\alpha \left(\left.\frac{\partial}{\partial u^\alpha}\right|_{j_p^1\phi} + \frac{1}{2} \left.\frac{\partial \Gamma_{ij}^\beta}{\partial u_i^\alpha}\right|_{j_p^1\phi} \left.\frac{\partial}{\partial u_j^\beta}\right|_{j_p^1\phi} \right).$$

The kernel of this endomorphism is $V\pi_{1,0}$, and defining its image to be $H_{1\Gamma}$ gives the required decomposition of $V\pi_1$. ∎

EXERCISES

5.4.1 Let Γ be a semi-holonomic second-order jet field on π. Show that if Γ is integrable, then it must be holonomic.

5.4.2 If Γ is a second-order jet field on π and $\xi \in TJ^1\pi$ has coordinate representation

$$\xi = \xi^i \left.\frac{\partial}{\partial x^i}\right|_{j_p^1\phi} + \xi^\alpha \left.\frac{\partial}{\partial u^\alpha}\right|_{j_p^1\phi} + \xi_j^\alpha \left.\frac{\partial}{\partial u_j^\alpha}\right|_{j_p^1\phi},$$

show that the component of ξ vertical over E under the decomposition of $TJ^1\pi$ by Γ has coordinate representation

$$\left(\xi^\alpha_j - \Gamma^\alpha_{ij}\xi^i - \tfrac{1}{2}\left.\frac{\partial\Gamma^\alpha_{jk}}{\partial u^\beta_k}\right|_{j^1_p\phi}(u^\beta_i\xi^i - \xi^\beta)\right)\left.\frac{\partial}{\partial u^\alpha_j}\right|_{j^1_p\phi}.$$

5.5 The Cartan Form

In this section, we shall continue our development of a jet-bundle description of the calculus of variations which we began in Section 4.4. So let $L \in C^\infty(J^1\pi)$ be a Lagrangian function; we have already seen that the local section ϕ of π is an extremal of L if, and only if,

$$\int_C (j^1\phi)^* d_{X^1} L\Omega = 0$$

whenever C is a compact m-dimensional submanifold of M satisfying $C \subset \operatorname{domain}(\phi)$, and whenever X is a vertical vector field on E satisfying $X|_{\pi^{-1}(\partial C)} = 0$. Our objectives here will be to show that ϕ must satisfy a family of partial differential equations called the Euler-Lagrange equations, and to find ways of representing these equations in a coordinate-free way.

To carry out this project, we shall need to generalise several of the objects described in Chapter 4 to involve 2-jets rather than 1-jets; many of these objects have already been constructed in exercises, and formal definitions of them will be given in Chapter 6, when we consider higher-order jet bundles. For instance, we shall need to use a generalisation of the horizontal vector-valued form on $J^1\pi$ which we described in Definition 4.5.3. This will be the vector-valued form along $\pi_{2,1}$ constructed as a vector-bundle endomorphism of $\pi^*_{2,1}(\tau_{J^1\pi})$ by mapping $(\xi, j^2_p\phi) \in \pi^*_{2,1}(TJ^1\pi)_{j^2_p\phi}$ to its horizontal component $((j^1\phi)_*(\pi_{1*}(\xi)), j^2_p\phi)$, and given in coordinates as

$$h = dx^i \otimes \frac{d}{dx^i},$$

where d/dx^i is now the vector field along $\pi_{2,1}$ given by

$$\frac{d}{dx^i} = \frac{\partial}{\partial x^i} + u^\alpha_i\frac{\partial}{\partial u^\alpha} + u^\alpha_{ij}\frac{\partial}{\partial u^\alpha_j}.$$

We shall also need to use the corresponding derivation of type d_*, which will be denoted d_h and will map r-forms on $J^1\pi$ to $(r+1)$-forms on $J^2\pi$. In coordinates,

$$\begin{aligned} d_h dx^i &= 0, \\ d_h du^\alpha &= dx^j \wedge du^\alpha_j, \\ d_h du^\alpha_i &= dx^j \wedge du^\alpha_{ij}, \\ \text{and} \quad d_h f &= \frac{df}{dx^j} dx^j; \end{aligned}$$

if ϕ is a local section of π then $(j^2\phi)^* \circ d_h = d \circ (j^1\phi)^*$. Finally, if X is a vertical vector field on E with coordinate representation $X = X^\alpha \partial / \partial u^\alpha$, then its second prolongation will be the vector field X^2 on $J^2\pi$ given by

$$X^2 = X^\alpha \frac{\partial}{\partial u^\alpha} + \frac{dX^\alpha}{dx^j} \frac{\partial}{\partial u^\alpha_j} + \frac{d}{dx^i}\left(\frac{dX^\alpha}{dx^j}\right)\frac{\partial}{\partial u^\alpha_{ij}}.$$

We shall start by obtaining a coordinate representation of the Euler-Lagrange equations.

Proposition 5.5.1 *Let $L \in C^\infty(J^1\pi)$ be a Lagrangian, and let C be a fixed compact m-dimensional submanifold of M lying within the domain of a single coordinate system. Suppose ϕ is a local section of π, where $C \subset$* domain (ϕ), *and where $\phi(C)$ lies within the domain of a single coordinate system on E. If ϕ is an extremal of L, then ϕ satisfies the Euler-Lagrange equations*

$$(j^2\phi)^* \left(\frac{\partial L}{\partial u^\alpha} - \frac{d}{dx^i} \frac{\partial L}{\partial u^\alpha_i} \right) = 0 \qquad 1 \le \alpha \le n$$

at every interior point of C.

Proof Let the vector field $X \in \mathcal{V}(\pi)$ satisfy $X|_{\pi^{-1}(\partial C)} = 0$. If the coordinate representation of X is $X^\alpha \partial/\partial u^\alpha$, then $X^\alpha(a) = 0$ whenever $\pi(a) \in \partial C$. Consequently

$$\int_{\partial C} (j^1\phi)^* \left(X^\alpha \frac{\partial L}{\partial u^\alpha_i} \left(\frac{\partial}{\partial x^i} \lrcorner \, \Omega \right) \right) = 0.$$

If we apply Stokes' Theorem to the integral in this equation, we find that

$$\begin{aligned} &\int_{\partial C} (j^1\phi)^* \left(X^\alpha \frac{\partial L}{\partial u^\alpha_i} \left(\frac{\partial}{\partial x^i} \lrcorner \, \Omega \right) \right) \\ = \; &\int_C d \left((j^1\phi)^* \left(X^\alpha \frac{\partial L}{\partial u^\alpha_i} \left(\frac{\partial}{\partial x^i} \lrcorner \, \Omega \right) \right) \right) \\ = \; &\int_C (j^2\phi)^* d_h \left(X^\alpha \frac{\partial L}{\partial u^\alpha_i} \left(\frac{\partial}{\partial x^i} \lrcorner \, \Omega \right) \right) \end{aligned}$$

$$= \int_C (j^2\phi)^* \left(\frac{\partial L}{\partial u_i^\alpha} d_h X^\alpha + X^\alpha d_h \left(\frac{\partial L}{\partial u_i^\alpha} \right) \right) \wedge \left(\frac{\partial}{\partial x^i} \lrcorner \Omega \right)$$

$$= \int_C (j^2\phi)^* \left(\frac{\partial L}{\partial u_i^\alpha} \frac{dX^\alpha}{dx^i} + X^\alpha \frac{d}{dx^i} \frac{\partial L}{\partial u_i^\alpha} \right) \Omega,$$

where as usual we have omitted the various projection maps, so that (for example) the symbol X^α represents three functions on the manifolds E, $J^1\pi$ and $J^2\pi$; it follows that the last integral also vanishes.

We may now apply this to the characterisation of extremals. We obtain

$$\begin{aligned} 0 &= \int_C (j^1\phi)^* d_{X^1} L\Omega \\ &= \int_C (j^1\phi)^* \left(X^\alpha \frac{\partial L}{\partial u^\alpha} + \frac{dX^\alpha}{dx^i} \frac{\partial L}{\partial u_i^\alpha} \right) \Omega \\ &= \int_C (j^2\phi)^* \left(X^\alpha \left(\frac{\partial L}{\partial u^\alpha} - \frac{d}{dx^i} \frac{\partial L}{\partial u_i^\alpha} \right) \right) \Omega. \end{aligned}$$

By taking a suitable variation field X, we can then show in the usual way that if p is any interior point of C then the vanishing of this integral implies that

$$(j^2\phi)^* \left(\frac{\partial L}{\partial u^\alpha} - \frac{d}{dx^i} \frac{\partial L}{\partial u_i^\alpha} \right) (p) = 0.$$

■

To demonstrate how this technique may be applied to a global construction of the Euler-Lagrange equations, we shall examine in more detail the operations carried out to the integrand in the last Proposition. Starting with the m-form $d_{X^1} L\Omega$ on $J^1\pi$, we lifted this form to $J^2\pi$, and then subtracted from it $d_h \Theta_L^X$, where Θ_L^X denotes the $(m-1)$-form on $J^1\pi$ written in coordinates as

$$\frac{\partial L}{\partial u_i^\alpha} X^\alpha \left(\frac{\partial}{\partial x^i} \lrcorner \Omega \right).$$

Since the variation field X vanished at points corresponding to the boundary of the region of integration, $d_h \Theta_L^X$ made no contribution to the integral; the purpose of the subtraction was to produce an integrand which did not involve derivatives of the coefficient functions X^α. If we write E_L^X for the m-form on $J^2\pi$ written in coordinates as

$$X^\alpha \left(\frac{\partial L}{\partial u^\alpha} - \frac{d}{dx^i} \frac{\partial L}{\partial u_i^\alpha} \right) \Omega,$$

then the operation on the integrand was in effect to write the equation

$$E_L^X = \pi_{2,1}^*(d_{X^1}L\Omega) - d_h\Theta_L^X$$

where for clarity we have reinstated the pull-back map $\pi_{2,1}^*$; the various differential forms involved satisfy

$$\begin{aligned} E_L^X &\in \textstyle\bigwedge_0^m \tilde{\pi}_2, \\ d_{X^1}L\Omega &\in \textstyle\bigwedge_0^m \tilde{\pi}_1, \\ \Theta_L^X &\in \textstyle\bigwedge_0^{m-1} \tilde{\pi}_1, \end{aligned}$$

where the tilde indicates the restriction of the jet projection to a suitably small portion of the appropriate jet manifold. This equation (or a closely related one) is commonly called the *equation of first variation*, and the key to preparing a global version of this construction is to note that each of the three m-forms involved may be regarded as the contraction of a suitable $(m+1)$-form with the second prolongation X^2 of the variation field:

$$\begin{aligned} E_L^X &= X^2 \lrcorner\, \delta L \\ \pi_{2,1}^*(d_{X^1}L\Omega) &= X^2 \lrcorner\, \pi_{2,1}^*(dL \wedge \Omega) \\ d_h\Theta_L^X &= -X^2 \lrcorner\, d_h\Theta_L, \end{aligned}$$

where we shall describe δL and Θ_L shortly (the choice of sign for Θ_L is purely conventional). Since the variation field X was chosen arbitrarily, the equation of first variation may then be written

$$\delta L = \pi_{2,1}^*(dL \wedge \Omega) + d_h\Theta_L,$$

where this equation is now required to hold globally on $J^2\pi$.

Since the $(m+1)$-form $\delta L \in \bigwedge_1^{m+1}\pi_2$ must have the property that $X^2 \lrcorner\, \delta L$ does not involve the derivatives of the functions X^α, δL must be horizontal over E; in coordinates

$$\delta L = \left(\frac{\partial L}{\partial u^\alpha} - \frac{d}{dx^i}\frac{\partial L}{\partial u_i^\alpha} \right) du^\alpha \wedge \Omega \in \textstyle\bigwedge_0^{m+1}\pi_{2,0}.$$

We shall call δL the *Euler-Lagrange form* of L. The m-form $\Theta_L \in \bigwedge_1^m \pi_1$ must then be chosen so that δL is horizontal over E. There are many possible choices of Θ_L which give this result; there is, however, only one such form which also has the same extremals as L, in the sense that $(j^1\phi)^*L\Omega = (j^1\phi)^*\Theta_L$ for every $\phi \in \Gamma_{loc}(\pi)$. This unique m-form on $J^1\pi$ will be called the *Cartan form* of L; in coordinates it is

$$\Theta_L = \frac{\partial L}{\partial u_i^\alpha}(du^\alpha - u_j^\alpha dx^j) \wedge \left(\frac{\partial}{\partial x^i} \lrcorner\, \Omega \right) + L\Omega.$$

The global construction is carried out readily with the aid of the canonical vector-valued m-form on $J^1\pi$ described in Section 4.7.

Theorem 5.5.2 *If $L \in C^\infty(J^1\pi)$, then the Cartan form of L may be defined globally by*

$$\Theta_L = d_{S_\Omega} L + L\Omega.$$

Proof It is clear from the coordinate representation

$$S_\Omega \lrcorner\, \sigma = \sigma^i_\alpha (du^\alpha - u^\alpha_j dx^j) \wedge \left(\frac{\partial}{\partial x^i} \lrcorner\, \Omega\right)$$

(where $\sigma \in \bigwedge^1 J^1\pi$) that the following properties are satisfied:

1. $(S_\Omega \lrcorner\, \sigma)_{j^1_p\phi}$ depends only upon the germ of σ at $j^1_p\phi$,
2. $\pi^*_{2,1}(\sigma \wedge \Omega) + d_h(S_\Omega \lrcorner\, \sigma) \in \bigwedge^{m+1}_1 \pi_2 \cap \bigwedge^{m+1}_0 \pi_{2,0}$, and
3. $(j^1\phi)^*(S_\Omega \lrcorner\, \sigma) = 0$ for every $\phi \in \Gamma_{loc}(\pi)$,

where these conditions have been selected so that they may be generalised later to higher-order systems. It is then immediate that Θ_L has the properties required of a Cartan form; the definition is global because S_Ω has been defined globally. ∎

Proposition 5.5.3 *If $S : \bigwedge^1 J^1\pi \longrightarrow \bigwedge^m_1 \pi_1 \cap \bigwedge^m_0 \pi_{1,0}$ satisfies*

$$\pi^*_{2,1}(\sigma \wedge \Omega) + d_h S(\sigma) \in \bigwedge^{m+1}_1 \pi_2 \cap \bigwedge^{m+1}_0 \pi_{2,0},$$

and if $(j^1\phi)^*(S(\sigma)) = 0$ for every $\phi \in \Gamma_{loc}(\pi)$, then for each $\sigma \in \bigwedge^1 J^1\pi$,

$$S(\sigma) = S_\Omega \lrcorner\, \sigma.$$

Proof This may readily be seen in local coordinates. The m-form $S_\Omega \lrcorner\, \sigma - S(\sigma)$ is an element of $\bigwedge^m_1 \pi_1 \cap \bigwedge^m_0 \pi_{1,0}$, and so it may be written locally as

$$S_\Omega \lrcorner\, \sigma - S(\sigma) = (\sigma^i)_\alpha du^\alpha \wedge \left(\frac{\partial}{\partial x^i} \lrcorner\, \Omega\right) + f\Omega$$

for some functions $(\sigma^i)_\alpha$, f on $J^1\pi$. Now $(j^1\phi)^*(S_\Omega \lrcorner\, \sigma - S(\sigma)) = 0$ for every $\phi \in \Gamma_{loc}(\pi)$, giving the relationship $f = -(\sigma^i)_\alpha u^\alpha_i$. Furthermore,

$$d_h(S_\Omega \lrcorner\, \sigma - S(\sigma)) = -\frac{d(\sigma^i)_\alpha}{dx^i} du^\alpha \wedge \Omega - (\sigma^i)_\alpha du^\alpha_i \wedge \Omega,$$

and from $d_h(S_\Omega \lrcorner\, \sigma - S(\sigma)) \in \bigwedge^{m+1}_0 \pi_{2,0}$ it follows that each $(\sigma^i)_\alpha = 0$. ∎

Corollary 5.5.4 *The Cartan form corresponding to a first-order Lagrangian is unique.* ∎

As we have seen, the Euler-Lagrange form δL incorporates the Euler-Lagrange equations in a global form. These equations may also be incorporated in a second-order jet field, in a way which generalises the construction of a second-order vector field in the traditional calculus of variations in a single independent variable. (These vector fields were introduced in Example 4.6.6.) Such a vector field is defined, in the time-dependent case, on the product manifold $\mathbf{R} \times TF$ ($\cong J^1\pi$ where $\pi = (\mathbf{R} \times F, pr_1, \mathbf{R})$). A time-dependent second-order vector field Γ is then required to satisfy two conditions:

1. $\Gamma \lrcorner \sigma = 0$ for every contact form σ on $J^1\pi$, or equivalently $\Gamma \lrcorner S_{dt} = 0$;
2. $d_\Gamma t = 1$.

In coordinates,

$$\Gamma = \frac{\partial}{\partial t} + \dot{q}^\alpha \frac{\partial}{\partial q^\alpha} + \Gamma^\alpha \frac{\partial}{\partial \dot{q}^\alpha}.$$

Starting with a time-dependent Lagrangian $L \in C^\infty(\mathbf{R} \times TF)$, the Euler-Lagrange field Γ_L (if it exists) is defined to satisfy

$$\Gamma_L \lrcorner d\Theta_L = 0,$$

where $\Theta_L = d_{S_{dt}} L + L\, dt$; such a vector field must have coordinate coefficients which satisfy

$$\Gamma^\alpha \frac{\partial^2 L}{\partial \dot{q}^\alpha \, \partial \dot{q}^\beta} = \left(\frac{\partial L}{\partial q^\beta} - \frac{\partial^2 L}{\partial t \, \partial \dot{q}^\beta} - \dot{q}^\alpha \frac{\partial^2 L}{\partial q^\alpha \, \partial \dot{q}^\beta} \right).$$

In the more general case, the association of a second-order jet field Γ with a Lagrangian $L \in C^\infty(J^1\pi)$ involves $d\Theta_L$, where Θ_L is the Cartan form of L; in coordinates

$$\begin{aligned} d\Theta_L &= \left(\frac{\partial L}{\partial u^\alpha} - \frac{\partial^2 L}{\partial x^i \, \partial u^\alpha_i} \right) du^\alpha \wedge \Omega \\ &\quad + \left(\frac{\partial^2 L}{\partial u^\beta \, \partial u^\alpha_i} du^\beta + \frac{\partial^2 L}{\partial u^\beta_j \, \partial u^\alpha_i} du^\beta_j \right) \wedge (du^\alpha - u^\alpha_k dx^k) \wedge \left(\frac{\partial}{\partial x^i} \lrcorner \Omega \right). \end{aligned}$$

Theorem 5.5.5 *Let Γ be a second-order jet field; then*

$$\tilde{\Gamma} \lrcorner d\Theta_L = (m-1) d\Theta_L + \Gamma^* \delta L.$$

If Γ is integrable then the integral sections of Γ are extremals of L if, and only if,

$$\widetilde{\Gamma} \lrcorner\, d\Theta_L = (m-1)d\Theta_L.$$

Proof From the coordinate representation of $d\Theta_L$ and $\widetilde{\Gamma}$,

$$\begin{aligned}\widetilde{\Gamma} \lrcorner\, d\Theta_L &= (m-1)d\Theta_L \\ &\quad + \left(\frac{\partial L}{\partial u^\alpha} - \frac{\partial^2 L}{\partial x^i \partial u^\alpha_i} - \frac{\partial^2 L}{\partial u^\beta \partial u^\alpha_i}u^\beta_i - \frac{\partial^2 L}{\partial u^\beta_j \partial u^\alpha_i}\Gamma^\beta_{ij}\right) du^\alpha \wedge \Omega \\ &= (m-1)d\Theta_L + \Gamma^* \delta L.\end{aligned}$$

Now suppose that Γ is integrable. Write $\delta L/\delta u^\alpha$ for the coefficient of $du^\alpha \wedge \Omega$ in the coordinate expression for δL. If every integral section of Γ is an extremal of L then for each integral section ϕ,

$$(j^1\phi)^* \Gamma^* \frac{\delta L}{\delta u^\alpha} = (j^2\phi)^* \frac{\delta L}{\delta u^\alpha} = 0.$$

But there is an integral section of Γ through each point of $J^1\pi$, so that $\Gamma^*(\delta L/\delta u^\alpha) = 0$. Conversely if $\widetilde{\Gamma} \lrcorner\, d\Theta_L = (m-1)d\Theta_L$ then $\Gamma^*(\delta L/\delta u^\alpha) = 0$, so if ϕ is an integral section of Γ then $(j^2\phi)^*(\delta L/\delta u^\alpha) = 0$. ∎

We shall call an integrable jet field Γ which satisfies these conditions an *Euler-Lagrange field* for L. When $m = 1$, the condition on $\widetilde{\Gamma}$ reduces to $\widetilde{\Gamma} \lrcorner\, d\Theta_L = 0$; writing $\overline{\Gamma}$ for the time-dependent vector field corresponding to $\widetilde{\Gamma}$ (so that $\widetilde{\Gamma} = dt \otimes \overline{\Gamma}$), the condition becomes $dt \wedge (\overline{\Gamma} \lrcorner\, d\Theta_L) = 0$, demonstrating the sense in which Theorem 5.5.5 generalises the one-dimensional result.

EXERCISE

5.5.1 Let π be the trivial bundle $(\mathbf{R}^2 \times \mathbf{R}, pr_1, \mathbf{R})$, with global coordinates (x^1, x^2, u). Let the Lagrangian $L : J^1\pi \longrightarrow \mathbf{R}$ be given by

$$L = \tfrac{1}{2}u_1 u_2 - \cos u.$$

Confirm that the Cartan form of L is

$$\Theta_L = \tfrac{1}{2}u_2 du \wedge dx^2 - \tfrac{1}{2}u_1 du \wedge dx^1 + \left(\tfrac{1}{2}u_1 u_2 - \cos u\right) dx^1 \wedge dx^2,$$

and that the Euler-Lagrange form is

$$\delta L = (u_{12} - \sin u) du \wedge dx^1 \wedge dx^2.$$

(The equation $u_{12} = \sin u$ is known as the *sine-Gordon* equation.)

REMARKS

Semi-holonomic jets and the Spencer operator (and their higher-order analogues, which are described in Chapter 6) may be used to investigate the "formal integrability" of partial differential equations: the idea is to construct a Taylor series solution to the equation by repeated differentiation, as in the proof of the Cauchy-Kowalewskaya theorem for analytic equations. The result is termed "formal" integrability because, in the C^∞ case, there is no guarantee that the resulting series will converge. This topic is examined in detail in [15] and [9]; note that, in the latter reference, jets are defined in an algebraic rather than a geometric context.

A discussion of the Cartan form in first-order field theories, and of its relationship to the affine structure of first-order jet bundles, may be found in [6].

Chapter 6

Higher-order Jet Bundles

In this chapter, we shall extend our definitions to encompass jets of arbitrary order, with the particular objective of studying the higher-order calculus of variations. Many of these extended definitions do not involve any significant new ideas, and so the proofs of our results will often be left as exercises. There is, however, a problem of notation: this was already beginning to appear in Chapter 5, where we saw that the derivative coordinates u^α_{ij} on $J^2\pi$ were symmetric in i and j. Continuing with the same notation would require coordinates denoted by $u^\alpha_{i_1 i_2 \ldots i_k}$ on $J^k\pi$, with symmetry in all the indices $i_1, \ldots, i_k$. In the remaining two chapters, we shall take advantage of this symmetry in order to use *multi-index notation*, where the derivative coordinates (and, indeed, the independent variables themselves) may be denoted simply by u^α_I.

6.1 Multi-index Notation

Let (E, π, M) be a bundle, with $\dim M = m$.

Definition 6.1.1 A *multi-index* is an m-tuple I of natural numbers. The *components* of I are denoted $I(j)$, where j is an ordinary index, $1 \leq j \leq m$. The multi-index 1_j is defined by $1_j(j) = 1$, $1_j(i) = 0$ for $i \neq j$. Addition and subtraction of multi-indexes are defined componentwise (although the result of a subtraction might not be a multi-index): $(I \pm J)(i) = I(i) \pm J(i)$. The length of a multi-index is $|I| = \sum_{i=1}^m I(i)$, and its factorial is $I! = \prod_{i=1}^m (I(i))!$. ■

It is important to be clear about the way we use multi-indexes, because there is an alternative way of using a single letter to represent a family of subscripts or superscripts which is sometimes found in the literature. Our convention is that the j-th component of the multi-index (a natural number, arbitrarily large) represents the number of occasions that an index with value j occurs in the ordinary representation.

Example 6.1.2 Let π be the trivial bundle $(\mathbf{R}^3 \times \mathbf{R}, pr_1, \mathbf{R})$, with coordinates $(x^1, x^2, x^3; u)$. The first derivative coordinates on $J^2\pi$ are

$$u_1, \ u_2, \ u_3,$$

and the second derivative coordinates are

$$u_{11}, \ u_{12}, \ u_{13}, \ u_{22}, \ u_{23}, \ u_{33}.$$

In multi-index notation, the first derivative coordinates would be written as

$$u_{(1,0,0)}, \ u_{(0,1,0)}, \ u_{(0,0,1)},$$

and the second derivative coordinates as

$$u_{(2,0,0)}, \ u_{(1,1,0)}, \ u_{(1,0,1)}, \ u_{(0,2,0)}, \ u_{(0,1,1)}, \ u_{(0,0,2)}.$$

■

Definition 6.1.3 The symbol $\partial^{|I|}/\partial x^I$ is defined by

$$\frac{\partial^{|I|}}{\partial x^I} = \prod_{i=1}^{m} \left(\frac{\partial}{\partial x^i} \right)^{I(i)}.$$

If $|I| = 0$ then $\partial^{|I|}/\partial x^I$ is the identity operator. ■

Typically, capital letters $I, J, K \ldots$ will denote multi-indexes. The summation convention will not extend to multi-indexes: any such such sum will always be indicated explicitly. However, the summation convention for ordinary indices will apply to the subscript of a multi-index such as 1_j.

As an example of the use of multi-indexes, we shall demonstrate the following useful result, a higher-order version of Leibniz' rule for partial derivatives.

Proposition 6.1.4 *If $f, g \in C^\infty(M)$ then*

$$\frac{\partial^{|I|}(fg)}{\partial x^I} = \sum_{J+K=I} \frac{I!}{J!\,K!} \frac{\partial^{|J|} f}{\partial x^J} \frac{\partial^{|K|} g}{\partial x^K}.$$

Proof By induction. The result is clearly true when $|I| = 0$; so suppose it is true whenever $|I| = r$. We shall show that it is then true for $I + 1_i$, and since every multi-index of length $r + 1$ may be written in such a form for some I and some i, the inductive step will follow.

Now

$$\begin{aligned}
\frac{\partial^{|I|+1}(fg)}{\partial x^{I+1_i}} &= \sum_{J+K=I} \frac{I!}{J!\,K!}\left(\frac{\partial^{|J|+1}f}{\partial x^{J+1_i}}\frac{\partial^{|K|}g}{\partial x^K} + \frac{\partial^{|J|}f}{\partial x^J}\frac{\partial^{|K|+1}g}{\partial x^{K+1_i}}\right) \\
&= \sum_{\substack{J+K=I+1_i \\ J(i)\geq 1}} \frac{I!}{(J-1_i)!\,K!}\frac{\partial^{|J|}f}{\partial x^J}\frac{\partial^{|K|}g}{\partial x^K} \\
&\quad + \sum_{\substack{J+K=1_i \\ K(i)\geq 1}} \frac{I!}{J!\,(K-1_i)!}\frac{\partial^{|J|}f}{\partial x^J}\frac{\partial^{|K|}g}{\partial x^K}.
\end{aligned}$$

We may combine the two separate sums by adopting the convention that if, for example, $J(i) = 0$, then the quantity $((J-1_i)!)^{-1}$ is deemed to be zero, even though $J - 1_i$ is not a *bona fide* multi-index. This convention is just the analogue of the usual convention $((-1)!)^{-1} = 0$. So then

$$\frac{\partial^{|I|+1}(fg)}{\partial x^{I+1_i}} = \sum_{J+K=I+1_i}\left(\frac{I!}{(J-1_i)!\,K!} + \frac{I!}{J!\,(K-1_i)!}\right)\frac{\partial^{|J|}f}{\partial x^J}\frac{\partial^{|K|}g}{\partial x^K}.$$

The result now follows by considering the coefficient in parentheses on the right-hand side. If $J(i) = 0$ then the first term of this coefficient is zero by convention; in the second term, $K(i) = I(i) + 1$, so that

$$\frac{I!}{J!\,(K-1_i)!} = \frac{(I+1_i)!}{J!\,K!}.$$

A similar result holds if $K(i) = 0$; note that $J(i) + K(i) = I(i) + 1 > 0$. Finally if $J(i), K(i)$ are both non-zero then

$$\begin{aligned}
I!\left(\frac{1}{(J-1_i)!K!} + \frac{1}{J!\,(K-1_i)!}\right) &= \frac{I!}{(J-1_i)!\,(K-1_i)!}\left(\frac{1}{J(i)} + \frac{1}{K(i)}\right) \\
&= \frac{I!}{(J-1_i)!\,(K-1_i)!}\frac{(J(i)+K(i))}{J(i)K(i)} \\
&= \frac{I!}{(J-1_i)!\,(K-1_i)!}\frac{(I(i)+1)}{J(i)K(i)} \\
&= \frac{(I+1_i)!}{J!\,K!}.
\end{aligned}$$

■

We shall also use multi-index notation when referring to *symmetric* covariant tensors and tensor fields; we shall need to use tensors of this kind when describing the affine structure of jet bundles. A section ξ of the bundle $(S^rT^*M, S^r\tau_M^*, M)$ of symmetric $(0,r)$ tensors over M will be written locally in coordinates as

$$\xi = \sum_{|I|=r} \xi_I dx^I,$$

where each dx^I is a symmetric product of the basis 1-forms dx^i.

EXERCISE

6.1.1 Let

$$n(ij) = \frac{|1_i + 1_j|!}{(1_i + 1_j)!}$$

so that $n(ij)$ is the number of *distinct* indices represented by i and j. If $L \in C^\infty(J^2\pi)$, show that the 1-form dL has coordinate representation

$$dL = \frac{\partial L}{\partial x^i}dx^i + \frac{\partial L}{\partial u^\alpha}du^\alpha + \frac{\partial L}{\partial u_i^\alpha}du_i^\alpha + \frac{1}{n(ij)}\frac{\partial L}{\partial u_{ij}^\alpha}du_{ij}^\alpha,$$

where the final term may also be written in multi-index notation as

$$\sum_{|I|=2} \frac{\partial L}{\partial u_I^\alpha}du_I^\alpha.$$

6.2 Higher-order Jets

We have defined 1-jets and 2-jets of local sections $\phi \in \Gamma_p(\pi)$ to be equivalence classes of local sections which have the same value and first (or first and second) partial derivatives at p. An obvious way to extend this idea is to define further equivalence relations where higher derivatives of the sections are required to be the same. The k-jet of a section will then be the equivalence class containing those sections with the same partial derivatives of order up to k. As before, we shall present the definition in terms of local coordinates, and our proof that the particular choice of coordinate system does not matter will be expressed in terms of multi-indexes.

Lemma 6.2.1 *Let (E, π, M) be a bundle and let $p \in M$. Suppose that $\phi, \psi \in \Gamma_p(\pi)$ satisfy $\phi(p) = \psi(p)$. Let (x^i, u^α) and (y^j, v^β) be two adapted coordinate systems around $\phi(p)$, and suppose also that*

$$\left.\frac{\partial^{|I|}(u^\alpha \circ \phi)}{\partial x^I}\right|_p = \left.\frac{\partial^{|I|}(u^\alpha \circ \psi)}{\partial x^I}\right|_p$$

for $1 \leq \alpha \leq n$, and for every multi-index I with $1 \leq |I| \leq k$. Then

$$\left.\frac{\partial^{|J|}(v^\beta \circ \phi)}{\partial y^J}\right|_p = \left.\frac{\partial^{|J|}(v^\beta \circ \psi)}{\partial y^J}\right|_p$$

for $1 \leq \beta \leq n$, and for every multi-index J with $1 \leq |J| \leq k$.

Proof The first part of this proof uses induction on the length of the multi-index J. Suppose we have shown that, in some neighbourhood of p,

$$\frac{\partial^{|J|}(v^\beta \circ \phi)}{\partial y^J} = F^\beta_J \circ \left(x^k; \frac{\partial^{|K|}(u^\alpha \circ \phi)}{\partial x^K}\right) \qquad 0 \leq |K| \leq |J|,$$

where the smooth function F^β_J is independent of the choice of section ϕ. Then, by the chain rule,

$$\begin{aligned}\frac{\partial^{|J|+1}(v^\beta \circ \phi)}{\partial y^{J+1_j}} &= \frac{\partial x^i}{\partial y^j}\left(\sum_{|L|=0}^{|J|} \frac{\partial^{|L|+1}(u^\gamma \circ \phi)}{\partial x^{L+1_i}} F^{\beta L}_{J\gamma} \circ \left(x^k; \frac{\partial^{|K|}(u^\alpha \circ \phi)}{\partial x^K}\right)\right. \\ &\qquad \left. + \frac{\partial F^\beta_J}{\partial x^i} \circ \left(x^k; \frac{\partial^{|K|}(u^\alpha \circ \phi)}{\partial x^K}\right)\right) \qquad 0 \leq |K| \leq |J|,\end{aligned}$$

where $F^{\beta L}_{J\gamma}$ denotes the partial derivative of F^β_J corresponding to the $\partial^{|L|}(u^\gamma \circ \phi)/\partial x^L$ coordinate; this equation is valid in the same neighbourhood of p. We may therefore certainly write

$$\frac{\partial^{|J|+1}(v^\beta \circ \phi)}{\partial y^{J+1_j}} = F^\beta_{J+1_j} \circ \left(x^k; \frac{\partial^{|K|}(u^\alpha \circ \phi)}{\partial x^K}\right) \qquad 0 \leq |K| \leq |J| + 1,$$

where $F^\beta_{J+1_j}$ is again independent of the choice of section ϕ. (We observe that $F^\beta_{J+1_j}$ is affine-linear in the coordinates corresponding to the highest order partial derivatives.)

Now every multi-index of length $|J| + 1$ may be written as the sum of a multi-index of length J and a multi-index of the form 1_j, so the induction step is valid. Furthermore,

$$\begin{aligned}\frac{\partial(v^\beta \circ \phi)}{\partial y^j} &= \frac{\partial x^i}{\partial y^j}\left(\left(\frac{\partial v^\beta}{\partial x^i} \circ \phi\right) + \left(\frac{\partial v^\beta}{\partial u^\alpha} \circ \phi\right)\frac{\partial(u^\alpha \circ \phi)}{\partial x^i}\right) \\ &= F^\beta_{1_j} \circ \left(x^k; \frac{\partial(u^\alpha \circ \phi)}{\partial x^k}\right),\end{aligned}$$

exactly as in the proof of Lemma 5.1.1. We therefore have in general that, for any multi-index J of arbitrary length,

$$\frac{\partial^{|J|}(v_\beta \circ \phi)}{\partial y^J} = F^\beta_J \circ \left(x^k; \frac{\partial^{|K|}(u^\alpha \circ \phi)}{\partial x^K}\right) \qquad 0 \leq |K| \leq |J|$$

in some suitably small neighbourhood of p, and thus that

$$\left.\frac{\partial^{|J|}(v^\beta \circ \phi)}{\partial y^J}\right|_p = F_J^\beta\left(x^k(p); \left.\frac{\partial^{|K|}(u^\alpha \circ \phi)}{\partial x^K}\right|_p\right) \qquad 0 \leq |K| \leq |J|.$$

The result now follows for $1 \leq |J| \leq k$ by applying the conditions of the lemma. ∎

Definition 6.2.2 Let (E, π, M) be a bundle and let $p \in M$. Define the local sections $\phi, \psi \in \Gamma_p(\pi)$ to be *k-equivalent at p* if $\phi(p) = \psi(p)$ and if, in some adapted coordinate system (x^i, u^α) around $\phi(p)$,

$$\left.\frac{\partial^{|I|}\phi^\alpha}{\partial x^I}\right|_p = \left.\frac{\partial^{|I|}\psi^\alpha}{\partial x^I}\right|_p$$

for $1 \leq |I| \leq k$ and $1 \leq \alpha \leq n$. The equivalence class containing ϕ is called the *k-jet of ϕ at p* and is denoted $j_p^k\phi$. ∎

The equivalence class $j_p^k\phi$ always contains a local section which, in coordinates (x^i, u^α), is a polynomial of degree not greater than k. This is, of course, the k-th Taylor polynomial of ϕ around p.

The set of all the k-jets of local sections of π has a natural structure as a differentiable manifold, and the construction of the atlas which describes this structure is a straightforward generalisation of the corresponding constructions on $J^1\pi$ and $J^2\pi$.

Definition 6.2.3 The *k-th jet manifold of* π is the set

$$\{j_p^k\phi : p \in M, \phi \in \Gamma_p(\pi)\}$$

and is denoted $J^k\pi$. The functions π_k and $\pi_{k,0}$, called the *source* and *target* projections respectively, are defined by

$$\begin{aligned} \pi_k : J^k\pi &\longrightarrow M \\ j_p^k\phi &\longmapsto p \end{aligned}$$

and

$$\begin{aligned} \pi_{k,0} : J^k\pi &\longrightarrow E \\ j_p^k\phi &\longmapsto \phi(p). \end{aligned}$$

If $1 \leq l < k$ then the *l-jet projection* is the function $\pi_{k,l}$ defined by

$$\begin{aligned} \pi_{k,l} : J^k\pi &\longrightarrow J^l\pi \\ j_p^k\phi &\longmapsto j_p^l\phi \end{aligned}$$

∎

It is clear from this definition that $\pi_k = \pi \circ \pi_{k,0}$, and that if $0 \leq m < l$ then $\pi_{k,m} = \pi_{l,m} \circ \pi_{k,l}$. It is conventional to regard $\pi_{k,k}$ as the identity map on $J^k\pi$, and to identify $J^0\pi$ with E.

$$\begin{array}{ccccccccc}
\cdots & J^k\pi & \xrightarrow{\pi_{k,k-1}} & J^{k-1}\pi & \cdots & J^1\pi & \xrightarrow{\pi_{1,0}} & E \\
& \downarrow \pi_k & & \downarrow \pi_{k-1} & & \downarrow \pi_1 & & \downarrow \pi \\
\cdots & M & \xrightarrow[id_M]{} & M & \cdots & M & \xrightarrow[id_M]{} & M
\end{array}$$

Definition 6.2.4 Let (E, π, M) be a bundle and let (U, u) be an adapted coordinate system on E, where $u = (x^i, u^\alpha)$. The *induced coordinate system* (U^k, u^k) on $J^k\pi$ is defined by

$$\begin{aligned} U^k &= \{j^k_p\phi : \phi(p) \in U\} \\ u^k &= (x^i, u^\alpha, u^\alpha_I), \end{aligned}$$

where $x^i(j^k_p\phi) = x^i(p)$, $u^\alpha(j^k_p\phi) = u^\alpha(\phi(p))$, and the $n({}^{m+k}C_k - 1)$ functions

$$u^\alpha_I : U^k \longrightarrow \mathbf{R}$$

are specified by

$$u^\alpha_I(j^k_p\phi) = \left.\frac{\partial^{|I|}\phi^\alpha}{\partial x^I}\right|_p$$

and are known as *derivative coordinates*. ∎

Note that if $|I| \leq l < k$ then the coordinate function u^α_I on $J^k\pi$ is the pullback by $\pi_{k,l}$ of the coordinate function u^α_I on $J^l\pi$.

Proposition 6.2.5 *Given an atlas of adapted charts (U, u) on E, the corresponding collection of charts (U^k, u^k) is a finite-dimensional C^∞ atlas on $J^k\pi$.*

Proof First, note that every k-jet $j^k_p\phi$ is in the domain of one such chart, namely any chart (U^k, u^k) where $\phi(p) \in U$. We must now show that, if (U, u) and (V, v) are two charts in the atlas on E such that $U^k \cap V^k$ is non-empty, then the transition function

$$v^k \circ (u^k)^{-1}$$

is smooth (where we have again omitted the explicit restriction of $(u^k)^{-1}$ to a subset of its domain).

As before, the component functions of $v^k \circ (u^k)^{-1}$ are $y^j \circ (u^k)^{-1}$, $v^\beta \circ (u^k)^{-1}$, and $v^\beta_J \circ (u^k)^{-1}$, but the domain of each of these functions is now an open subset of $\mathbf{R}^{m+n} \times \mathbf{R}^N$, where $N = n({}^{m+k}C_k - 1)$. Since $pr_1 \circ u^k = u \circ \pi_{k,0}$, we have

$$\begin{aligned} y^j \circ (u^k)^{-1} &= y^j \circ u^{-1} \circ pr_1 \\ v^\beta \circ (u^k)^{-1} &= v^\beta \circ u^{-1} \circ pr_1, \end{aligned}$$

so that the first two sets of component functions are smooth. As far as the third set is concerned,

$$\begin{aligned} v^\beta_J(j^k_p\phi) &= \left.\frac{\partial^{|J|}(v^\beta \circ \phi)}{\partial y^J}\right|_p \\ &= F^\beta_J\left(x^k(p); \left.\frac{\partial^{|K|}(u^\alpha \circ \phi)}{\partial x^K}\right|_p\right) \qquad 0 \le |K| \le |J|, \end{aligned}$$

where F^β_J are the functions introduced in the proof of Lemma 6.2.1 and shown there to be smooth. ∎

Once again, if we show that $J^k\pi$ is the total space of a bundle, then Proposition 1.1.14 will imply that it satisfies the topological conditions we require for it to be a manifold. There will now, however, be $k+1$ different bundles of which it is the total space.

Lemma 6.2.6 *The function $\pi_{k,k-1} : J^k\pi \longrightarrow J^{k-1}\pi$ is a smooth surjective submersion.* ∎

Corollary 6.2.7 *The functions $\pi_{k,l} : J^k\pi \longrightarrow J^l\pi$ (where $1 \le l < k$), $\pi_{k,0} : J^k\pi \longrightarrow E$ and $\pi_k : J^k\pi \longrightarrow M$ are smooth surjective submersions.* ∎

Given for the moment that the atlas on each $J^k\pi$ defines a manifold, we now see that the triples $(J^k\pi, \pi_{k,l}, J^l\pi)$, $(J^k\pi, \pi_{k,0}, E)$ and $(J^k\pi, \pi_k, M)$ all become fibred manifolds. The proof that π_k is a bundle will again involve the local trivialisations of π, whereas the remaining triples are still be bundles even if π is only a fibred manifold. Furthermore, our results that $\pi_{1,0}$ and $\pi_{2,1}$ are affine bundles may be generalised to an assertion that $\pi_{k,k-1}$ is an affine bundle. However, the remaining $\pi_{k,l}$ are *not* affine bundles: for example, in $\pi_{k,k-2}$ the transition functions are *quadratic* functions of the fibre coordinates. This is another manifestation of our observation in the proof of Lemma 6.2.1 concerning the affine-linear appearance of the highest-order derivatives in the expression of the chain rule.

Proposition 6.2.8 *If $0 \le l < k$ then $J^k\pi$ is a manifold, and $(J^k\pi, \pi_{k,l}, J^l\pi)$ is a bundle.*

Proof Let $j^l_p\phi \in J^l\pi$, and let (U, u) be an adapted coordinate system around $\phi(p) \in E$. Then $U^k = \pi^{-1}_{k,l}(U^l)$, and the map

$$\begin{array}{rcl} t^{k,l}_p : U^k & \longrightarrow & U^l \times \mathbf{R}^N \\ j^k_q\psi & \longmapsto & (j^l_q\psi; u^\alpha_I(j^k_q\psi)) \quad l+1 \le |I| \le k \end{array}$$

(where $N = n\sum_{r=l+1}^{k} {}^{m+r-1}C_r$) is a diffeomorphism, because it is the composite $((u^l)^{-1} \times id_{\mathbf{R}^N}) \circ u^k$; clearly $pr_1 \circ t^{k,l}_p = \pi_{k,l}|_{U^k}$. By taking $l = 0$, it follows from Proposition 1.1.14 that $J^k\pi$ is a manifold. The maps $t^{k,l}_p$ then become local trivialisations, and each $\pi_{k,l}$ becomes a bundle. ∎

To demonstrate the affine structure of the bundle $\pi_{k,k-1}$, we should indicate a corresponding vector bundle over $J^{k-1}\pi$. This will be the bundle with total space $\pi^*_{k-1}(S^kT^*M) \otimes \pi^*_{k-1,0}(V\pi)$: formally, it is the bundle

$$\Big(\pi^*_{k-1}(S^kT^*M) \otimes \pi^*_{k-1,0}(V\pi),$$
$$\left(S^k\tau^*_{J^{k-1}\pi}\Big|_{\pi^*_{k-1}(S^kT^*M)}\right) \otimes \pi^*_{k-1,0}\left(\tau_E|_{V\pi}\right), J^{k-1}\pi\Big).$$

The proof is just a straightforward generalisation of the proof of Theorem 5.1.7.

Theorem 6.2.9 *The triple $(J^k\pi, \pi_{k,k-1}, J^{k-1}\pi)$ may be given the structure of an affine bundle modelled on the vector bundle*

$$\left(S^k\tau^*_{J^{k-1}\pi}\Big|_{\pi^*_{k-1}(S^kT^*M)}\right) \otimes \pi^*_{k-1,0}\left(\tau_E|_{V\pi}\right)$$

in such a way that, for each adapted chart (U, u) on E, the map

$$\begin{array}{rcl} t_u : \pi^{-1}_{k,k-1}(U^{k-1}) & \longrightarrow & U^{k-1} \times \mathbf{R}^N \\ j^k_p\phi & \longmapsto & (j^{k-1}_p\phi, u^\alpha_I(j^k_p\phi)) \qquad |I| = k, \end{array}$$

where $N = n\,{}^{m+k-1}C_k$, is an affine local trivialisation.

Proof In coordinates, if $a \in J^{k-1}\pi$, then a typical element

$$\xi \in (\pi^*_{k-1}(S^kT^*M) \otimes \pi^*_{k-1,0}(V\pi))_a$$

may be written as

$$\xi = \sum_{|I|=k} \xi_I^\alpha \left(dx^I \otimes \frac{\partial}{\partial u^\alpha} \right)_a .$$

If the image of $j^k_{\pi_{k-1}(a)}\phi$ under the action of ξ is denoted by $\xi[j^k_{\pi_{k-1}(a)}\phi]$, then

$$u_I^\alpha(\xi[j^k_{\pi_{k-1}(a)}\phi]) = u_I^\alpha(j^k_{\pi_{k-1}(a)}\phi) + \xi_I^\alpha \qquad |I| = k,$$

which is independent of the choice of coordinate system. ■

Lemma 6.2.10 *If $W \subset M$ is an open submanifold then*

$$J^k\left(\pi|_{\pi^{-1}(W)}\right) \cong \pi_k^{-1}(W).$$

■

Definition 6.2.11 If $p \in M$ then the fibre $\pi_k^{-1}(p)$ is denoted $J^k_p\pi$ rather than $(J^k\pi)_p$. ■

Lemma 6.2.12 *Let π be the trivial bundle $(\mathbf{R}^m \times F, pr_1, \mathbf{R}^m)$. Then the fibred manifold $(J^k\pi, \pi_k, \mathbf{R}^m)$ is trivial.*

Proof Similar to the proof of Lemma 4.1.20. ■

Proposition 6.2.13 *Let (E, π, M) be a bundle. Then $(J^k\pi, \pi_k, M)$ is a bundle.*

Proof Similar to the proof of Proposition 4.1.21. ■

Example 6.2.14 Let π be the trivial bundle $(\mathbf{R} \times F, pr_1, \mathbf{R})$. We have already seen that $J^1\pi \cong \mathbf{R} \times TF$ and $J^2\pi \cong \mathbf{R} \times T^2F$ (the latter in Exercise 5.1.3); in general, $J^k\pi \cong \mathbf{R} \times T^kF$, where T^kF is the k-th order tangent manifold to F. The elements of T^kF are equivalence classes of curves through each point in F, where curves through the same point are equivalent if they have the same derivatives of order up to k. There is an obvious identification of T^kF with $J^k_0\pi$. In mechanics, where coordinates (t, q^α) are used on $\mathbf{R} \times F$, the coordinates on T^kF are often written as $(q^\alpha_{(r)})$, where the subscript r indicates the number of dots above the q: of course, $(r) \in \mathbf{N}^1$ is really just a multi-index. ■

As we might expect, some of the local sections of $(J^k\pi, \pi_k, M)$ may be characterised as arising from sections of π.

Definition 6.2.15 Let ϕ be a local section of π with domain $W \subset M$. The *k-th prolongation of* ϕ is the map $j^k\phi : W \longrightarrow J^k\pi$ defined by

$$j^k\phi(p) = j^k_p\phi.$$

■

Note that $\pi_k \circ j^k\phi = id_W$, so that $j^k\phi$ really is a section; also, if $k > l$, then $\pi_{k,l} \circ j^k\phi = j^l\phi$. In local coordinates, $j^k\phi$ is given by

$$\left(\phi^\alpha, \frac{\partial^{|I|}\phi^\alpha}{\partial x^I}\right) \qquad 1 \leq |I| \leq k.$$

Using the identification of $J^0\pi$ with E, we may also identify $j^0\phi$ with ϕ.

Lemma 6.2.16 *If* $\psi \in \Gamma_{loc}(\pi_k)$ *then* ψ *is the k-th prolongation of some* $\phi \in \Gamma_{loc}(\pi)$ *if, and only if,* $j^k(\pi_{k,0} \circ \psi) = \psi$. ■

We may also define the prolongation of bundle morphisms.

Definition 6.2.17 Let (E, π, M) and (F, ρ, N) be bundles, and let $(f, \overline{f}) : \pi \longrightarrow \rho$ be a bundle morphism, where $\overline{f}$ is a diffeomorphism. The *k-th prolongation of* f is the map $j^k(f, \overline{f}) : J^k\pi \longrightarrow J^k\rho$ defined by

$$j^k(f, \overline{f})(j^k_p\phi) = j^k_{\overline{f}(p)}\tilde{f}(\phi).$$

We also write j^kf instead of $j^k(f, \overline{f})$ where there is no ambiguity. ■

As before, this definition does not depend on the particular choice of ϕ, and we may rewrite the formula as

$$\widetilde{j^kf}(j^k\phi) = j^k\tilde{f}(\phi).$$

Lemma 6.2.18 *Both* $(j^kf, f) : \pi_{k,0} \longrightarrow \rho_{k,0}$ *and* $(j^kf, \overline{f}) : \pi_k \longrightarrow \rho_k$ *are bundle morphisms.* ■

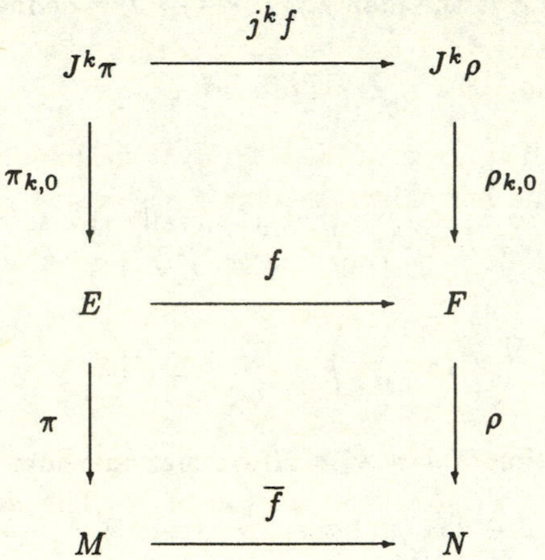

Lemma 6.2.19 *Let $f : \pi \longrightarrow \rho$ and $g : \rho \longrightarrow \sigma$ be bundle morphisms which project to diffeomorphisms; then $j^k(g \circ f, \overline{g \circ f}) = j^k(g, \overline{g}) \circ j^k(f, \overline{f})$. In addition, $j^k(id_E, id_M) = id_{J^k\pi}$.* ■

We may also define differential equations on π, in a way which directly generalises our earlier definition of a first-order differential equation.

Definition 6.2.20 Let (E, π, M) be a bundle. A *differential equation* on π is a closed embedded submanifold S of the jet manifold $J^k\pi$. The *order* of S is the largest natural number r satisfying

$$\pi_{r,r-1}^{-1}(\pi_{k,r-1}S) \neq \pi_{k,r}S.$$

■

This description of the order of a differential equation is intended to concentrate attention on the case $r = k$; the point of the definition is that the additional derivative variables u_I^α, where $|I| > r$, do not provide any further information about S. It follows from this that a first-order differential equation, as specified in Definition 4.1.24, might have order zero—but of course the result then isn't really a differential equation at all. We shall normally regard a differential equation of order k as being defined on the bundle π_k.

Definition 6.2.21 A *solution* of the differential equation S is a local section $\phi \in \Gamma_W(\pi)$ satisfying $j_p^k\phi \in S$ for every $p \in W$. ■

An alternative description of a solution is that it is a local section ϕ whose prolongation $j^k\phi$ takes its values in $S \subset J^k\pi$.

It is often the case that a differential equation is defined by a bundle morphism whose domain is a jet bundle.

Definition 6.2.22 Let (E, π, M) and (H, ρ, M) be bundles, and let $(f, id_M) : \pi_k \longrightarrow \rho$ be a bundle morphism, so that $f : J^k\pi \longrightarrow H$. The *differential operator determined by* f is the map $\mathcal{D}_f : \Gamma_{loc}(\pi) \longrightarrow \Gamma_{loc}(\rho)$ given by

$$(\mathcal{D}_f(\phi))(p) = f(j^k_p\phi).$$

■

Definition 6.2.23 Let $\mathcal{D}_f$ be the differential operator determined by $f : J^k\pi \longrightarrow H$, and let χ be a local section of ρ. The *differential equation determined by* $\mathcal{D}_f$ *and* χ is the submanifold

$$S_{f;\chi} = \{j^k_p\phi : f(j^k_p\phi) = \chi(p)\} \subset J^k\pi.$$

■

A solution of a differential equation determined in this manner is then nothing but a local section $\phi \in \Gamma_W(\pi)$ satisfying $\mathcal{D}_f(\phi) = \chi|_W$. Frequently, of course, ρ is a trivial vector bundle and χ is its zero section, but this is not a requirement of our definition.

Example 6.2.24 Let π and ρ both be the trivial bundles $(\mathbf{R}^2 \times \mathbf{R}, pr_1, \mathbf{R})$, with coordinates $(x^1, x^2; u)$. Let $f : J^2\pi \longrightarrow \mathbf{R}^2 \times \mathbf{R}$ be given by

$$f(j^2_p\phi) = (p^1, p^2; u_{12}(j^2_p\phi) - \sin(\phi(p))),$$

where $p = (p^1, p^2) \in \mathbf{R}^2$, and let $z \in \Gamma(\rho)$ be the zero section. Then

$$S_{f,z} = \{j^2_p\phi : (u_{12} - \sin u)(j^2_p\phi) = 0\},$$

so that solutions of $S_{f,z}$ satisfy the sine-Gordon equation

$$\frac{\partial^2\phi}{\partial x^1\,\partial x^2} = \sin\phi.$$

■

As an application of the process of prolonging a bundle morphism, we may define a *symmetry* of the differential equation $S \subset J^k\pi$ to be a bundle isomorphism $(f, \overline{f})$ of π with itself, such that $\tilde{f}(\phi)$ is a solution of S exactly when ϕ is a solution. We may express this requirement by demanding that

$j^k f(S) = S$.

We may also use the fact that $(J^k\pi, \pi_k, M)$ is a bundle to consider repeated jets, as in Section 5.2. The l-jet manifold of π_k will be denoted $J^l\pi_k$, and will contain l-jets of all the local sections of π_k:

$$J^l\pi_k = \{j^l_p\psi : \psi \in \Gamma_p(\pi_k)\}.$$

If local coordinates on E are (x^i, u^α), and on $J^k\pi$ are (x^i, u^α_I), $0 \le |I| \le k$, then coordinates on $J^l\pi_k$ are

$$(x^i, u^\alpha_{I;J}) \qquad 0 \le |I| \le k, \quad 0 \le |J| \le l,$$

where the functions $u^\alpha_{I;J}$ are defined by

$$u^\alpha_{I;J}(j^l_p\psi) = \left.\frac{\partial^{|J|}\psi^\alpha_I}{\partial x^J}\right|_p$$

using the standard coordinate representation $\psi^\alpha_I = u^\alpha_I \circ \psi$. Here, too, there is a distinguished subset containing those elements $j^l_p\psi$ where the local section ψ is itself the prolongation $j^k\phi$ of a local section of π, and we may define a map $\iota_{l,k}$ which generalises the map $\iota_{1,1}$ described in Section 5.2.

Definition 6.2.25 The map $\iota_{l,k} : J^{k+l}\pi \longrightarrow J^l\pi_k$ is defined by

$$\iota_{l,k}(j^{k+l}_p\phi) = j^l_p(j^k\phi).$$

■

In coordinates,

$$\begin{aligned} u^\alpha_{I;J}(\iota_{l,k}(j^{k+l}_p\phi)) &= u^\alpha_{I;J}(j^l_p(j^k\phi)) \\ &= \left.\frac{\partial^{|J|}}{\partial x^J}\right|_p (u^\alpha_I \circ j^k\phi) \\ &= \left.\frac{\partial^{|J|}}{\partial x^J}\right|_p \left(\frac{\partial^{|I|}\phi^\alpha}{\partial x^I}\right) \\ &= \left.\frac{\partial^{|I+J|}\phi^\alpha}{\partial x^{I+J}}\right|_p \\ &= u^\alpha_{I+J}(j^{k+l}_p\phi). \end{aligned}$$

It follows that $\iota_{l,k}(J^{k+l}\pi)$ is the subset of $J^l\pi_k$ where, for every local coordinate system $(x^i, u^\alpha_{I;J})$, if $I_1 + J_1 = I_2 + J_2$ then $u^\alpha_{I_1;J_1}(j^l_p\psi) = u^\alpha_{I_2;J_2}(j^l_p\psi)$.

Since, in these coordinates, $\iota_{l,k}$ is represented by a linear injection, it is an embedding. We may, of course, continue this procedure and use the fact that $(\pi_k)_l$ is a bundle to define *its* r-jet bundle, and so on: in fact, we shall only need to use several levels of repeated jets when $k = l = r = \ldots = 1$, and we shall write π_1^k for $((\ldots(\pi_1)_1\ldots)_1)_1$.

As with $\iota_{1,1}$, we may use $\iota_{l,k}$ to define the prolongation of bundle morphisms $(f, \overline{f}) : \pi_k \longrightarrow \rho$ as maps $J^{k+l}\pi \longrightarrow J^l\rho$ rather than $J^l\pi_k \longrightarrow J^l\rho$: we simply consider $j^l(f, \overline{f}) \circ \iota_{l,k}$. We may also prolong differential equations by this method, for if $M = N$, if χ is a section of ρ, and if $S \subset J^k\pi$ is the differential equation determined by f and χ, then

$$j^l S = \{j_p^{k+l}\phi : j^l(f, \overline{f})(j_p^{k+l}\phi) = j_p^l\chi\} \subset J^{k+l}\pi$$

is also a differential equation, and is called the l-th prolongation of S. In classical notation, $j^l S$ describes the family of partial differential equations obtained by differentiating the original equations $0, 1, 2, \ldots, l$ times with respect to the independent variables x^i.

The way these maps fit together may be summarised in the following two commutative diagrams, which generalise the diagrams given for second-order jets and repeated jets in Chapter 5:

$$\begin{array}{ccc}
J^{k+l+m}\pi & \xrightarrow{\ \iota_{l+m,k}\ } & J^{l+m}\pi_k \\
\Big\downarrow{\scriptstyle \pi_{k+l+m,k+l}} & & \Big\downarrow{\scriptstyle (\pi_k)_{l+m,l}} \\
J^{k+l}\pi & \xrightarrow[\ \iota_{l,k}\ ]{} & J^l\pi_k
\end{array}$$

and

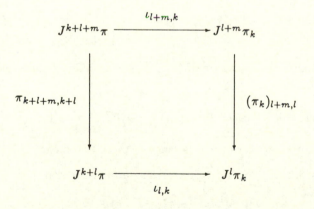

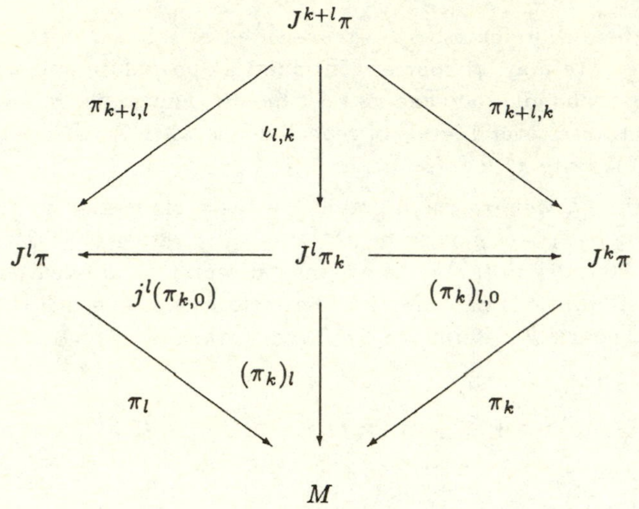

Finally in this section we shall define semi-holonomic jets. As with semi-holonomic 2-jets, these may be constructed by considering two different maps between the same pair of jet manifolds; this time the two manifolds are $J^1\pi_k$ and $J^1\pi_{k-1}$. First, the maps $(\pi_k)_{1,0} : J^1\pi_k \longrightarrow J^k\pi$ and $\iota_{1,k-1} : J^k\pi \longrightarrow J^1\pi_{k-1}$ may be composed to give the map $\iota_{1,k-1} \circ (\pi_k)_{1,0}$; secondly, the map $\pi_{k,k-1} : J^k\pi \longrightarrow J^{k-1}\pi$ may be regarded as a bundle morphism $(\pi_{k,k-1}, id_M) : \pi_k \longrightarrow \pi_{k-1}$, and so may be prolonged to give a map $j^1(\pi_{k,k-1})$. Note that if $j^1_p\psi \in J^1\pi_k$, then its images under these two maps will be in the same fibre of $J^1\pi_{k-1}$ over $J^{k-1}\pi$, because

$$\begin{aligned}(\pi_{k-1})_{1,0}(\iota_{1,k-1}((\pi_k)_{1,0}(j^1_p\psi))) &= (\pi_{k-1})_{1,0}(j^1(\pi_{k,k-1})(j^1_p\psi)) \\ &= \pi_{k,k-1}(\psi(p)).\end{aligned}$$

This means that the structure of $(\pi_{k-1})_{1,0}$ as an affine bundle modelled on the vector bundle

$$\left(\pi^*_{k-1}(T^*M) \otimes V\pi_{k-1}, (\tau^*_{J^{k-1}\pi}|_{\pi^*_{k-1}(T^*M)}) \otimes (\tau_{J^{k-1}\pi}|_{V\pi_{k-1}}), J^{k-1}\pi\right)$$

may be used to construct the difference of these two maps.

Definition 6.2.26 The *k-jet Spencer operator* is the map

$$D_k : J^1\pi_k \longrightarrow \pi^*_{k-1}(T^*M) \otimes V\pi_{k-1}$$

defined by requiring $D_k(j^1_p\psi)$ to be the unique element of $\pi^*_{k-1}(T^*M)\otimes V\pi_{k-1}$ whose affine action on $J^1\pi_{k-1}$ maps $\iota_{1,k-1}((\pi_k)_{1,0}(j^1_p\psi))$ to $j^1(\pi_{k,k-1})(j^1_p\psi)$. ∎

In local coordinates,

$$D_k(j^1_p\psi) = \sum_{|I|=0}^{k-1} (u^\alpha_{I;1_i}(j^1_p\psi) - u^\alpha_{I+1_i;}(j^1_p\psi)) \left(dx^i \otimes \frac{\partial}{\partial u^\alpha_I}\right)_{\pi_{k,k-1}(\psi(p))}$$

Definition 6.2.27 The *semi-holonomic $(k+1)$-jet manifold* $\widehat{J}^{k+1}\pi$ is the submanifold $D_k^{-1}(0)$ of $J^1\pi_k$. ∎

We now have the inclusions $\iota_{1,k}(J^{k+1}\pi) \subset \widehat{J}^{k+1}\pi \subset J^1\pi_k$; in terms of coordinates, we may say that $\iota_{1,k}(J^{k+1}\pi)$ is the submanifold of $J^1\pi_k$ where the derivative coordinates are totally symmetric, whereas $\widehat{J}^{k+1}\pi$ is the submanifold where all except the highest order derivative coordinates are totally symmetric (so that we may take $(x^i, u^\alpha_{I;}, u^\alpha_{J;1_i})$ as a coordinate system on $\widehat{J}^{k+1}\pi$, where $0 \leq |I| \leq k$ and $|J| = k$).

EXERCISES

6.2.1 Complete the proofs of Lemma 6.2.12, and of Proposition 6.2.13.

6.2.2 Let π be the trivial bundle $(\mathbf{R}^2 \times \mathbf{R}, pr_1, \mathbf{R})$, with global coordinates $(x_1, x_2; u)$. Let $S \subset J^2\pi$ be the differential equation $u_{12} = \sin u$ described in Exercise 5.5.1. Show that the first prolongation j^1S is the subset of $J^3\pi$ described by the equations

$$\begin{aligned} u_{12} &= \sin u, \\ u_{112} &= u_1 \cos u, \\ u_{122} &= u_2 \cos u. \end{aligned}$$

6.3 The Contact Structure

In Section 4.3, we explained that the intrinsic structure of the affine bundle $(J^1\pi, \pi_{1,0}, E)$ could be captured by certain vector fields along $\pi_{1,0}$ and the dual differential forms on $J^1\pi$. In Section 4.5, we saw further how this information could be summarised in the vector-valued forms h and v which we termed the *contact structure on* π_1. These ideas may be generalised without difficulty to the affine bundle $(J^{k+1}\pi, \pi_{k+1,k}, J^k\pi)$, and yield two families of d_*-derivations called the *horizontal* and *vertical differentials*. We have already glimpsed the horizontal differential d_h in Section 5.5, and we shall see in these last two chapters that d_h is an operator of fundamental importance in the calculus of variations.

Definition 6.3.1 Let (E, π, M) be a bundle, and suppose that $p \in M$, $\phi \in \Gamma_p(\pi)$, and $\zeta \in T_pM$. The kth *holonomic lift* of ζ by ϕ is defined to be

$$((j^k\phi)_*(\zeta), j_p^{k+1}\phi) \in \pi_{k+1,k}^*(TJ^k\pi).$$

■

A word about nomenclature is necessary here: we have chosen to call the resulting element of $\pi_{k+1,k}^*(TJ^k\pi)$ the k-th rather than the $(k+1)$-th holonomic lift, because it involves a tangent vector to $J^k\pi$, and so the original holonomic lift to $\pi_{1,0}^*(TE)$ described in Definition 4.3.1 should properly be called the *zeroth* holonomic lift.

Theorem 6.3.2 *Let (E, π, M) be a bundle and let $j_p^{k+1}\phi \in J^{k+1}\pi$. There is then a canonical decomposition of the vector space $\pi_{k+1,k}^*(TJ^k\pi)_{j_p^{k+1}\phi}$ as a direct sum of two subspaces*

$$\pi_{k+1,k}^*(V\pi_k)_{j_p^{k+1}\phi} \oplus (j^k\phi)_*(T_pM),$$

where $(j^k\phi)_*(T_pM)$ denotes the collection of k-th holonomic lifts of tangent vectors in T_pM by ϕ.

Proof Similar to the proof of Theorem 4.3.2; of course we must check that the k-th holonomic lift is well-defined for different choices of ϕ with the same $(k+1)$-jet at p, but if $\gamma : \mathbf{R} \longrightarrow M$ is a curve with $\gamma(0) = p$, $[\gamma] = \zeta$, then $(j^k\phi)_*\zeta = [j^k\phi \circ \gamma]$, and

$$\left.\frac{d}{dt}\right|_{t=0} (u_I^\alpha \circ j^k\phi \circ \gamma) = \left.\frac{\partial^{|I|+1}\phi^\alpha}{\partial x^{I+1_i}}\right|_p \left.\frac{d\gamma^i}{dt}\right|_{t=0}$$

depends only on the derivatives of ϕ of order $\leq k+1$. ■

Corollary 6.3.3 *The vector bundle $(\pi_{k+1,k}^*(TJ^k\pi), \pi_{k+1,k}^*(\tau_{J^k\pi}), J^k\pi)$ may be written as the direct sum of two sub-bundles*

$$(\pi_{k+1,k}^*(V\pi_k) \oplus H\pi_{k+1,k}, \pi_{k+1,k}^*(\tau_{J^k\pi}), J^k\pi),$$

where $H\pi_{k+1,k}$ is the union of the fibres $(j^k\phi)_*(T_pM)$ for $p \in M$. ■

In coordinates, the k-th holonomic lift of

$$\zeta = \zeta^i \left.\frac{\partial}{\partial x^i}\right|_p$$

is given by

$$\begin{aligned}(j^k\phi)_*(\zeta) &= \zeta^i (j^k\phi)_* \left(\left.\frac{\partial}{\partial x^i}\right|_p \right) \\ &= \zeta^i \left(\left.\frac{\partial}{\partial x^i}\right|_{j^k_p\phi} + \sum_{|I|=0}^{k} \left.\frac{\partial(u^\alpha_I \circ \phi)}{\partial x^i}\right|_{j^k_p\phi} \left.\frac{\partial}{\partial u^\alpha_I}\right|_{j^k_p\phi} \right) \\ &= \zeta^i \left(\left.\frac{\partial}{\partial x^i}\right|_{j^k_p\phi} + \sum_{|I|=0}^{k} u^\alpha_{I+1_i}(j^{k+1}_p\phi) \left.\frac{\partial}{\partial u^\alpha_I}\right|_{j^k_p\phi} \right).\end{aligned}$$

Dual to the construction of holonomic lifts is the specification of contact cotangent vectors; these are contained in the kernels of prolongations.

Definition 6.3.4 An element $(\eta, j^{k+1}_p\phi) \in \pi^*_{k+1,k}(T^*J^k\pi)$ is called a *contact cotangent vector* if $(j^k\phi)^*(\eta) = 0$. ∎

As before, this definition does not depend on the particular choice of the local section ϕ, because $(j^k\phi)^*$ depends only on the derivatives of ϕ of order up to $(k+1)$, and so is completely determined by $j^{k+1}_p\phi$.

Proposition 6.3.5 *Let (E, π, M) be a bundle, and let $j^{k+1}_p\phi \in J^{k+1}\pi$. Then*

$$\pi^*_{k+1,k}(\pi^*_k(T^*M))_{j^{k+1}_p\phi} = \left(\pi^*_{k+1,k}(V\pi_k)_{j^{k+1}_p\phi}\right)^\circ$$

and

$$\pi^*_{k+1,k}(\ker (j^k\phi)^*) = \left((j^k\phi)_*(T_pM)\right)^\circ,$$

where $\pi^*_k(T^*M)$ is regarded as a submanifold of $T^*J^k\pi$, and $\pi^*_{k+1,k}(\ker (j^k\phi)^*)$ denotes the set of contact cotangent vectors in $\pi^*_{k+1,k}(T^*J^k\pi)_{j^{k+1}_p\phi}$. ∎

Theorem 6.3.6 *Let (E, π, M) be a bundle, and let $j^{k+1}_p\phi \in J^{k+1}\pi$. There is then a canonical decomposition of the vector space $\pi^*_{k+1,k}(T^*J^k\pi)_{j^{k+1}_p\phi}$ as a direct sum*

$$\pi^*_{k+1,k}(\pi^*_k(T^*M))_{j^{k+1}_p\phi} \oplus \pi^*_{k+1,k}(\ker (j^k\phi)^*).$$

∎

Corollary 6.3.7 *The vector bundle $(\pi^*_{k+1,k}(T^*J^k\pi), \pi^*_{k+1,k}(\tau^*_{J^k\pi}), J^{k+1}\pi)$ may be written as the direct sum of two sub-bundles*

$$(\pi^*_{k+1,k}(\pi^*_k(T^*M)) \oplus C^*\pi_{k+1,k}, \pi^*_{k+1,k}(\tau^*_{J^k\pi}), J^{k+1}\pi),$$

where $C^*\pi_{k+1,k}$ is the union of the fibres $\pi^*_{k+1,k}(\ker (j^k\phi)^*)$ for $p \in M$. ∎

In coordinates, if

$$\eta = \sum_{|I|=0}^{k} \eta_\alpha^I \, du_I^\alpha|_{j_p^{k+1}\phi} + \eta_i \, dx^i\Big|_{j_p^{k+1}\phi} \in \pi_{k+1,k}^*(\ker(j^k\phi)^*),$$

then

$$\sum_{|I|=0}^{k} \eta_\alpha^I \, d\phi_I^\alpha|_{j_p^{k+1}\phi} + \eta_i \, dx^i\Big|_{j_p^{k+1}\phi} = 0,$$

so that

$$\sum_{|I|=0}^{k} \eta_\alpha^I \left.\frac{\partial \phi_I^\alpha}{\partial x^i}\right|_{j_p^k\phi} + \eta_i = 0$$

for each index i. Consequently

$$\eta = \sum_{|I|=0}^{k} \eta_\alpha^I (du_I^\alpha - u_{I+1_i}^\alpha dx^i)_{j_p^{k+1}\phi}.$$

We shall, of course, be interested in sections of these bundles of tangent and cotangent vectors. Our notation for the sections will be a straightforward generalisation of the notation of Chapter 4.

Definition 6.3.8 The submodule of $\mathcal{X}(\pi_{k+1,k})$ corresponding to sections of $\pi_{k+1,k}^*(\tau_{J^k\pi})\Big|_{\pi_{k+1,k}^*(V\pi_k)}$ will be denoted by $\mathcal{X}^v(\pi_{k+1,k})$, and the submodule corresponding to sections of $\pi_{k+1,k}^*(\tau_{J^k\pi})\Big|_{H\pi_{k+1,k}}$ will be denoted by $\mathcal{X}^h(\pi_{k+1,k})$. An element of the submodule $\mathcal{X}^h(\pi_{k+1,k})$ will be called a *total derivative*. ∎

It follows from Corollary 6.3.3 that we may write

$$\mathcal{X}(\pi_{k+1,k}) = \mathcal{X}^v(\pi_{k+1,k}) \oplus \mathcal{X}^h(\pi_{k+1,k}).$$

Definition 6.3.9 Each vector field $X \in \mathcal{X}(M)$ corresponds to a total derivative, its *k-th holonomic lift* $X^k \in \mathcal{X}^h(\pi_{k+1,k})$, according to the rule

$$X^k_{j_p^{k+1}\phi} = (j^k\phi)_*(X_p).$$

∎

We should expect different holonomic lifts X^k, X^l to be π-related in the sense of Definition 3.4.11, and this is indeed the case: if $k > l$ then

$$\begin{aligned}\pi_{k,l*}\left(X^k_{j_p^{k+1}\phi}\right) &= \pi_{k,l*}((j^k\phi)_*(X_p)) \\ &= (j^l\phi)_*(X_p) \\ &= X^l_{j_p^{l+1}\phi} \\ &= X^l_{\pi_{k+1,l+1}(j_p^{k+1}\phi)}.\end{aligned}$$

Alternatively, we may consider the coordinate representations of X^k and X^l. To do this, we shall obtain a characterisation of X^k as a derivation, by taking $f \in C^\infty(J^k\pi)$ and unwinding the definitions:

$$\begin{aligned}(d_{X^k}f)(j_p^{k+1}\phi) &= X^k_{j_p^{k+1}\phi}(f) \\ &= (j^k\phi)_*(X_p(f)) \\ &= X_p(f \circ j^k\phi) \\ &= d_X(f \circ j^k\phi)(p).\end{aligned}$$

Another way of writing this is

$$(d_{X^k}f)(j^{k+1}\phi(p)) = d_X(f \circ j^k\phi)(p),$$

which gives

$$(j^{k+1}\phi)^*(d_{X^k}(f)) = d_X((j^k\phi)^*(f)),$$

so that

$$(j^{k+1}\phi)^* \circ d_{X^k} = d_X \circ (j^k\phi)^*$$

for every $\phi \in \Gamma_{loc}(\pi)$.

In coordinates, if $X = X^i\partial/\partial x^i$ then the coordinate expression of its k-th holonomic lift is

$$X^k = X^i\left(\frac{\partial}{\partial x^i} + \sum_{|I|=0}^{k} u^\alpha_{I+1_i}\frac{\partial}{\partial u^\alpha_I}\right).$$

(Of course, X^k on the left of this equation is a vector field along $\pi_{k+1,k}$, whereas X^i on the right is a function lifted from M to $J^{k+1}\pi$, but it will be clear from the context which type of object is intended.) The *coordinate total derivatives* are the holonomic lifts of the local vector fields $\partial/\partial x^i$; their coordinate representations are

$$\frac{\partial}{\partial x^i} + \sum_{|I|=0}^{k} u^\alpha_{I+1_i}\frac{\partial}{\partial u^\alpha_I},$$

and they are normally written as d/dx^i without any specific indication of the degree of holonomic lift involved. As derivations of type d_*, we see immediately that, when acting on the coordinate functions u_I^α on $J^k\pi$,

$$\frac{du_I^\alpha}{dx^i} = u_{I+1_i}^\alpha$$

as functions on $J^{k+1}\pi$.

Example 6.3.10 If π is the trivial bundle $(\mathbf{R} \times F, pr_1, \mathbf{R})$ with coordinates (t, q^α), then the coordinate representation of the total time derivative $d/dt \in \mathcal{X}(\pi_{k+1,k})$ is

$$\frac{d}{dt} = \frac{\partial}{\partial t} + \sum_{r=0}^{k} q_{(r+1)}^\alpha \frac{\partial}{\partial q_{(r)}^\alpha}.$$

If we use the identification $J^k\pi \cong \mathbf{R} \times T^kF$, and we denote by $\tau_F^{k+1,k}$ the unique map satisfying

$$\begin{array}{ccc} \mathbf{R} \times T^{k+1}F & \xrightarrow{\;pr_2\;} & T^{k+1}F \\ \Big\downarrow{\scriptstyle \pi_{k+1,k}} & & \Big\downarrow{\scriptstyle \tau_F^{k+1,k}} \\ \mathbf{R} \times T^kF & \xrightarrow[\;pr_2\;]{} & T^kF \end{array}$$

then the total time derivative induces an operator $T \in \mathcal{X}(\tau_F^{k+1,k})$ which is also called a total time derivative operator, and which satisfies

$$T_{pr_2(j_p^{k+1}\phi)} = (pr_2)_* \left(\left. \frac{d}{dt} \right|_{j_p^{k+1}\phi} \right).$$

In coordinates,

$$T = \sum_{r=0}^{k} q_{(r+1)}^\alpha \frac{\partial}{\partial q_{(r)}^\alpha}.$$

The derivation of type d_* corresponding to T will be denoted d_T. ∎

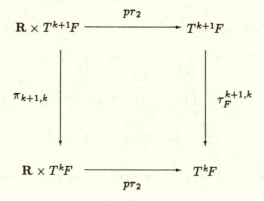

With the introduction of holonomic lifts of different orders, we can see how the Lie algebra structure of the vector fields on M is reflected in their lifts. Of course a holonomic lift is a vector field along $\pi_{k+1,k}$ (or $\pi_{k+2,k+1}$), and so the natural bracket operation is the Frölicher-Nijenhuis bracket. The result is then a vector field along $\pi_{k+2,k}$, and so is not in itself a holonomic lift: the difference, however, simply involves the jet projection $\pi_{k+2,k+1}$. To see this, we shall first prove a lemma about the characterisation of vector fields along $\pi_{k,l}$.

Lemma 6.3.11 *Suppose $X, Y \in \mathcal{X}(\pi_{k,l})$. If, for every $\phi \in \Gamma_{loc}(\pi)$,*

$$(j^k\phi)^* \circ d_X = (j^k\phi)^* \circ d_Y,$$

then $d_X = d_Y$ (so that $X = Y$).

Proof This follows directly from the definitions. For every $f \in C^\infty(J^l\pi)$,

$$(j^k\phi)^*(d_X(f)) = (j^k\phi)^*(d_Y(f)) \in C^\infty(M),$$

giving, for every $p \in M$,

$$((j^k\phi)^*(d_X(f)))(p) = ((j^k\phi)^*(d_Y(f)))(p),$$

so that

$$d_X f(j^k_p\phi) = d_Y f(j^k_p\phi).$$

Since this is true for every $j^k_p\phi \in J^k\pi$, it follows that $d_X f = d_Y f$, so that $d_X = d_Y$ and hence $X = Y$. ∎

Proposition 6.3.12 *If X, Y are vector fields on M, then*

$$\pi^*_{k+2,k+1} \circ d_{[X,Y]^k} = d_{[X^{k+1},Y^{k+1}]}.$$

Proof The operator $d_{[X^{k+1},Y^{k+1}]} = d_{X^{k+1}} \circ d_{Y^k} - d_{Y^{k+1}} \circ d_{X^k}$ represents the Lie derivative action of a vector field along $\pi_{k+2,k}$. It satisfies, for any $\phi \in \Gamma_{loc}(\pi)$,

$$\begin{aligned}
(j^{k+2}\phi)^* \circ (d_{X^{k+1}} \circ d_{Y^k} - d_{Y^{k+1}} \circ d_{X^k}) &= (d_X \circ d_Y - d_Y \circ d_X) \circ (j^k\phi)^* \\
&= d_{[X,Y]} \circ (j^k\phi)^* \\
&= (j^{k+1}\phi)^* \circ d_{[X,Y]^k} \\
&= (j^{k+2}\phi)^* \circ \pi^*_{k+2,k+1} \circ d_{[X,Y]^k},
\end{aligned}$$

using the properties of holonomic lifts and prolongations. The result then follows from Lemma 6.3.11. ∎

The corresponding decomposition of the module $\bigwedge_0^1 \pi_{k+1,k}$ of differential forms requires slightly less new notation, because the submodule corresponding to sections of $\pi_{k+1,k}^*(\tau_{J^k\pi}^*)\Big|_{\pi_{k+1,k}^*(\pi_k^*(T^*M))}$ may be identified with $\bigwedge_0^1 \pi_{k+1}$.

Definition 6.3.13 The submodule of $\bigwedge_0^1 \pi_{k+1,k}$ corresponding to sections of $\pi_{k+1,k}^*(\tau_{J^k\pi}^*)\Big|_{C^*\pi_{k+1,k}}$ will be denoted by $\bigwedge_C^1 \pi_{k+1,k}$, and its elements will be called *contact forms.* ∎

We may therefore write

$$\bigwedge\nolimits_0^1 \pi_{k+1,k} = \bigwedge\nolimits_0^1 \pi_{k+1} \oplus \bigwedge\nolimits_C^1 \pi_{k+1,k};$$

it is clear that $\bigwedge_C^1 \pi_{k+1,k}$ and $\mathcal{X}^h(\pi_{k+1,k})$ annihilate each other, as do $\bigwedge_0^1 \pi_{k+1}$ and $\mathcal{X}^v(\pi_{k+1,k})$. In coordinates, a contact form may be written as

$$\sigma = \sum_{|I|=0}^{k} \sigma_\alpha^I (du_I^\alpha - u_{I+1_i}^\alpha dx^i).$$

As with contact forms on $J^1\pi$, the contact forms on $J^{k+1}\pi$ may be characterised (among all the 1-forms on $J^{k+1}\pi$) as those which are pulled back to the zero form on M by prolongations.

Theorem 6.3.14 *If $\sigma \in \bigwedge^1 J^{k+1}\pi$ then $\sigma \in \bigwedge_C^1 \pi_{k+1,k}$ if, and only if, for every open submanifold $W \subset M$ and every $\phi \in \Gamma_W(\pi)$,*

$$(j^{k+1}\phi)^*(\sigma|_W) = 0.$$

∎

It is often convenient to encapsulate information about the decomposition of the bundle $\pi_{k+1,k}^*(\tau_{J^k\pi})$ in pairs of vector bundle endomorphisms or in a pair of vector-valued 1-forms, and this may be done in just the same way as in Section 4.5.

Definition 6.3.15 The two vector bundle endomorphisms $(h, id_{J^k\pi})$ and $(v, id_{J^k\pi})$ of $\pi_{k+1,k}^*(\tau_{J^k\pi})$ are defined by

$$\begin{aligned} h(\xi^h + \xi^v) &= \xi^h \\ v(\xi^h + \xi^v) &= \xi^v, \end{aligned}$$

where $\xi^h \in H\pi_{k+1,k}$ and $\xi^v \in \pi_{k+1,k}^*(V\pi_k)$. ∎

Clearly $h + v = \pi_{k+1,k*}$; we shall not normally indicate the particular map along which h or v is defined, because if $\xi \in \pi^*_{k+1,k}(TJ^k\pi)_{j^{k+1}_p\phi}$ and if $0 \leq l < k$, then

$$\pi_{k,l*}(h_{j^{k+1}_p\phi}(\xi)) = h_{j^{l+1}_p\phi}(\pi_{k+1,l+1*}(\xi)),$$

and similarly for v.

Definition 6.3.16 The two vector bundle endomorphisms $(h, id_{J^k\pi})$ and $(v, id_{J^k\pi})$ of $\pi^*_{k+1,k}(\tau^*_{J^k\pi})$ are defined by

$$\begin{aligned} h(\eta^h + \eta^v) &= \eta^h \\ v(\eta^h + \eta^v) &= \eta^v, \end{aligned}$$

where $\eta^h \in \pi^*_{k+1,k}(\pi^*_k(T^*M))$ and $\eta^v \in C^*\pi_{k+1,k}$. ■

Definition 6.3.17 The two vector-valued 1-forms h, v are the sections of the bundle $\pi^*_{k+1,k}(\tau^*_{J^k\pi}) \otimes \pi^*_{k+1,k}(\tau_{J^k\pi})$ defined by

$$\begin{aligned} h_{j^{k+1}_p\phi}(\xi, \eta) &= \eta(h(\xi)) \\ v_{j^{k+1}_p\phi}(\xi, \eta) &= \eta(v(\xi)), \end{aligned}$$

where $\xi \in \pi^*_{k+1,k}(TJ^k\pi)_{j^{k+1}_p\phi}$ and $\eta \in \pi^*_{k+1,k}(T^*J^k\pi)_{j^{k+1}_p\phi}$. ■

If we consider $\pi^*_{k+1,k}(\tau^*_{J^k\pi}) \otimes \pi^*_{k+1,k}(\tau_{J^k\pi})$ to be a sub-bundle of $\tau^*_{J^{k+1}\pi} \otimes \pi^*_{k+1,k}(\tau_{J^k\pi})$, we may regard h and v as vector-valued 1-forms along $\pi_{k+1,k}$ in the sense of Section 3.3; in coordinates, these vector-valued forms may be written as

$$\begin{aligned} h &= dx^i \otimes \frac{d}{dx^i} \\ v &= \sum_{|I|=0}^{k} (du^\alpha_I - u^\alpha_{I+1_i} dx^i) \otimes \frac{\partial}{\partial u^\alpha_I}. \end{aligned}$$

As in Section 4.5, we could now go on to define the Cartan distribution on $J^{k+1}\pi$, and use it to characterise prolongations. Rather than do this, however, we shall investigate some further properties of h and v which take account of the fact that these symbols really represent two families of π-related vector-valued forms. These properties will involve the corresponding derivations of type i_* and d_*, whose actions on the coordinate functions and coordinate 1-forms may be summarised as follows:

$$\begin{aligned} i_h(dx^i) &= d_h x^i &&= dx^i \\ i_h(du^\alpha_I) &= d_h u^\alpha_I &&= u^\alpha_{I+1_i} dx^i \\ & \quad\; d_h dx^i &&= 0 \\ & \quad\; d_h du^\alpha_I &&= dx^i \wedge du^\alpha_{I+1_i} \end{aligned}$$

$$\begin{aligned} i_v(dx^i) &= d_v x^i &&= 0 \\ i_v(du^\alpha_I) &= d_v u^\alpha_I &&= du^\alpha_I - u^\alpha_{I+1_i} dx^i \\ & \quad\; d_v dx^i &&= 0 \\ & \quad\; d_v du^\alpha_I &&= du^\alpha_{I+1_i} \wedge dx^i . \end{aligned}$$

We shall call d_h and d_v the horizontal and vertical differentials. Notice that, whereas i_h and i_v map $\bigwedge_0^r \pi_{k+1,k}$ to itself, d_h and d_v map $\bigwedge_0^r \pi_{k+1,k}$ to $\bigwedge_0^{r+1} \pi_{k+2,k+1}$.

One consequence of the relationship $h + v = \pi_{k+1,k*}$ is that $d_h + d_v = \pi_{k+1,k*} \circ d$; this yields the following lemma, which shows what happens when the exterior derivative d is taken to the other side of a prolongation.

Lemma 6.3.18 *For every $\phi \in \Gamma_{loc}(\pi)$, $d \circ (j^k\phi)^* = (j^{k+1}\phi)^* \circ d_h$.*

Proof The exterior derivative d commutes with pull-backs, so

$$\begin{aligned} d \circ (j^k\phi)^* &= (j^k\phi)^* \circ d \\ &= (j^{k+1}\phi)^* \circ \pi^*_{k+1,k} \circ d \\ &= (j^{k+1}\phi)^* \circ (d_h + d_v). \end{aligned}$$

But for any 1-form σ, $v \lrcorner \sigma$ is a contact form, so that $(j^{k+1}\phi)^*(v \lrcorner \sigma) = 0$; therefore $(j^{k+1}\phi)^* \circ i_v = 0$, and so $(j^{k+1}\phi)^* \circ d_v = 0$. ∎

Another important property of the horizontal and vertical differentials is the following.

Lemma 6.3.19 *The horizontal and vertical differentials satisfy $d_h^2 = d_v^2 = 0$.*

Proof The derivations d_h^2, d_v^2 are of type d_* and degree 2 along $\pi_{k+2,k}$. We shall show, using coordinates, that they both vanish on $C^\infty(J^k\pi)$. For d_h^2,

$$\begin{aligned} d_h^2 f &= d_h \left(\frac{df}{dx^i} dx^i \right) \\ &= \frac{d^2 f}{dx^j\, dx^i} dx^j \wedge dx^i \\ &= 0, \end{aligned}$$

so that $d_h^2 = 0$. For d_v^2, we shall first consider $d_v^2 u_I^\alpha$:

$$\begin{aligned} d_v^2 u_I^\alpha &= d_v(du_I^\alpha - u_{I+1_i}^\alpha dx^i) \\ &= du_{I+1_i}^\alpha \wedge dx^i - (du_{I+1_i}^\alpha - u_{I+1_i+1_j}^\alpha dx^j) \wedge dx^i \\ &= 0. \end{aligned}$$

We then see that, in general,

$$\begin{aligned} d_v^2 f &= d_v\left(\sum_{|I|=0}^{k} \frac{\partial f}{\partial u_I^\alpha}(du_I^\alpha - u_{I+1_i}^\alpha dx^i)\right) \\ &= \sum_{|I|=0}^{k} \sum_{|J|=0}^{k} \frac{\partial^2 f}{\partial u_J^\beta \partial u_I^\alpha}(du_J^\beta - u_{J+1_j}^\beta dx^j) \wedge (du_I^\alpha - u_{I+1_i}^\alpha dx^i) \\ &= 0. \end{aligned}$$

■

Corollary 6.3.20 *The horizontal and vertical differentials satisfy*

$$d_h \circ d_v + d_v \circ d_h = 0.$$

Proof This follows immediately from $(d_h + d_v)^2 = \pi_{k+2,k}^* \circ d^2$. ■

These results show that, if we consider the spaces $\bigwedge_0^r \pi_{k+1,k}$ of r-forms on $J^{k+1}\pi$ which are totally horizontal over $J^k\pi$, then d_h and d_v may be considered as coboundary operators. To show how they fit together, we shall define a canonical splitting of $\bigwedge_0^r \pi_{k+1,k}$ which generalises the splitting $\bigwedge_0^1 \pi_{k+1,k} = \bigwedge_0^1 \pi_{k+1} \oplus \bigwedge_C^1 \pi_{k+1,k}$. To do this, we shall denote the p-fold composite $i_h \circ \ldots \circ i_h$ by i_h^p, where $0 \leq p \leq r$. We shall also let A denote the $(r+1) \times (r+1)$ matrix

$$\begin{pmatrix} 1 & \ldots & 1 & 1 & 1 \\ r & \ldots & 2 & 1 & 0 \\ r^2 & \ldots & 4 & 1 & 0 \\ \vdots & & \vdots & \vdots & \vdots \\ r^r & \ldots & 2^r & 1 & 0 \end{pmatrix},$$

so that $A_{pq} = (r-q)^p$ for $0 \leq p, q \leq r$, and we shall put $B = A^{-1}$.

Definition 6.3.21 The map $\Phi_s : \bigwedge_0^r \pi_{k+1,k} \longrightarrow \bigwedge_0^r \pi_{k+1,k}$ is defined by

$$\Phi_s(\theta) = \sum_{p=0}^{r} B_{ps} i_h^p(\theta),$$

and the image $\Phi_s(\bigwedge_0^r \pi_{k+1,k})$ is denoted by $\Phi_s^{r-s}(\pi_{k+1})$. ■

It is immediate from the construction that Φ_s is a homomorphism of $C^\infty(J^{k+1}\pi)$ modules, and therefore that $\Phi_s^{r-s}(\pi_{k+1})$ is a module. Since

$$\sum_{s=0}^{r} A_{sq}\Phi_s(\theta) = i_h^q(\theta),$$

it follows that

$$\theta = \sum_{s=0}^{r} A_{s0}\Phi_s(\theta) = \sum_{s=0}^{r} \Phi_s(\theta).$$

It is also true that

$$\Phi_{s_1}^{r-s_1}(\pi_{k+1}) \cap \Phi_{s_2}^{r-s_2}(\pi_{k+1}) = 0$$

for $s_1 \neq s_2$, and we may see this by using coordinates. Suppose first that, locally,

$$\theta = (du_{I_1}^{\alpha_1} - u_{I_1+1_j}^{\alpha_1}dx^j) \wedge \ldots \wedge (du_{I_q}^{\alpha_q} - u_{I_q+1_j}^{\alpha_q}dx^j) \wedge dx^{i_{q+1}} \wedge \ldots \wedge dx^{i_r},$$

so that θ has q contact factors and $(r-q)$ factors which are horizontal over M. Then $i_h^p(\theta) = (r-q)^p\theta$, so that

$$\begin{aligned} \Phi_s(\theta) &= \sum_{p=0}^{r} B_{ps}(r-q)^p\theta \\ &= \delta_{sq}\theta, \end{aligned}$$

because the numbers $(r-q)^p$ for $0 \leq p \leq r$ constitute the q-th column of the matrix A. Now any element of $\bigwedge_0^r \pi_{k+1,k}$ may be written locally as a unique combination of elements of the form of θ, so it follows that Φ_s picks out those terms with s contact factors and r horizontal factors. We have therefore obtained the following result.

Proposition 6.3.22 $\bigwedge_0^r \pi_{k+1,k} = \bigoplus_{s=0}^r \Phi_s^{r-s}(\pi_{k+1})$. ∎

As a special case of this construction, note that when $r = 0$ we have

$$\Phi_0^0(\pi_{k+1}) = \bigwedge^0 J^{k+1}\pi,$$

and when $r = 1$ we have

$$\Phi_0^1(\pi_{k+1}) = \bigwedge_0^1 \pi_{k+1},$$

$$\Phi_1^0(\pi_{k+1}) = \bigwedge_C^1 \pi_{k+1,k}.$$

We may now construct the following large commutative diagram, and this will be the framework for the "variational bicomplex" to be introduced in Chapter 7.

$$
\begin{array}{ccccccc}
\Phi_0^0(\pi_k) & \xrightarrow{d_v} & \Phi_1^0(\pi_{k+1}) & \cdots & \Phi_s^0(\pi_{k+s}) & \xrightarrow{d_v} & \Phi_{s+1}^0(\pi_{k+s+1}) \\
\downarrow{\scriptstyle d_h} & & \downarrow{\scriptstyle -d_h} & & \downarrow{\scriptstyle (-1)^s d_h} & & \downarrow{\scriptstyle (-1)^{s+1} d_h} \\
\Phi_0^1(\pi_{k+1}) & \xrightarrow{d_v} & \Phi_1^1(\pi_{k+2}) & \cdots & \Phi_s^1(\pi_{k+s+1}) & \xrightarrow{d_v} & \Phi_{s+1}^1(\pi_{k+s+2}) \\
\downarrow{\scriptstyle -d_h} & & \downarrow{\scriptstyle d_h} & & \downarrow{\scriptstyle (-1)^{s+1} d_h} & & \downarrow{\scriptstyle (-1)^s d_h} \\
\Phi_0^2(\pi_{k+2}) & \xrightarrow{d_v} & \Phi_1^2(\pi_{k+3}) & \cdots & \Phi_s^2(\pi_{k+s+2}) & \xrightarrow{d_v} & \Phi_{s+1}^2(\pi_{k+s+3}) \\
\vdots & & \vdots & & \vdots & & \vdots \\
\Phi_0^{m-1}(\pi_{k+m-1}) & \xrightarrow{d_v} & \Phi_1^{m-1}(\pi_{k+m}) & \cdots & \Phi_s^{m-1}(\pi_{k+m+s-1}) & \xrightarrow{d_v} & \Phi_{s+1}^{m-1}(\pi_{k+m+s}) \\
\downarrow{\scriptstyle (-1)^{m-1} d_h} & & \downarrow{\scriptstyle (-1)^m d_h} & & \downarrow{\scriptstyle (-1)^{m+s-1} d_h} & & \downarrow{\scriptstyle (-1)^{m+s} d_h} \\
\Phi_0^m(\pi_{k+m}) & \xrightarrow{d_v} & \Phi_1^m(\pi_{k+m+1}) & \cdots & \Phi_s^m(\pi_{k+m+s}) & \xrightarrow{d_v} & \Phi_{s+1}^m(\pi_{k+m+s+1})
\end{array}
$$

We may choose any $k \in \mathbf{Z}$ when constructing this diagram, but if $k < 0$ then only part of the diagram exists, namely the part containing the spaces $\Phi_s^r(\pi_{k+r+s})$ where $k + r + s \geq 0$.

We shall call an element of $\Phi_s^{r-s}(\pi_{k+1})$ an *s-contact r-form on $J^{k+1}\pi$*. A word of caution is necessary here: when $r \leq \dim M$, it would seem natural to try to characterise a contact r-form on $J^{k+1}\pi$ as an r-form $\theta \in \bigwedge^r J^{k+1}\pi$ satisfying $(j^{k+1}\phi)^*\theta = 0$ for every $\phi \in \Gamma_{loc}(\pi)$, by analogy with Theorem 4.3.14. When $r = 1$ this is just a contact form as previously defined, and when $\dim M = 1$ it is indeed a sum of s-contact r-forms for $1 \leq s \leq r$. However, when $\dim M \geq 2$ and $r \geq 2$ there are contact r-forms which cannot be written as sums of elements of $\Phi_s^{r-s}(\pi_{k+1})$, because they are not even elements of $\bigwedge_0^r \pi_{k+1,k}$.

Example 6.3.23 Let π be the trivial bundle $(\mathbf{R}^2 \times \mathbf{R}, pr_1, \mathbf{R})$, with coordinates $(x^1, x^2; u)$. Then the 2-form $\theta = d_v du = du_1 \wedge dx^1 + du_2 \wedge dx^2 \in \bigwedge^2 J^1\pi$

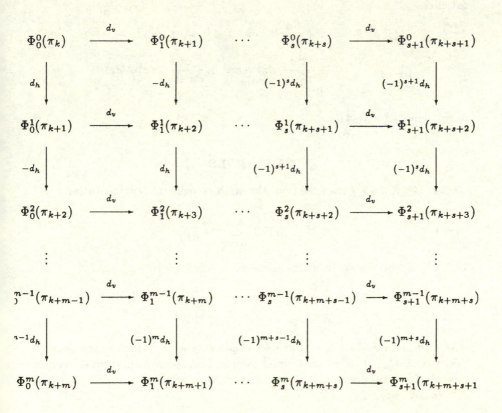

satisfies

$$\begin{aligned}(j^1\phi)^*\theta &= d\left(\frac{\partial\phi}{\partial x^1}\right)\wedge dx^1 + d\left(\frac{\partial\phi}{\partial x^2}\right)\wedge dx^2 \\ &= \frac{\partial^2\phi}{\partial x^2\,\partial x^1}dx^2\wedge dx^1 + \frac{\partial^2\phi}{\partial x^1\,\partial x^2}dx^1\wedge dx^2 \\ &= 0;\end{aligned}$$

nevertheless, $\theta \notin \bigwedge_0^2\pi_{1,0}$. ∎

EXERCISES

6.3.1 Let X be a vector field on $J^k\pi$ with coordinate representation

$$X = X^i\frac{\partial}{\partial x^i} + \sum_{|I|=0}^{k} X_I^\alpha\frac{\partial}{\partial u_I^\alpha}.$$

Let X^h, X^v be the vector fields along $\pi_{k+1,k}$ defined by

$$\begin{aligned}X^h_{j_p^{k+1}\phi} &= (X_{j_p^k\phi}, j_p^{k+1}\phi)^h \\ X^v_{j_p^{k+1}\phi} &= (X_{j_p^k\phi}, j_p^{k+1}\phi)^v,\end{aligned}$$

where the superscripts h, v indicate components with respect to the decomposition of $\pi_{k+1,k}^*(TJ^k\pi)$ as a direct sum of holonomic and vertical vectors described in Corollary 6.3.3. Confirm that the coordinate representations of X^h and X^v are

$$\begin{aligned}X^h &= X^i\frac{d}{dx^i}, \\ X^v &= \sum_{|I|=0}^{k}(X_I^\alpha - X^i u_{I+1_i}^\alpha)\frac{\partial}{\partial u_I^\alpha}.\end{aligned}$$

6.3.2 Confirm the validity of Proposition 6.3.5 by an argument using coordinates.

6.3.3 Supply a proof of Theorem 6.3.14 by adapting the coordinate proof of Theorem 4.3.14 to use multi-index notation.

6.3.4 If $f, g \in C^\infty(J^k\pi)$ and I is a multi-index, show that the repeated total derivative of a product satisfies Leibniz' rule in the following form:

$$\begin{aligned}\frac{d^{|I|}(fg)}{dx^I} &= \sum_{J+K=I}\frac{I!}{J!\,K!}\left(\pi^*_{k+|I|,k+|J|}\frac{d^{|J|}f}{dx^J}\right)\left(\pi^*_{k+|I|,k+|K|}\frac{d^{|K|}g}{dx^K}\right) \\ &\in C^\infty(J^{k+|I|}\pi).\end{aligned}$$

(We normally omit the pullback maps preceding the various derivatives in the above formula and simply write it as

$$\frac{d^{|I|}(fg)}{dx^I} = \sum_{J+K=I} \frac{I!}{J!\,K!} \frac{d^{|J|}f}{dx^J} \frac{d^{|K|}g}{dx^K}.\Bigg)$$

6.3.5 Show by induction that, if $f \in C^\infty(J^k\pi)$ and $\phi \in \Gamma_W(\pi)$, then

$$\left.\frac{d^{|I|}f}{dx^I}\right|_{j_p^{k+|I|}\phi} = \left.\frac{\partial^{|I|}(f \circ j^k\phi)}{\partial x^I}\right|_p .$$

6.3.6 If R, R' are π-related vector-valued r-forms along the projections $\pi_{k+m,k}$ and $\pi_{k+m+1,k+1}$ respectively, define the *derivations along* $\pi_{k+m+1,k}$ *of type* h_* *and* v_* *determined by* R to be

$$\begin{aligned} h_R &= i_R \circ d_h + (-1)^r d_h \circ i_{R'} \\ v_R &= i_R \circ d_v + (-1)^r d_v \circ i_{R'}, \end{aligned}$$

by analogy with the definition of derivations of type d_*. Show that

$$v_R = d_{R \,\lrcorner\, v} + i_{[v,R]}$$

and

$$h_R = d_{R \,\lrcorner\, h} + i_{[h,R]}.$$

6.3.7 Suppose that R^k is a family of π-related vector-valued r-forms along $\pi_{k+m,k}$, and that S^k is a family of π-related vector-valued s-forms along $\pi_{k+l,k}$. Show that the operator

$$v_{R^{k+l+1}} \circ v_{S^k} - (-1)^{rs} v_{S^{k+m+1}} \circ v_{R^k}$$

is a derivation along $\pi_{k+l+m+2,k}$ of type v_*. (The corresponding vector-valued $(r+s)$-form along $\pi_{k+l+m+1,k}$ may be denoted $[R,S]^v$, and called the *vertical bracket* of R and S.

6.4 Vector Fields and their Prolongations

In Section 4.4, we saw how a vector field X on E could be prolonged to a vector field on $J^1\pi$. For a vertical vector field, this was comparatively straightforward: in coordinates, the coefficients of the basis fields in the derivative variables were obtained by differentiating the coefficients of the basis fields in the independent variables. Where the vector field was not vertical, the result was rather more complicated, although if the vector field happened to be projectable then its flow could be prolonged to give the flow of the prolonged field. In both cases, however, we started by describing a suitable bundle morphism which we then composed with j^1X to give the required section of $\tau_{J^1\pi}$.

All this generalises easily to give higher-order prolongations of a vector field on E. We shall, however, choose to extend our generalisation in a slightly different direction, by starting with vector fields along $\pi_{k,0}$ rather than on E. These will be called *generalised vector fields*, and the reason for including them in our discussion is that they provide yet another means of representing certain types of differential equation.

Definition 6.4.1 A *generalised vector field* is a section X of the pull-back bundle $(\pi^*_{k,0}(TE), \pi^*_{k,0}(\tau_E), J^k\pi)$; X is a *vertical generalised vector field* if it is also a section of the sub-bundle $(\pi^*_{k,0}(V\pi), \pi^*_{k,0}(\tau_E|_{V\pi}), J^k\pi)$. ∎

We shall, as usual, regard generalised vector fields as maps $J^k\pi \longrightarrow TE$; in coordinates, we have

$$X = X^i \frac{\partial}{\partial x^i} + X^\alpha \frac{\partial}{\partial u^\alpha},$$

where the functions X^i, X^α are defined locally on $J^k\pi$; more explicitly, we have

$$X_{j^k\phi} = X^i(j^k\phi)\left.\frac{\partial}{\partial x^i}\right|_{\phi(p)} + X^\alpha(j^k\phi)\left.\frac{\partial}{\partial u^\alpha}\right|_{\phi(p)}.$$

If X is vertical (and so may be regarded as a map $J^k\pi \longrightarrow V\pi$) then its coordinate expression is just

$$X = X^\alpha \frac{\partial}{\partial u^\alpha}.$$

We shall also regard vector fields on E as generalised vector fields, using the identification $J^0\pi \cong E$.

Example 6.4.2 The zeroth-order holonomic lift of a vector field on M is a generalised vector field; in coordinates,

$$X = X^i \frac{\partial}{\partial x^i} + X^i u^\alpha_i \frac{\partial}{\partial u^\alpha}.$$

Higher-order holonomic lifts are *not* generalised vector fields. ∎

Any generalised vector field $X : J^k\pi \longrightarrow TE$ gives rise to a family of generalised vector fields defined on higher-order jet manifolds, namely $X \circ \pi_{m,k} : J^m\pi \longrightarrow TE$. Equally, X itself may have arisen by the same process from a generalised vector field on a lower-order jet manifold. We may use this idea to define the *order* of X.

Definition 6.4.3 If $X : J^k\pi \longrightarrow TE$ is a generalised vector field, then the *order* of X is the smallest natural number l such that there is a generalised vector field $Y : J^l\pi \longrightarrow TE$ satisfying $X = Y \circ \pi_{k,l}$. ■

A vector field on E is therefore a generalised vector field of order zero, and a zeroth-order holonomic lift is a generalised vector field of order one. It will often be convenient to adopt the convention that a generalised vector field of order k is defined on $J^k\pi$ (rather than on a higher-order jet manifold).

As usual, we shall be particularly interested in vertical generalised vector fields, and in fact every generalised vector field has a vertical representative. This applies even to generalised vector fields of order zero: in this case, however, the vertical representative is a generalised vector field of order one; we have already pointed out that a connection is required to yield the vertical representative as a vector field on E.

Definition 6.4.4 If $X : J^k\pi \longrightarrow TE$ is a generalised vector field of order k, the *vertical representative of* X is the generalised vector field X^v of order $\max\{k, 1\}$ defined by

$$X^v_{j^k_p\phi} = X_{j^k_p\phi} - \phi_*(\pi_*(X_{j^k_p\phi}))$$

when $k > 0$, and by

$$X^v_{j^1_p\phi} = X_{\phi(p)} - \phi_*(\pi_*(X_{\phi(p)}))$$

when $k = 0$. ■

In coordinates, if

$$X = X^i \frac{\partial}{\partial x^i} + X^\alpha \frac{\partial}{\partial u^\alpha},$$

then

$$X^v = (X^\alpha - X^i u^\alpha_i)\frac{\partial}{\partial u^\alpha}.$$

When $k = 1$, the vertical representative of X is just the vector field along $\pi_{1,0}$ which would be obtained from the canonical decomposition of the bundle $(\pi^*_{1,0}(TE), \pi^*_{1,0}(\tau_E), J^1\pi)$ into its sub-bundles of vertical and holonomic tangent vectors.

The relationship between generalised vector fields and differential equations arises when we consider the action of a vertical generalised vector field upon local sections of the bundle π.

Definition 6.4.5 If $X : J^k\pi \longrightarrow V\pi$ is a vertical generalised vector field and if $\phi \in \Gamma_W(\pi)$, then the local section $\tilde{X}(\phi) \in \Gamma_W(\nu_\pi)$ is defined by

$$\tilde{X}(\phi) = X \circ j^k\phi.$$

■

So far, this is just a generalisation of the action of a vertical vector field on E, as described in Lemma 3.2.18. We may, however, take the idea further by considering maps $\gamma : W \times I \longrightarrow E$, where $W \subset M$ and where $I \subset \mathbf{R}$ is a non-empty open interval. If $\gamma_t : W \longrightarrow E$ is defined as usual by $\gamma_t(p) = \gamma(p,t)$, and if $\pi \circ \gamma_t = id_W$, then we may construct the local section $\tilde{X}(\gamma_t) \in \Gamma_W(\nu_\pi)$; for each $p \in W$, $\tilde{X}(\gamma_t)(p) \in V_{\gamma(p,t)}\pi$. On the other hand, we may obtain an element of $V_{\gamma(p,t)}\pi$ directly from γ by considering the tangent vector $[s \longmapsto \gamma(p, s+t)]$.

Definition 6.4.6 A *solution* of the vertical generalised vector field X is a map $\gamma : W \times I \longrightarrow E$ which satisfies $\pi \circ \gamma_t = id_W$ for each $t \in I$, and

$$\tilde{X}(\gamma_t(p)) = [s \longmapsto \gamma(p, s+t)]$$

for each $(p,t) \in W \times I$. ■

In coordinates, if $X = X^\alpha \partial/\partial u^\alpha$ then γ is a solution of X if

$$X^\alpha(j^k_p\gamma_t) \left.\frac{\partial}{\partial u^\alpha}\right|_{\gamma(p,t)} = \left.\frac{\partial}{\partial t}\right|_{t=0} (t \longmapsto \gamma^\alpha(p, s+t)) \left.\frac{\partial}{\partial u^\alpha}\right|_{\gamma(p,t)}$$

or, in more traditional language, if

$$X^\alpha \left(x^i, \gamma^\beta, \frac{\partial \gamma^\beta}{\partial x^i}, \ldots, \frac{\partial^{|I|}\gamma^\beta}{\partial x^I} \right) = \frac{\partial \gamma^\alpha}{\partial t}$$

for $1 \leq \alpha \leq n$: a general set of n evolution equations for the functions γ. If X has order zero, then these are ordinary differential equations whose solutions are given by the flow of X. If the order of X is greater than zero, then there is no concept of a flow unless "infinite jets" are used, and indeed there is no general existence theorem for the solutions of these partial differential equations. Of course, we may also express these equations in the language of Definition 6.2.20; to do this, we would need to consider the bundle $(E \times \mathbf{R}, \pi \times id_{\mathbf{R}}, M \times \mathbf{R})$ in order to include the time coordinate explicitly as an independent variable. We shall not go into the details of this relationship here.

Example 6.4.7 An evolution equation which it is often convenient to consider in this form is the Korteweg-de Vries equation, given (with a suitable choice of scaling factors) as

$$u_t = u_{xxx} + 6uu_x.$$

On the bundle $(\mathbf{R} \times \mathbf{R}, pr_1, \mathbf{R})$ with coordinates (x, u), the corresponding vertical generalised vector field is just

$$(u_{(3)} + 6uu_{(1)})\frac{\partial}{\partial u}.$$

■

We shall now consider how to prolong vector fields and generalised vector fields. This may be done by generalising Theorem 4.4.1, Definition 4.4.7 and Proposition 4.4.8.

Theorem 6.4.8 *There is a canonical diffeomorphism* $i_l : J^l\nu_\pi \longrightarrow V\pi_l$ *which projects onto the identity of* M.

Proof As with Theorem 4.4.1, this diffeomorphism may be constructed by considering maps $\gamma : W \times \mathbf{R} \longrightarrow E$ which satisfy $\pi(\gamma(p,t)) = p$. We may then define a map

$$\begin{aligned} j^l_p\gamma : \mathbf{R} &\longrightarrow J^l\pi \\ t &\longmapsto j^l_p\gamma_t, \end{aligned}$$

where $\gamma_t : W \longrightarrow E$ satisfies $\gamma t(q) = \gamma(q,t)$, and so we obtain the tangent vector $[j^l_p\gamma] \in V\pi_l$. On the other hand, we may also define a map

$$\begin{aligned} [\gamma] : W &\longrightarrow V\pi \\ q &\longmapsto [\gamma_q], \end{aligned}$$

where $\gamma_q : \mathbf{R} \longrightarrow E$ satisfies $\gamma_q(t) = \gamma(q,t)$, and so obtain the l-jet $j^l_p[\gamma] \in J^l\nu_\pi$. The map $i_l : J^l\nu_\pi \longrightarrow V\pi_l$ is then given by the correspondence $j^l_p[\gamma] \longmapsto [j^l_p\gamma]$. ■

In terms of coordinates, the diffeomorphism i_l is represented by a simple rearrangement: the coordinates on $J^l\nu_\pi$ are

$$(x^i, u^\alpha; \dot{u}^\alpha; u^\alpha_I, \dot{u}^\alpha_I) \qquad 1 \leq |I| \leq l$$

whereas those on $V\pi_l$ are

$$(x^i, u^\alpha; u^\alpha_I; \dot{u}^\alpha, \dot{u}^\alpha_I) \qquad 1 \leq |I| \leq l.$$

Definition 6.4.9 If $X : J^k\pi \longrightarrow V\pi$ is a vertical generalised vector field, then its *l-th prolongation* is the vector field X^l along $\pi_{k+l,l}$ defined by

$$X^l = i_l \circ j^l X \circ \iota_{l,k} : J^{k+l}\pi \longrightarrow V\pi_l.$$

■

As we might expect, prolongations of different orders are related by the jet projections: if $l > m$ then $\pi_{l,m*}(X^l_{j^{k+l}_p\phi}) = X^m_{j^{k+m}_p\phi}$, so that X^l is $\pi_{l,m}$-related to X^m. In coordinates, if

$$X^l = \sum_{|I|=0}^{l} X^\alpha_I \frac{\partial}{\partial u^\alpha_I},$$

then

$$\begin{aligned} X^\alpha_I(j^{k+l}_p\phi) &= \left.\frac{\partial^{|I|}(X^\alpha \circ j^k\phi)}{\partial x^I}\right|_p \\ &= \left.\frac{d^{|I|}X^\alpha}{dx^I}\right|_{j^{k+|I|}_p\phi}. \end{aligned}$$

We shall, as usual, just write

$$X^l = \sum_{|I|=0}^{l} \frac{d^{|I|}X^\alpha}{dx^I}\frac{\partial}{\partial u^\alpha_I},$$

by pulling all the coefficient functions back to $J^{k+l}\pi$.

Proposition 6.4.10 *Let $X : J^k\pi \longrightarrow V\pi$ and $Y : J^m\pi \longrightarrow V\pi$ be vertical generalised vector fields. Then the Frölicher-Nijenhuis bracket $[X^m, Y^k] : J^{k+m}\pi \longrightarrow V\pi$ is a vertical generalised vector field.*

Proof The vector fields X^m and X are π-related in the sense of Definition 3.4.11, as are Y^k and Y, so that $[X^m, Y^k]$ is defined as a vector field along $\pi_{k+m,0}$. If $f \in \pi^*(C^\infty(M))$ then $d_X f = d_Y f = 0$, so that $d_{[X^m,Y^k]}f = 0$. It follows that $[X^m, Y^k]$ is vertical over M, and therefore defines a vertical generalised vector field. ■

On the other hand, the Frölicher-Nijenhuis bracket of a vertical generalised vector field and a holonomic lift will always vanish. In the following result, note that, despite the similarity in notation, X^{l+1} is a holonomic lift, whereas Y^{k+l} is a prolongation.

Proposition 6.4.11 *If $X : J^k\pi \longrightarrow V\pi$ is a vertical generalised vector field on E, and Y is a vector field on M, then for any natural number l,*

$$[Y^{k+l}, X^{l+1}] = 0$$

Proof In local coordinates. Suppose that $X = X^\alpha \partial/\partial u^\alpha$ and $Y = Y^i \partial/\partial x^i$. Then for $f \in C^\infty(J^l\pi)$,

$$d_{Y^{k+l}} d_{X^l} f = Y^i \frac{d}{dx^i}\left(\sum_{|I|=0}^{l} \frac{d^{|I|}X^\alpha}{dx^I} \frac{\partial f}{\partial u^\alpha_I} \right),$$

so that

$$\begin{aligned} d_{Y^{k+l}} d_{X^l} x^j &= 0, \\ d_{Y^{k+l}} d_{X^l} u^\beta_J &= Y^i \frac{d^{|J|+1} X^\beta}{dx^{J+1_i}}. \end{aligned}$$

On the other hand,

$$\begin{aligned} d_{X^{l+1}} d_{Y^l} f &= \sum_{|I|=0}^{l+1} \frac{d^{|I|}X^\alpha}{dx^I} \frac{\partial}{\partial u^\alpha_I}\left(Y^i \frac{df}{dx^i} \right) \\ &= Y^i \sum_{|I|=0}^{l+1} \frac{d^{|I|}X^\alpha}{dx^I} \frac{\partial}{\partial u^\alpha_I}\left(\frac{df}{dx^i} \right), \end{aligned}$$

because the functions Y^i have been pulled back from M, so that

$$\begin{aligned} d_{X^{l+1}} d_{Y^l} x^j &= 0, \\ d_{X^{l+1}} d_{Y^l} u^\beta_J &= Y^i \frac{d^{|J|+1} X^\beta}{dx^{J+1_i}}. \end{aligned}$$

Consequently $d_{Y^{k+l}} \circ d_{X^l} = d_{X^{l+1}} \circ d_{Y^l}$. ∎

The converse of this result gives a characterisation of those vertical vector fields along $\pi_{k+l,l}$ which are l-prolongations of vertical generalised vector fields.

Proposition 6.4.12 *Suppose that the vector fields $X'' \in \mathcal{X}^v(\pi_{k+l+1,l+1})$ and $X' \in \mathcal{X}^v(\pi_{k+l,l})$ are π-related. If, for every vector field $Y \in \mathcal{X}(M)$,*

$$[Y^{k+l}, X''] = 0,$$

then there is a vertical generalised vector field $X \in \mathcal{X}(\pi_{k,0})$ such that $X' = X^l$, $X'' = X^{l+1}$.

Proof It is clear that we shall need to define $X \in \mathcal{X}(\pi_{k,0})$ by $X_{j^k_p\phi} = \pi_{l,0*}(X'_{j^{k+l}_p\phi})$; we shall use coordinates to show that $X^l = X'$ and $X^{l+1} = X''$. So take, for Y, vector fields on M which may be expressed locally as the coordinate vector fields $\partial/\partial x^i$. Suppose also that

$$X' = \sum_{|I|=0}^{l} X^\alpha_I \frac{\partial}{\partial u^\alpha_I}$$

and

$$X'' = \sum_{|I|=0}^{l+1} X^\alpha_I \frac{\partial}{\partial u^\alpha_I},$$

where the functions X^α_I are defined locally on $J^{k+l}\pi$ for $0 \leq |I| \leq l$, and are defined locally on $J^{k+l+1}\pi$ for $|I| = l+1$. Then

$$d_{Y^{k+l}} \circ d_{X'} - d_{X''} \circ d_{Y^l} = 0,$$

so that

$$\begin{aligned} \sum_{|I|=0}^{l} \frac{dX^\alpha_I}{dx^i} \frac{\partial}{\partial u^\alpha_I} &= \sum_{|I|=0}^{l+1} X^\alpha_I \sum_{|J|=0}^{l} \frac{\partial u^\beta_{J+1_i}}{\partial u^\alpha_I} \frac{\partial}{\partial u^\beta_J} \\ &= \sum_{|J|=0}^{l} X^\alpha_{J+1_i} \frac{\partial}{\partial u^\alpha_J}. \end{aligned}$$

Hence, equating coefficients,

$$X^\alpha_{I+1_i} = \frac{dX^\alpha_I}{dx^i} \qquad 0 \leq |I| \leq l,$$

so that

$$X^\alpha_I = \frac{d^{|I|}X^\alpha}{dx^I} \qquad 0 \leq |I| \leq l+1,$$

as required. ∎

The result dual to that of the last two propositions is that the prolongations of vertical generalised vector fields are characterised as those vector fields along $\pi_{k+l,l}$ which, as derivations of type d_*, map contact forms to contact forms.

Proposition 6.4.13 *If $X' \in \mathcal{X}^v(\pi_{k+l,l})$ is the prolongation of a vertical generalised vector field and $\sigma \in \bigwedge^1 J^l\pi$ is a contact form then $d_{X'}\sigma \in \bigwedge^1 J^{k+l}\pi$ is a contact form. Conversely, if $X' \in \mathcal{X}^v(\pi_{k+l,l})$ has the property that $d_{X'}\sigma$ is a contact form on $J^{k+l}\pi$ whenever σ is a contact form on $J^l\pi$, then X' is the prolongation of a vertical generalised vector field.*

Proof Suppose first that $X' = X^l$, where X is a vertical generalised vector field, and that σ is a contact form. Then $\sigma \in \bigwedge_0^1 \pi_{l,l-1}$, so for each point $j_p^l\phi \in J^l\pi$ there is a cotangent vector $\eta \in T^*J^{l-1}\pi$ such that $\sigma_{j_p^l\phi} = \pi_{l,l-1}^*(\eta)$. Then

$$\begin{aligned}(i_{X^l}\sigma)_{j_p^{k+l}\phi} &= \sigma_{j_p^l\phi}(X^l_{j_p^{k+l}\phi}) \\ &= \pi_{l,l-1}^*(\eta)(X^l_{j_p^{k+l}\phi}) \\ &= \eta(\pi_{l,l-1*}(X^l_{j_p^{k+l}\phi})) \\ &= \eta(X^{l-1}_{j_p^{k+l-1}\phi}),\end{aligned}$$

so that $i_{X^l}\sigma \in \pi_{k+l,k+l-1}^*(C^\infty(J^{k+l-1}\pi))$, and consequently

$$d(i_{X^l}\sigma) \in \textstyle\bigwedge_0^1 \pi_{k+l,k+l-1}.$$

By a similar argument, we may show that

$$i_{X^l}(d\sigma) \in \textstyle\bigwedge_0^1 \pi_{k+l,k+l-1},$$

and therefore it follows that $d_{X^l}\sigma \in \bigwedge_0^1 \pi_{k+l,k+l-1}$. If now Y is an arbitrary vector field on M, then

$$d_{X^l}(Y^{l-1} \lrcorner\, \sigma) = [X^l, Y^{k+l-1}] \lrcorner\, \sigma + Y^{k+l-1} \lrcorner\, d_{X^l}\sigma.$$

Now $Y^{l-1} \lrcorner\, \sigma = 0$ and $[X^l, Y^{k+l-1}] = 0$, so that $Y^{k+l-1} \lrcorner\, d_{X^l}\sigma = 0$; it follows that $d_{X^l}\sigma$ is a contact form.

Conversely, suppose $X' \in \mathcal{X}^v(\pi_{k+l,l})$ satisfies the property described in the statement of the proposition. Then, for each contact form σ on $J^l\pi$, $d_{X'}\sigma$ is a contact form on $J^{k+l}\pi$, and so is an element of $\bigwedge_0^1 \pi_{k+l,k+l-1}$. An argument in coordinates then shows that X' must be π-related to a vector field along $\pi_{k+l-1,l-1}$. If Y is an arbitrary vector field on M, then

$$d_{X'}(Y^{l-1} \lrcorner\, \sigma) = [X', Y^{k+l-1}] \lrcorner\, \sigma + Y^{k+l-1} \lrcorner\, d_{X'}\sigma,$$

but both $Y^{l-1} \lrcorner\, \sigma$ and $Y^{k+l-1} \lrcorner\, d_{X'}\sigma$ are zero, so that $[X', Y^{k+l-1}] \lrcorner\, \sigma$ is zero. Since σ is arbitrary, $[X', Y^{k+l-1}] \in \mathcal{X}^h(\pi_{k+l,l-1})$, and since X' is vertical and Y^{k+l-1} is projectable, $[X', Y^{k+l-1}]$ is also vertical, and hence is zero. Since Y is arbitrary, X is therefore the prolongation of a vertical generalised vector field. ∎

As in Section 4.4, we may now move on to construct the l-th prolongation of an arbitrary vector field on E (or, indeed, of an arbitrary generalised vector field) by defining a map $r_l : J^l(\pi \circ \tau_E) \longrightarrow TJ^l\pi$. The new definition is a direct generalisation of the earlier one.

Definition 6.4.14 The map $r_l : J^l(\pi \circ \tau_E) \longrightarrow TJ^l\pi$ is defined by

$$r_l(j^l_p\psi) = i_l(j^l_p(\psi - \phi_* \circ X)) + (j^l\phi)_*(X_p),$$

where $\phi = \tau_E \circ \psi$ and $X = \pi_* \circ \psi$. ■

Proposition 6.4.15 *The pair (r_l, id_{TE}) is a bundle morphism from $(J^l(\pi \circ \tau_E), (\pi \circ \tau_E)_{l,0}, TE)$ to $(TJ^l\pi, (\pi_{l,0})_*, TE)$. If $\psi \in \Gamma_W(\pi \circ \tau_E)$ satisfies*

$$\psi(p) = \psi^i(p) \left.\frac{\partial}{\partial x^i}\right|_{\phi(p)} + \psi^\alpha(p) \left.\frac{\partial}{\partial u^\alpha}\right|_{\phi(p)},$$

where $\phi = \tau_E \circ \psi$, $\psi^i = \dot{x}^i \circ \psi$ and $\psi^\alpha = \dot{u}^\alpha \circ \psi$, then

$$r_l(j^l_p\psi) = \psi^i(p) \left.\frac{\partial}{\partial x^i}\right|_{j^l_p\phi} + \psi^\alpha(p) \left.\frac{\partial}{\partial u^\alpha}\right|_{j^l_p\phi}$$

$$+ \sum_{|I|=0}^{l} \left(\dot{u}^\alpha_I - \sum_{\substack{J+K=I \\ J\neq 0}} \frac{I!}{J!\,K!} \dot{x}^j_J u^\alpha_{K+1_j} \right) (j^l_p\psi) \left.\frac{\partial}{\partial u^\alpha_I}\right|_{j^l_p\phi},$$

so that $r_l(j^l_p\psi)$ does not depend upon the particular choice of ψ to represent the jet $j^l_p\psi$.

Proof The coordinate expression is derived from that given in Proposition 4.4.8 by an application of Leibniz' rule. ■

Definition 6.4.16 If $X : J^k\pi \longrightarrow TE$ is a generalised vector field, then the *l-th prolongation of X* is the vector field X^l along $\pi_{k+l,l}$ defined by

$$X^l = r_l \circ j^l X \circ \iota_{l,k} : J^{k+l}\pi \longrightarrow TJ^l\pi.$$

■

In coordinates, if $X = X^i\partial/\partial x^i + X^\alpha\partial/\partial u^\alpha$ then

$$X^l = X^i \frac{\partial}{\partial x^i} + \sum_{|I|=0}^{l} \left(\frac{d^{|I|}X^\alpha}{dx^I} - \sum_{\substack{J+K=I \\ J\neq 0}} \frac{I!}{J!\,K!} \frac{d^{|J|}X^j}{dx^J} u^\alpha_{K+1_j} \right) \frac{\partial}{\partial u^\alpha_I}.$$

EXERCISES

6.4.1 If X is a generalised vector field of order $k > 0$, show by using coordinates that the l-th prolongation X^l satisfies

$$(X^l)_{j_p^{k+l}\phi} = (X^v)^l_{j_p^{k+l}\phi} + (j^l\phi)_*(\pi_*(X_{j_p^k\phi})),$$

where X^v is the vertical representative of X described in Definition 6.4.4.

6.4.2 If X is a vector field on E, show that its l-th prolongation X^l satisfies

$$(X^l)_{j_p^l\phi} = (X^v)^l_{j_p^{l+1}\phi} + (j^l\phi)_*(\pi_*(X_{\phi(p)})).$$

Explain why the expression on the right-hand side is well-defined, despite the appearance of an $(l+1)$-jet.

6.4.3 If $X : J^k\pi \longrightarrow TE$ is a generalised vector field and if Y is a vector field on M, show that, for any natural number l and any point $j_p^{k+l+1}\phi \in J^{k+l+1}\pi$,

$$[Y^{k+l}, X^{l+1}]_{j_p^{k+l+1}\phi}$$

is the holonomic lift of a tangent vector in T_pM.

6.4.4 Suppose the vector fields $X'' \in \mathcal{X}(\pi_{k+l+1,l+1})$ and $X' \in \mathcal{X}(\pi_{k+l,l})$ are π-related. If, for every vector field Y on M and every point $j_p^{k+l+1}\phi \in J^{k+l+1}\pi$,

$$[Y^{k+l}, X'']_{j_p^{k+l+1}\phi} \in TJ^l\pi$$

is the holonomic lift of a tangent vector in T_pM, then there is a generalised vector field X such that $X' = X^l$ and $X'' = X^{l+1}$.

6.4.5 If $X^l \in \mathcal{X}(\pi_{k+l,l})$ is the prolongation of a generalised vector field X and $\sigma \in \bigwedge^1 J^l\pi$ is a contact form, show that $d_{X^l}\sigma \in \bigwedge^1 J^{k+l}\pi$ is also a contact form.

6.4.6 If $X' \in \mathcal{X}(\pi_{k+l,l})$ has the property that $d_{X'}\sigma$ is a contact form on $J^{k+l}\pi$ whenever σ is a contact form on $J^l\pi$, show that X' is the prolongation of a generalised vector field $X : J^k\pi \longrightarrow TE$.

6.4.7 Suppose that the vector field $X \in \mathcal{X}(E)$ is projectable onto M, and let ψ_t be the flow of X in a neighbourhood of $\phi(p) \in E$ for given $\phi \in \Gamma_W(\pi)$ and $p \in W$. Show that the prolonged vector field X^l is related to the prolonged flow $j^l\psi_t$ by

$$X^l_{j_p^l\phi} = [t \longmapsto j^l\psi_t(j_p^l\phi)].$$

6.5 The Higher-order Cartan Form

In Section 5.5 we saw how a problem in the calculus of variations could be reformulated in terms of the Euler-Lagrange equations, initially in terms of coordinate representations, and then subsequently in a global context. The variational problem was first-order, in the sense that the Lagrangian function L was defined on the first jet manifold $J^1\pi$, but the resulting Euler-Lagrange equations were second-order: the Euler-Lagrange form δL was defined on $J^2\pi$. The coordinate-free version of "integration by parts" used to construct δL involved the Cartan form Θ_L; this had been obtained from the vector-valued m-form S_Ω, and therefore from the vector-valued 1-forms S_ω on $J^1\pi$.

The generalisation of this procedure to the higher-order calculus of variations starts as we might expect.

Definition 6.5.1 A *k-th order Lagrangian (density) on* π is a function $L \in C^\infty(J^k\pi)$. ∎

Definition 6.5.2 The local section $\phi \in \Gamma_W(\pi)$ is an *extremal of* L if

$$\left.\frac{d}{dt}\right|_{t=0} \int_C (j^k(\psi_t \circ \phi))^* L\Omega = 0,$$

whenever C is a compact m-dimensional submanifold of M with $C \subset W$, and whenever $X \in \mathcal{V}(\pi)$ has flow ψ_t and satisfies $X|_{\pi^{-1}(\partial C)} = 0$. ∎

Lemma 6.5.3 *The local section ϕ is an extremal of L if, and only if,*

$$\int_C (j^k\phi)^* d_{X^k} L\Omega = 0.$$

∎

It turns out, however, that subsequent stages of the procedure involve unexpected difficulties. Although it is always possible to find a unique globally-defined Euler-Lagrange form on $J^{2k}\pi$, there is a degree of arbitrariness in the Cartan form employed in the construction: if $k > 1$ and $m = \dim M > 1$, then there will be different Cartan forms which carry out the same function. The reason for this is to do with the commutativity of repeated partial differentiation; the problem can only arise when there are two or more independent variables, and when the Lagrangian involves second (or higher) derivatives. It is possible, by imposing a condition on the Cartan form, to regain uniqueness for second-order Lagrangians, but we shall see that this condition is inadequate for third-order Lagrangians.

We shall start our investigation of this problem by extending the vertical lift operators of Section 4.7 to higher-order jet manifolds. Formerly, we were able to construct a tensor S which, when acting by contraction, yielded a map

$$\left(\omega_i dx^i, \xi^\alpha \frac{\partial}{\partial u^\alpha}\right) \longmapsto \xi^\alpha \omega_i \frac{\partial}{\partial u^\alpha_i}.$$

For higher-order jet manifolds, the operator is more complicated: on $J^2\pi$, for instance, the corresponding map has the property that

$$\left(\omega_i dx^i, \xi^\alpha \frac{\partial}{\partial u^\alpha}\right) \longmapsto \xi^\alpha \left(\omega_i \frac{\partial}{\partial u^\alpha_i} + \frac{\partial \omega_i}{\partial x^j} \frac{\partial}{\partial u^\alpha_{ij}}\right),$$

and so no longer represents a tensor, because it involves derivatives of the coefficients of the 1-form. This complication is necessary for the operator to behave properly under coordinate changes, and it arises naturally in the generalisation of the construction in Theorem 4.7.1.

Theorem 6.5.4 *Suppose given a point $j^k_p\phi \in J^k\pi$, a closed 1-form $\omega \in \bigwedge^1 M$, and a tangent vector $\zeta \in V_{j^{k-1}_p\phi}\pi_{k-1}$. Let W be a neighbourhood of $p \in M$ and let $\gamma : W \times \mathbf{R} \longrightarrow E$ satisfy $\pi \circ \gamma = pr_1$ and $[t \longmapsto j^{k-1}_p \gamma_t] = \zeta$, where $\gamma_t : W \longrightarrow E$ is given by $\gamma_t(q) = \gamma(q,t)$; suppose also that $j^k_p\gamma_0 = j^k_p\phi$. Let $f \in C^\infty(M)$ satisfy $f(p) = 0$, $df|_{W'} = \omega|_{W'}$ for some neighbourhood $W' \subset W$ of p. Then the new tangent vector*

$$[t \longmapsto j^k_p(q \longmapsto \gamma(q, tf(q)))],$$

denoted by the symbol $\omega \otimes_{j^k_p\phi} \zeta$, is an element of $V_{j^k_p\phi}\pi_{k,0}$ which is independent of the choices of γ and f.

Proof We must first establish the existence of suitable maps γ, and this may be done using coordinates. So let (x^i, u^α) be coordinates around $\phi(p) \in E$, and let

$$\zeta = \sum_{|I|=0}^{k-1} \zeta^\alpha_I \left.\frac{\partial}{\partial u^\alpha_I}\right|_{j^{k-1}_p\phi}.$$

We may find a map γ such that, whenever t sufficiently small,

$$\left.\frac{\partial^{|I|}\gamma^\alpha}{\partial x^I}\right|_{p,t} = t\zeta^\alpha_I + u^\alpha_I(j^{k-1}_p\phi)$$

for $0 \leq |I| \leq k-1$, by choosing γ to be a $(k-1)$-th degree polynomial in the coordinates x^i in a neighbourhood of $(p, 0)$. By adding a k-th degree polynomial in these coordinates, we may also ensure that

$$\left.\frac{\partial^{|J|}\gamma^\alpha}{\partial x^J}\right|_{p;t=0} = u^\alpha_J(j^k_p\phi)$$

for $|J| = k$, as required.

We may now use γ and the function f to construct a new map $\chi : W \times \mathbf{R} \longrightarrow E$ by the rule $\chi(q,t) = \gamma(q, tf(q))$. We then have

$$\left.\frac{\partial^{|I|}\chi^\alpha}{\partial x^I}\right|_{p;t=0} = \left.\frac{\partial^{|I|}\gamma^\alpha}{\partial x^I}\right|_{p;t=0}$$

for $0 \leq |I| \leq k$, so that $j^k_p\chi_0 = j^k_p\phi$ (where $\chi_t : W \longrightarrow E$ satisfies $\chi_t(q) = \chi(q,t)$), and so that $\omega \circledvee_{j^k_p\phi} \zeta = [t \longmapsto j^k_p\chi_t]$ is indeed a tangent vector to $J^k\pi$ at $j^k_p\phi$.

We also have

$$\left.\frac{\partial \chi^\alpha}{\partial t}\right|_{t=0}(q) = f(q)\left.\frac{\partial\gamma^\alpha}{\partial t}\right|_{t=0}(q)$$

for $q \in W$, and if we apply the higher-order version of Leibniz' rule to this equation, we obtain

$$\left.\frac{\partial^{|I|+1}\chi^\alpha}{\partial t\,\partial x^I}\right|_{p;t=0} = \sum_{J+K=I}\frac{I!}{J!\,K!}\left.\frac{\partial^{|J|}f}{\partial x^J}\right|_p\left.\frac{\partial^{|K|+1}\gamma^\alpha}{\partial t\,\partial x^K}\right|_{p;t=0}$$

for $0 \leq |I| \leq k-1$, showing that the coordinates of the tangent vector $\omega \circledvee_{j^k_p\phi}\zeta$ depend on ω and ζ, rather than the maps f and γ chosen to represent them. Since, in particular,

$$\left.\frac{\partial\chi^\alpha}{\partial t}\right|_{p;t=0} = f(p)\left.\frac{\partial\gamma^\alpha}{\partial t}\right|_{p;t=0} = 0,$$

we see that $\omega \circledvee_{j^k_p\phi}\zeta$ is vertical over E. ■

We may find the coordinate representation of $\omega \circledvee_{j^k_p\phi}\zeta$ by the following calculation:

$$\begin{aligned}
\omega \circledvee_{j^k_p\phi}\zeta &= \sum_{|I|=0}^{k}\left.\frac{\partial^{|I|+1}\chi^\alpha}{\partial t\,\partial x^I}\right|_{t=0;p}\left.\frac{\partial}{\partial u^\alpha_I}\right|_{j^k_p\phi} \\
&= \sum_{|I|=0}^{k}\sum_{J+K=I}\frac{I!}{J!\,K!}\left.\frac{\partial^{|K|+1}\gamma^\alpha}{\partial t\,\partial x^K}\right|_{t=0;p}\left.\frac{\partial^{|J|}f}{\partial x^J}\right|_p\left.\frac{\partial}{\partial u^\alpha_I}\right|_{j^k_p\phi} \\
&= \sum_{|J+K|=0}^{k}\frac{(J+K)!}{J!\,K!}\left.\frac{\partial^{|K|+1}\gamma^\alpha}{\partial t\,\partial x^K}\right|_{t=0;p}\left.\frac{\partial^{|J|}f}{\partial x^J}\right|_p\left.\frac{\partial}{\partial u^\alpha_{J+K}}\right|_{j^k_p\phi} \\
&= \sum_{|J+K|=0}^{k-1}\frac{(J+K+1_i)!}{(J+1_i)!\,K!}\left.\frac{\partial^{|K|+1}\gamma^\alpha}{\partial t\,\partial x^K}\right|_{t=0;p}\left.\frac{\partial^{|J|+1}f}{\partial x^{J+1_i}}\right|_p\left.\frac{\partial}{\partial u^\alpha_{J+K+1_i}}\right|_{j^k_p\phi} \\
&= \sum_{|J+K|=0}^{k-1}\frac{(J+K+1_i)!}{(J+1_i)!\,K!}\zeta^\alpha_K\left.\frac{\partial^{|J|}\omega_i}{\partial x^J}\right|_p\left.\frac{\partial}{\partial u^\alpha_{J+K+1_i}}\right|_{j^k_p\phi}.
\end{aligned}$$

It is clear from this coordinate representation that different vertical lifts are related correctly by the jet bundle structure, so that if $\zeta \in V_{j_p^{k-1}\phi}\pi_{k-1}$ and $0 < l < k$ then $\pi_{k,l*}(\omega \circledvee_{j_p^k\phi} \zeta) = \omega \circledvee_{j_p^l\phi} (\pi_{k-1,l-1*}(\zeta))$.

Example 6.5.5 Let π be the trivial bundle $(\mathbf{R}\times F, pr_1, \mathbf{R})$ with coordinates (t, q^α), and let $\zeta \in V_{j_p^{k-1}\phi}\pi_{k-1}$ have coordinate representation

$$\zeta = \sum_{r=0}^{k-1} \zeta_{(r)}^{\alpha} \left.\frac{\partial}{\partial q_{(r)}^{\alpha}}\right|_{j_p^{k-1}\phi}$$

Then $dt \circledvee_{j_p^k\phi} \zeta$ has coordinate representation

$$dt \circledvee_{j_p^k\phi} \zeta = \sum_{r=0}^{k-1}(r+1)\zeta_{(r)}^{\alpha} \left.\frac{\partial}{\partial q_{(r+1)}^{\alpha}}\right|_{j_p^k\phi}.$$

■

We may now combine the operation of the vertical lift of tangent vectors with the vertical vector-valued form along $\pi_{k,k-1}$, to define a vector-valued 1-form $S_\omega^{(k)}$ on $J^k\pi$. This will be a direct generalisation of the corresponding object S_ω introduced on $J^1\pi$ in Chapter 4.

Definition 6.5.6 If $\omega \in \bigwedge^1 M$ satisfies $d\omega = 0$, then the vector-valued 1-form $S_\omega^{(k)} \in \bigwedge_C^1 \pi_{k+1,k} \otimes \mathcal{V}(\pi_{k,0})$ is defined by

$$(S_\omega^{(k)})_{j_p^k\phi}(\xi) = \omega \circledvee_{j_p^k\phi} pr_1(v(\pi_{k,k-1*}(\xi), j_p^k\phi))$$

where $\xi \in T_{j_p^k\phi}(J^k\pi)$. ■

In coordinates,

$$S_\omega^{(k)} = \sum_{|J+K|=0}^{k-1} \frac{(J+K+1_i)!}{(J+1_i)!\,K!} \frac{\partial^{|J|}\omega_i}{\partial x^J}(du_K^\alpha - u_{K+1_j}^\alpha dx^j) \otimes \frac{\partial}{\partial u_{J+K+1_i}^\alpha}.$$

For a given 1-form ω, the vector-valued 1-forms $S_\omega^{(k)}$ on different jet manifolds are compatible with the bundle structure; this follows from the corresponding property of vertical lifts.

Lemma 6.5.7 *If $X \in \mathcal{X}(J^k\pi)$ is $\pi_{k,l}$-related to $Y \in \mathcal{X}(J^l\pi)$, then $X \lrcorner S_\omega^{(k)}$ is $\pi_{k,l}$-related to $Y \lrcorner S_\omega^{(l)}$. If $\sigma \in \bigwedge^1 J^l\pi$, then $S_\omega^{(k)} \lrcorner \pi_{k,l}^*(\sigma) = \pi_{k,l}^*(S_\omega^{(l)} \lrcorner \sigma)$.*

Proof For each $j^k_p\phi \in J^k\pi$,

$$\begin{aligned}
\pi_{k,l*}((S^{(k)}_\omega)_{j^k_p\phi}(X_{j^k_p\phi})) &= \pi_{k,l*}(\omega \circledvee_{j^k_p\phi} pr_1(v(\pi_{k,k-1*}(X_{j^k_p\phi}), j^k_p\phi))) \\
&= \omega \circledvee_{j^l_p\phi} (\pi_{k-1,l-1*}(pr_1(v(\pi_{k,k-1*}(X_{j^k_p\phi}), j^k_p\phi)))) \\
&= \omega \circledvee_{j^l_p\phi} (pr_1(v(\pi_{l,l-1*}(Y_{j^l_p\phi}), j^l_p\phi))) \\
&= (S^{(l)}_\omega)_{j^l\phi}(Y_{j^l_p\phi}).
\end{aligned}$$

If now $\xi \in T_{j^k_p\phi}J^k\pi$ then

$$\begin{aligned}
(S^{(k)}_\omega \lrcorner\, \pi^*_{k,l}(\sigma))_{j^k_p\phi}(\xi) &= \pi^*_{k,l}(\sigma_{j^l_p\phi})((S^{(k)}_\omega)_{j^k_p\phi}(\xi)) \\
&= \sigma_{j^l_p\phi}(\pi_{k,l*}((S^{(k)}_\omega)_{j^k_p\phi}(\xi))) \\
&= \sigma_{j^l_p\phi}((S^{(l)}_\omega)_{j^l_p\phi}(\pi_{k,l*}(\xi))) \\
&= (S^{(l)}_\omega \lrcorner\, \sigma)_{j^l_p\phi}(\pi_{k,l*}(\xi)) \\
&= (\pi^*_{k,l}(S^{(l)}_\omega \lrcorner\, \sigma))_{j^k_p\phi}(\xi).
\end{aligned}$$

■

We may also consider the contraction of two of these vector-valued 1-forms corresponding to different 1-forms on M, and we shall see that they commute. In this proposition, as on other occasions when we are considering only a single jet manifold $J^k\pi$, we shall omit the superscript k and refer simply to S_ω.

Proposition 6.5.8 *If $\omega^1, \omega^2 \in \bigwedge^1 M$ satisfy $d\omega^1 = d\omega^2 = 0$, then*

$$S_{\omega^1} \lrcorner\, S_{\omega^2} = S_{\omega^2} \lrcorner\, S_{\omega^1}.$$

Proof If $k = 1$ then, as we remarked in Section 4.7, $S_{\omega^1} \lrcorner\, S_{\omega^2} = 0$, and so there is nothing to prove. We may therefore assume that $k > 1$.

Let $\zeta \in T_{j^{k-2}_p\phi}(J^{k-2}\pi)$. In Theorem 6.5.4, let ζ be represented by a map γ which still satisfies $j^k_p\gamma_0 = j^k_p\phi$, even though now

$$\zeta = [t \longmapsto j^{k-2}_p\gamma_t].$$

If locally $\omega^1 = df^1$ and $\omega^2 = df^2$ then

$$\omega^1 \circledvee_{j^{k-1}_p\phi} \zeta = [t \longmapsto j^{k-1}_p(q \longmapsto \gamma(q, tf^1(q)))],$$

and from the additional restriction on γ we obtain

$$\omega^2 \circledvee_{j^k_p\phi} (\omega^1 \circledvee_{j^{k-1}_p\phi} \zeta) = [t \longmapsto j^k_p(q \longmapsto \gamma(q, tf^1(q)f^2(q)))],$$

which is clearly also equal to $\omega^1 \circledvee_{j_p^k\phi} (\omega^2 \circledvee_{j_p^{k-1}\phi} \zeta)$.

Now let $\xi \in T_{j_p^k\phi}(J^k\pi)$. Then

$$\begin{aligned}
&(S_{\omega^1} \lrcorner S_{\omega^2})_{j_p^k\phi}(\xi) \\
&= (S_{\omega^2})_{j_p^k\phi}((S_{\omega^1})_{j_p^k\phi}(\xi)) \\
&= \omega^2 \circledvee_{j_p^k\phi} pr_1(v(\pi_{k,k-1*}(\omega^1 \circledvee_{j_p^k\phi} pr_1(v(\pi_{k,k-1*}(\xi), j_p^k\phi)), j_p^k\phi))).
\end{aligned}$$

Now $\omega^1 \circledvee_{j_p^k\phi} pr_1(v(\pi_{k,k-1*}(\xi), j_p^k\phi)) \in V_{j_p^k\phi}\pi_{k,0} \subset V_{j_p^k\phi}\pi_k$, so that

$$\begin{aligned}
&pr_1(v(\pi_{k,k-1*}(\omega^1 \circledvee_{j_p^k\phi} pr_1(v(\pi_{k,k-1*}(\xi), j_p^k\phi))), j_p^k\phi)) \\
&= \pi_{k,k-1*}(\omega^1 \circledvee_{j_p^k\phi} pr_1(v(\pi_{k,k-1*}(\xi), j_p^k\phi))) \\
&= \omega^1 \circledvee_{j_p^{k-1}\phi} \pi_{k-1,k-2*}(pr_1(v(\pi_{k,k-1*}(\xi), j_p^k\phi))).
\end{aligned}$$

It therefore follows that

$$\begin{aligned}
&(S_{\omega^1} \lrcorner S_{\omega^2})_{j_p^k\phi}(\xi) \\
&= \omega^2 \circledvee_{j_p^k\phi} (\omega^1 \circledvee_{j_p^{k-1}\phi} \pi_{k-1,k-2*}(pr_1(v(\pi_{k,k-1*}(\xi), j_p^k\phi)))) \\
&= \omega^1 \circledvee_{j_p^k\phi} (\omega^2 \circledvee_{j_p^{k-1}\phi} \pi_{k-1,k-2*}(pr_1(v(\pi_{k,k-1*}(\xi), j_p^k\phi)))) \\
&= (S_{\omega^2} \lrcorner S_{\omega^1})_{j_p^k\phi}(\xi).
\end{aligned}$$

■

By virtue of this lemma, we will be justified in using a multi-index notation for the contraction of several of these vector-valued forms. If $(\omega^1, \ldots, \omega^m)$ is a family of m closed 1-forms on M, we may define S_{ω^I} by

$$S_{\omega^{I+1_i}} = S_{\omega^I} \lrcorner S_{\omega^i}.$$

Of course, the idea is that $(\omega^1, \ldots, \omega^m)$ should form a basis of closed 1-forms: however, the topological nature of M may prohibit this, and so we shall also allow the use of a family which only forms a basis on some open submanifold of M.

The vector-valued 1-forms S_ω are of importance in showing that the horizontal differential d_h is locally exact: in fact, for any $\sigma \in \bigwedge^1 J^k\pi$, the relationship

$$i_{S_\omega^{(k+1)}} d_h\sigma - d_h i_{S_\omega^{(k)}}\sigma = \pi_{k+1}^*(\omega) \wedge i_v \pi_{k+1,k}^*(\sigma)$$

may be obtained from a calculation in local coordinates. We shall examine this question in more detail in Chapter 7. For the moment, however, the most important feature of these vector-valued forms will be that the map $\omega \longmapsto S_\omega$ depends on the derivatives of the coefficients of ω, and so cannot be used directly to define a vector-valued m-form S_Ω on $J^k\pi$. We shall therefore adopt a rather more roundabout technique, which we shall illustrate by an example before giving a general proof.

Example 6.5.9 Let π be the trivial bundle $(\mathbf{R}^m \times \mathbf{R}^n, pr_1, \mathbf{R}^m)$ with global coordinates (x^i, u^α), and let σ be a 1-form on $J^2\pi$. (In the context of a variational problem we would take $\sigma = dL$, where $L \in C^\infty(J^2\pi)$ was a second-order Lagrangian.) In coordinates, we have

$$\sigma = \sigma_i dx^i + \sigma_\alpha du^\alpha + \sigma^i_\alpha du^\alpha_i + \sigma^{ij}_\alpha du^\alpha_{ij},$$

where, for this example, we have reverted to ordinary subscript notation for the derivative coordinates on $J^2\pi$, and where $\sigma^{ij}_\alpha = \sigma^{ji}_\alpha$. Note that the sum in the final term is over *all* pairs of indices i, j with $1 \leq i, j \leq m$: if $\sigma = dL$ then we have

$$\sigma^{ij}_\alpha = \frac{1}{n(ij)} \frac{\partial L}{\partial u^\alpha_{ij}},$$

where $n(ij)$ is the number of distinct indices represented by i and j, as in Exercise 6.1.1.

In order to use our first-order theory on this second-order example, we shall use the relationship between $J^2\pi$ and $J^1\pi_1$ described in Section 5.2. Now $\iota_{1,1} : J^2\pi \longrightarrow J^1\pi_1$ is an embedding, but to extend the 1-form σ from $J^2\pi$ to $J^1\pi_1$ we require a projection $\tau : J^1\pi_1 \longrightarrow J^2\pi$, and for this example we may use the projection given by the coordinate system:

$$\begin{aligned} u^\alpha_i(\tau(j^1_p\psi)) &= \tfrac{1}{2}(u^\alpha_i(j^1_p\psi) + u^\alpha_{;i}(j^1_p\psi)) \\ u^\alpha_{ij}(\tau(j^1_p\psi)) &= \tfrac{1}{2}(u^\alpha_{i;j}(j^1_p\psi) + u^\alpha_{j;i}(j^1_p\psi)). \end{aligned}$$

The resulting 1-form $\tau^*(\sigma)$ on $J^1\pi_1$ has coordinate representation

$$\tau^*(\sigma) = \sigma_i dx^i + \sigma_\alpha du^\alpha + \tfrac{1}{2}\sigma^i_\alpha(du^\alpha_i + du^\alpha_{;i}) + \sigma^{ij}_\alpha du^\alpha_{i;j},$$

where we have omitted the pull-back maps τ^* in front of the coefficient functions, and where we have made use of the symmetry $\sigma^{ij}_\alpha = \sigma^{ji}_\alpha$.

We may now apply the first-order theory on $J^1\pi_1$. Here, of course, the operator S'_Ω takes the form

$$\begin{aligned} S'_\Omega &= (du^\alpha - u^\alpha_{;k}dx^k) \wedge \left(\frac{\partial}{\partial x^j} \lrcorner \Omega\right) \otimes \frac{\partial}{\partial u^\alpha_{;j}} \\ &\quad + (du^\alpha_i - u^\alpha_{i;k}dx^k) \wedge \left(\frac{\partial}{\partial x^j} \lrcorner \Omega\right) \otimes \frac{\partial}{\partial u^\alpha_{i;j}}, \end{aligned}$$

because the functions u^α and u^α_i are all regarded as independent coordinates. Similarly, the horizontal differential d'_h is here to be regarded as a map $\bigwedge^r J^1\pi \longrightarrow \bigwedge^{r+1} J^1\pi_1$ or $\bigwedge^r J^1\pi_1 \longrightarrow \bigwedge^{r+1} J^2\pi_1$; the composition $\iota^*_{1,1} \circ d'_h : \bigwedge^r J^1\pi \longrightarrow \bigwedge^{r+1} J^2\pi$ is then the original horizontal differential d_h. Taking account of this, we find that

$$\begin{aligned} d'_h S'_\Omega(\tau^*(\sigma)) &= -\tfrac{1}{2}\frac{d\sigma^i_\alpha}{dx^i} du^\alpha \wedge \Omega - \tfrac{1}{2}\sigma^i_\alpha du^\alpha_{;i} \wedge \Omega \\ &\quad -\frac{d\sigma^{ij}_\alpha}{dx^i} du^\alpha_j \wedge \Omega - \sigma^{ij}_\alpha du^\alpha_{j;i} \wedge \Omega, \end{aligned}$$

so that the "Euler-Lagrange form" on $J^1\pi$ is then

$$\begin{aligned} E\sigma &= (\pi_1)^*_{2,1}(\tau^*(\sigma) \wedge \Omega) + d'_h S'_\Omega(\tau^*(\sigma)) \\ &= \left(\sigma_\alpha - \tfrac{1}{2}\frac{d\sigma^i_\alpha}{dx^i}\right) du^\alpha \wedge \Omega + \left(\tfrac{1}{2}\sigma^j_\alpha - \frac{d\sigma^{ij}_\alpha}{dx^i}\right) du^\alpha_j \wedge \Omega. \end{aligned}$$

Our integration by parts has given us an $(m+1)$-form on $J^2\pi_1$ which is horizontal over $J^1\pi$: it is an element of $\bigwedge^{m+1}_1(\pi_1)_2 \cap \bigwedge^{m+1}_0(\pi_1)_{2,0}$ as described in Theorem 5.5.2. Our target, however, is an $(m+1)$-form horizontal over E, and so we must integrate by parts again. We can do this by using the injection $\iota_{2,1} : J^3\pi \longrightarrow J^2\pi_1$ to obtain an $(m+1)$-form $\iota^*_{2,1}(E\sigma)$ on $J^3\pi$ horizontal over $J^1\pi$, and employing a variant of the first-order theory. The two features we must deal with are that we now have an $(m+1)$-form rather than a 1-form, and that the coefficient functions are defined on $J^3\pi$ rather than on $J^1\pi$. Neither of these factors presents any problem, and we can apply S_Ω and d_h (in their original forms on $J^1\pi$ rather than on $J^1\pi_1$) to obtain the horizontal differential of the "Cartan form",

$$\begin{aligned} d_h S_\Omega(\iota^*_{2,1}(E\sigma)) &= \left(-\tfrac{1}{2}\frac{d\sigma^j_\alpha}{dx^j} + \frac{d^2\sigma^{ij}_\alpha}{dx^i\,dx^j}\right) du^\alpha \wedge \Omega \\ &\quad - \left(\tfrac{1}{2}\sigma^j_\alpha - \frac{d\sigma^{ij}_\alpha}{dx^i}\right) du^\alpha_j \wedge \Omega. \end{aligned}$$

Finally, therefore, we obtain

$$\begin{aligned} E^2\sigma &= \pi^*_{4,3}(\iota^*_{2,1}(E\sigma)) + d_h S_\Omega(\iota^*_{2,1}(E\sigma)) \\ &= \left(\sigma_\alpha - \frac{d\sigma^j_\alpha}{dx^j} + \frac{d^2\sigma^{ij}_\alpha}{dx^i\,dx^j}\right) du^\alpha \wedge \Omega \end{aligned}$$

as an $(m+1)$-form on $J^4\pi$ horizontal over E; it is an element of $\bigwedge^{m+1}_1\pi_4 \cap \bigwedge^{m+1}_0\pi_{4,0}$. If $L \in C^\infty(J^2\pi)$ is a second-order Lagrangian and $\sigma = dL$, then the result is the Euler-Lagrange form for L,

$$\begin{aligned}\delta L &= E^2(dL) \\ &= \left(\frac{\partial L}{\partial u^\alpha} - \frac{d}{dx^j}\left(\frac{\partial L}{\partial u^\alpha_j}\right) + \frac{1}{n(ij)}\frac{d^2}{dx^i\,dx^j}\left(\frac{\partial L}{\partial u^\alpha_{ij}}\right)\right) du^\alpha \wedge \Omega.\end{aligned}$$

■

This example suggests that it might be possible to find a Cartan form and an Euler-Lagrange form for a Lagrangian defined on a jet manifold of arbitrary order: if $L \in C^\infty(J^k\pi)$ then we would need to perform k integrations by parts to obtain the Euler-Lagrange form, of which the first $(k-1)$ integrations would yield the corresponding Cartan form. The problem, of course, is that we need to use projections from repeated jet manifolds to their holonomic submanifolds, and these projections may be defined in many different ways. In our example, we used the global coordinate system on the base manifold to define a suitable projection. More generally, we shall construct such a projection by using tubular neighbourhoods.

Definition 6.5.10 Let M be an embedded closed submanifold of the manifold H, and let $(N_H M, \nu, M)$ be the normal bundle of M in H. A *tubular neighbourhood of M in H* is a neighbourhood U of M in H, a neighbourhood V of the image of the zero section $z(M)$ in $N_H M$, and a diffeomorphism $f : U \longrightarrow V$ satisfying $f|_M = z$. The map $\nu \circ f : U \longrightarrow M$ is called the *projection* of the tubular neighbourhood. ■

It may be shown that tubular neighbourhoods always exist, and we may therefore use the projection of such a neighbourhood to "spread out" the value of a differential form.

Before proving the general result, we shall dispose of the two technical features mentioned in the example, by using the vector-valued m-form S_Ω on $J^1\pi$ to construct a map from $(m+1)$-forms to m-forms on $J^1\pi$, and hence a map from $(m+1)$-forms to m-forms on $J^s\pi$. We shall use the same notation S_Ω for these new maps as for the original vector-valued m-forms.

Definition 6.5.11 The map $S_\Omega : \bigwedge_1^{m+1}\pi_1 \longrightarrow \bigwedge^m J^1\pi$ is defined by the rule that, for $\theta \in \bigwedge_1^{m+1}\pi_1$,

$$S_\Omega(\theta) = S_\Omega \lrcorner\, \sigma,$$

where $\sigma \in \bigwedge^1 J^1\pi$ satisfies $\theta = \sigma \wedge \Omega$. ■

Of course, we need to check that this definition makes sense. To see this, note that the vector-valued m-form S_Ω, when regarded as an alternating m-linear map

$$\mathcal{X}(J^1\pi) \times \ldots \times \mathcal{X}(J^1\pi) \longrightarrow \mathcal{X}(J^1\pi)$$

is vertical over M: it takes its values in $\mathcal{V}(\pi_1) \subset \mathcal{X}(J^1\pi)$. The transposed map $\bigwedge^1 J^1\pi \longrightarrow \bigwedge^m J^1\pi$ may therefore be defined on the quotient space $\mathcal{V}^*(\pi_1)$ of vertical 1-forms described in Definition 3.3.10. By Proposition 3.3.11, $\mathcal{V}^*(\pi_1)$ is isomorphic to $\bigwedge_1^{m+1}\pi_1$, and the isomorphism is given by

$$[\sigma] \longmapsto \sigma \wedge \Omega,$$

so that our new map S_Ω is really no more than a reformulation of the old one.

Our second definition uses the fact that the module over $C^\infty(J^s\pi)$ generated by $\pi_{s,1}^*(\bigwedge_1^{m+1}\pi_1)$ may be written as $\bigwedge_0^{m+1}\pi_{s,1} \cap \bigwedge_1^{m+1}\pi_s$: the $(m+1)$-forms in this module are those which are not only m-horizontal over M, but are also completely horizontal over $J^1\pi$.

Definition 6.5.12 For any $s > 0$, the map $S_\Omega : \bigwedge_0^{m+1}\pi_{s,1} \cap \bigwedge_1^{m+1}\pi_s \longrightarrow \bigwedge^m J^s\pi$ is defined by considering the map

$$\begin{array}{rcl} \pi_{s,1}^*(\bigwedge_1^{m+1}\pi_1) & \longrightarrow & \bigwedge^m J^s\pi \\ \pi_{s,1}^*(\theta) & \longmapsto & \pi_{s,1}^*(S_\Omega(\theta)), \end{array}$$

which is well-defined because $\pi_{s,1}^*$ is injective, and extending to $\bigwedge_0^{m+1}\pi_{s,1} \cap \bigwedge_1^{m+1}\pi_s$ by $C^\infty(J^s\pi)$-linearity. ∎

Note that, as a consequence of the contact properties of the original vector-valued m-form, this last operator S_Ω takes its values in a sub-module of $\bigwedge^m J^s\pi$: it is always the case that $S_\Omega(\theta) \in \bigwedge_0^m \pi_{s,0} \cap \bigwedge_1^m \pi_s$, and indeed that $(j^s\phi)^*(S_\Omega(\theta)) = 0$ for every $\phi \in \Gamma_{loc}(\pi)$.

We are now in a position to apply the induction argument. We shall use the notation $\pi_1^k = ((\ldots\pi_1)_1\ldots)_1)_1$ for the bundle of k-th repeated 1-jets as described in Section 6.2. The induction step will be that, if the result is true for $J^r\nu$ for an arbitrary bundle ν, then it is also true for $J^{r+1}\pi$; we shall, of course, let ν be the bundle π_1. We shall formulate the result by supposing that the necessary tubular neighbourhoods have been specified, and by demonstrating the existence of an operator with properties which generalise those given in Theorem 5.5.2 for the first-order Cartan form.

Theorem 6.5.13 *Suppose given, for $0 \leq r \leq k-2$, a family of tubular neighbourhoods of $J^{k-r}\pi_1^r$ in $J^{k-r-1}\pi_1^{r+1}$. Corresponding to this family, there is then an* $\mathbf{R}$*-linear operator $S_\Omega^{(k)} : \bigwedge^1 J^k\pi \longrightarrow \bigwedge_0^m \pi_{2k-1,k-1} \cap \bigwedge_1^m \pi_{2k-1}$ satisfying the conditions*

1. *$(S_\Omega^{(k)}(\sigma))_{j_p^{2k-1}\phi}$ depends only on the germ of σ at $j_p^k\phi$;*
2. *$\pi_{2k,k}^*(\sigma \wedge \Omega) + d_h(S_\Omega^{(k)}(\sigma)) \in \bigwedge_0^{m+1}\pi_{2k,0} \cap \bigwedge_1^{m+1}\pi_{2k}$; and*

3. $(j^{2k-1}\phi)^*(S_\Omega^{(k)}(\sigma)) = 0$ *for every* $\phi \in \Gamma_{loc}(\pi)$.

Proof The proof is by induction on k. When $k = 1$, no tubular neighbourhoods are needed and the operator is just S_Ω as previously defined; so suppose $k > 1$. The induction hypothesis is that, for every bundle $\nu : F \longrightarrow M$ and family of tubular neighbourhoods of $J^{k-r-1}\nu_1^r$ in $J^{k-r-2}\nu_1^{r+1}$, where $0 \leq r \leq k-3$, there is an **R**-linear operator $S_\Omega^{(k-1)} : \bigwedge^1 J^{k-1}\nu \longrightarrow \bigwedge_0^m \nu_{2k-3,k-2} \cap \bigwedge_1^m \nu_{2k-3}$ such that, for $\tilde{\sigma} \in \bigwedge^1 J^{k-1}\nu$, the three properties

1. $(S_\Omega^{(k-1)}(\tilde{\sigma}))_{j_p^{2k-3}\psi}$ depends only on the germ of $\tilde{\sigma}$ at $j_p^{k-1}\psi$,
2. $\nu_{2k-2,k-1}^*(\tilde{\sigma} \wedge \Omega) + d_h(S_\Omega^{(k-1)}(\tilde{\sigma})) \in \bigwedge_0^{m+1}\nu_{2k-2,0} \cap \bigwedge_1^{m+1}\nu_{2k-2}$, where d_h in this expression is the horizontal differential on the jet bundles of ν, and
3. $(j^{2k-3}\psi)^*(S_\Omega^{(k-1)}(\tilde{\sigma})) = 0$ for $\psi \in \Gamma_{loc}(\nu)$,

are satisfied.

Choose ν to be $\pi_1 : J^1\pi \longrightarrow M$. Given $\sigma \in \bigwedge^1 J^k\pi$, transfer σ to $\iota_{k-1,1}(J^k\pi)$ along $\iota_{k-1,1}$; extend it to the tubular neighbourhood using the neighbourhood's projection; and then extend it in an arbitrary manner as a smooth 1-form $\tilde{\sigma}$ over the whole of $J^{k-1}\pi_1$.

By the induction hypothesis,

$$\begin{aligned} S_\Omega^{(k-1)}(\tilde{\sigma}) &\in \textstyle\bigwedge_0^m(\pi_1)_{2k-3,k-2} \cap \bigwedge_1^m(\pi_1)_{2k-3} \\ &\subset \textstyle\bigwedge^m J^{2k-3}\pi_1, \end{aligned}$$

and if we write $E^{(k-1)}(\tilde{\sigma})$ for the result of the next integration by parts,

$$E^{(k-1)}\tilde{\sigma} = (\pi_1)_{2k-2,k-1}^*(\tilde{\sigma} \wedge \Omega) + d_h(S_\Omega^{(k-1)}(\tilde{\sigma})),$$

then

$$\begin{aligned} E^{(k-1)}\tilde{\sigma} &\in \textstyle\bigwedge_0^{m+1}(\pi_1)_{2k-2,0} \cap \bigwedge_1^{m+1}(\pi_1)_{2k-2} \\ &\subset \textstyle\bigwedge^{m+1} J^{2k-2}\pi_1, \end{aligned}$$

so that $\iota_{2k-2,1}^*(E^{(k-1)}\tilde{\sigma})$ is an $(m+1)$-form on $J^{2k-1}\pi$.

We may now use the basic relationship for repeated jets illustrated in Section 6.2 by a commutative diagram,

$$(\pi_j)_{l+m,l} \circ \iota_{l+m,j} = \iota_{l,j} \circ \pi_{j+l+m,j+l}.$$

This relationship yields $(\pi_1)_{2k-2,0} \circ \iota_{2k-2,1} = \pi_{2k-1,1}$, so that

$$\iota_{2k-2,1}^*(\textstyle\bigwedge_0^{m+1}(\pi_1)_{2k-2,0}) \subset \bigwedge_0^{m+1}\pi_{2k-1,1},$$

and consequently $\pi_1 \circ (\pi_1)_{2k-2,0} \circ \iota_{2k-2,1} = \pi_{2k-1}$, so that

$$\iota^*_{2k-2,1}(\bigwedge{}_1^{m+1}(\pi_1)_{2k-2}) \subset \bigwedge{}_1^{m+1}\pi_{2k-1}.$$

Therefore

$$\iota^*_{2k-2,1}(E^{(k-1)}\tilde{\sigma}) \in \bigwedge{}_0^{m+1}\pi_{2k-1,1} \cap \bigwedge{}_1^{m+1}\pi_{2k-1},$$

so by using Definition 6.5.12 we may apply S_Ω to this $(m+1)$-form to obtain the m-form

$$S_\Omega(\iota^*_{2k-2,1}(E^{(k-1)}\tilde{\sigma})) \in \bigwedge{}_0^m\pi_{2k-1,0} \cap \bigwedge{}_1^m\pi_{2k-1}$$

on $J^{2k-1}\pi$.

Now this m-form will be one term of the m-form $S_\Omega^{(k)}(\sigma)$; the other term will be one which, when its horizontal differential is taken and the result added to $\sigma \wedge \Omega$, yields the $(m+1)$-form $\iota^*_{2k-2,1}(E^{(k-1)}\tilde{\sigma})$. More formally, we will let

$$S_\Omega^{(k)}(\sigma) = \theta + S_\Omega(\iota^*_{2k-2,1}(E^{(k-1)}\tilde{\sigma})),$$

where the m-form θ on $J^{2k-1}\pi$ will be chosen so that the $(m+1)$-form

$$\pi^*_{2k,k}(\sigma \wedge \Omega) + d_h(S_\Omega^{(k)}(\sigma)) = \pi^*_{2k,k}(\sigma \wedge \Omega) + d_h\theta + d_h S_\Omega(\iota^*_{2k-2,1}(E^{(k-1)}\tilde{\sigma}))$$

has the property of being totally horizontal over E: we shall therefore be able to regard the latter as the Euler-Lagrange form $E^{(k)}\sigma$. If we require θ to satisfy the equation

$$\pi^*_{2k,k}(\sigma \wedge \Omega) + d_h\theta = \pi^*_{2k,2k-1}(\iota^*_{2k-2,1}(E^{(k-1)}\tilde{\sigma})),$$

then we will be sure that

$$E^{(k)}\sigma = \pi^*_{2k,2k-1}(\iota^*_{2k-2,1}(E^{(k-1)}\tilde{\sigma})) + d_h S_\Omega(\iota^*_{2k-2,1}(E^{(k-1)}\tilde{\sigma}))$$

will have the appropriate property. We shall therefore set θ to equal

$$\pi^*_{2k-1,2k-2}(\iota^*_{2k-3,1}(S_\Omega^{(k-1)}(\tilde{\sigma})),$$

which is an element of $\bigwedge{}_0^m\pi_{2k-2,k-1} \cap \bigwedge{}_1^m\pi_{2k-2}$ by virtue of the relationship $(\pi_1)_{2k-3,k-2} \circ \iota_{2k-3,1} = \iota_{k-2,1} \circ \pi_{2k-2,k-1}$. The definition of $E^{(k-1)}\tilde{\sigma}$ then shows that θ will satisfy the required equation.

Our definition of $S_\Omega^{(k)}$ is therefore

$$\begin{aligned} S_\Omega^{(k)}(\sigma) &= \pi^*_{2k-1,2k-2}(\iota^*_{2k-3,1}(S_\Omega^{(k-1)}(\tilde{\sigma}))) + S_\Omega(\iota^*_{2k-2,1}(E^{(k-1)}\tilde{\sigma})) \\ &\in \bigwedge{}_0^m\pi_{2k-1,k-1} \cap \bigwedge{}_1^m\pi_{2k-1}, \end{aligned}$$

and clearly $(S_\Omega^{(k)}(\sigma))_{j_p^{2k-1}\phi}$ depends only on the germ of σ at $j_p^k\phi$. The operator $S_\Omega^{(k)}$ satisfies the second required property by construction. As far as the third property is concerned, if $\phi \in \Gamma_{loc}(\pi)$, then

$$\begin{aligned} & (j^{2k-1}\phi)^*(S_\Omega^{(k)}(\sigma)) \\ = \; & (j^{2k-3}(j^1\phi))^* S_\Omega^{(k-1)}(\tilde{\sigma}) + (j^{2k-1}\phi)^*(S_\Omega(\iota_{2k-2,1}^*(E^{(k-1)}\tilde{\sigma}))), \end{aligned}$$

where the first term vanishes by the induction hypothesis because $j^1\phi$ is a local section of π_1, and the second term vanishes by virtue of the properties of S_Ω. ■

Corollary 6.5.14 *If $L \in C^\infty(J^k\pi)$ where $k \geq 1$, then a Cartan form for L may be constructed globally by*

$$\Theta_L = S_\Omega^{(k)}(dL) + \pi_{2k-1,k}^*(L\Omega).$$

■

The preceding argument provides a satisfactory demonstration of the existence of a suitable Cartan form; as we have already remarked, however, the uniqueness of such a form is a rather more complicated affair. Nevertheless, the Euler-Lagrange form which is constructed from the Cartan form by the equation of first variation is always unique (so that when two distinct Cartan forms may be found, their difference will necessarily be annihilated by the horizontal differential d_h). Although our construction of a Cartan form gave an m-form which was totally horizontal over $J^{k-1}\pi$, it is *a priori* possible that an m-form with suitable properties could be found in $\bigwedge_0^m \pi_{2k-1,k} \cap \bigwedge_1^m \pi_{2k-1}$; we shall therefore express the following result in slightly more general terms.

Proposition 6.5.15 *If $L \in C^\infty(J^k\pi)$, and if $\Theta_1, \Theta_2 \in \bigwedge_0^m \pi_{2k-1,k} \cap \bigwedge_1^m \pi_{2k-1}$ have the property that both the $(m+1)$-forms*

$$\begin{aligned} \delta L_1 &= \pi_{2k,k}^*(dL \wedge \Omega) + d_h\Theta_1 \\ \delta L_2 &= \pi_{2k,k}^*(dL \wedge \Omega) + d_h\Theta_2 \end{aligned}$$

are elements of $\bigwedge_0^{m+1} \pi_{2k,0} \cap \bigwedge_1^{m+1} \pi_{2k}$, then $\delta L_1 = \delta L_2$.

Proof We shall use coordinates to show that $d_h(\Theta_1 - \Theta_2) = 0$. First, because both Θ_1 and Θ_2 are elements of $\bigwedge_0^m \pi_{2k-1,k} \cap \bigwedge_1^m \pi_{2k-1}$, it follows that their difference $\Theta_1 - \Theta_2$ may be expressed locally as

$$\Theta_1 - \Theta_2 = \sigma^i \wedge \left(\frac{\partial}{\partial x^i} \lrcorner \, \Omega \right),$$

where the 1-forms σ^i are elements of $\bigwedge_0^1 \pi_{2k-1,k}$. If the coordinate representation of each σ^i is

$$(\sigma^i)_j dx^j + \sum_{|I|=0}^{k} (\sigma^i)_\alpha^I du_I^\alpha,$$

then

$$d_h \sigma^i = \frac{d(\sigma^i)_j}{dx^m} dx^m \wedge dx^j + \sum_{|I|=0}^{k} \left(\frac{d(\sigma^i)_\alpha^I}{dx^m} dx^m \wedge du_I^\alpha + (\sigma^i)_\alpha^I dx^m \wedge du_{I+1_m}^\alpha \right),$$

and so

$$d_h(\Theta_1 - \Theta_2) = -\sum_{|I|=0}^{k} \left(\frac{d(\sigma^i)_\alpha^I}{dx^i} du_I^\alpha + (\sigma^i)_\alpha^I du_{I+1_i}^\alpha \right) \wedge \Omega.$$

Since $d_h(\Theta_1 - \Theta_2) \in \bigwedge_0^{m+1} \pi_{2k,0}$, the only non-zero terms in this expression are those in $du^\alpha \wedge \Omega$, with coefficients $-d/dx^i((\sigma^i)_\alpha)$. From the vanishing of the other terms, we may calculate recursively that these coefficients equal

$$\begin{aligned} &\sum_{|I|=k} (-1)^{k+1} \frac{d^{|I|+1}((\sigma^i)_\alpha^I)}{dx^{I+1_i}} \\ = &\sum_{|I|=k} (-1)^{k+1} \sum_{J=I+1_i} \frac{d^{|J|}((\sigma^i)_\alpha^I)}{dx^J}. \end{aligned}$$

But for each fixed multi-index J with $|J| = k+1$, the sum

$$\sum_{I+1_i=J} ((\sigma^i)_\alpha^I)$$

equals the coefficient of $du_J^\alpha \wedge \Omega$, which is zero. The coefficient of $du^\alpha \wedge \Omega$ is then a sum of derivatives of the coefficients of the $du_J^\alpha \wedge \Omega$ where $|J| = k+1$, and so itself is zero. ∎

To obtain the coordinate representation of the unique Euler-Lagrange form, we shall make a particular choice of tubular neighbourhood which yields a particular choice of Cartan form. While the coordinate description of the Cartan form may only be valid for this coordinate system, Proposition 6.5.15 implies that the representation of δL is valid in an arbitrary coordinate system. So suppose (x^i, u^α) is a coordinate system on $U \subset E$ and that, for each s with $1 \leq s \leq k$, (x^i, u_I^α) and $(x^i, u_{;J}^\alpha, u_{1_i;J}^\alpha)$ are the corresponding coordinate systems on $U^s \subset J^s\pi$ and $U_1^{s-1} \subset J^{s-1}\pi_1$ respectively, where $|I| \leq s$, $|J| \leq s-1$. Then a projection $\tau_s : U_1^{s-1} \longrightarrow U^s$ may be defined by the rule

$$\begin{aligned} x^i(\tau_s(j_p^{s-1}\psi)) &= x^i(j_p^{s-1}\psi); \\ u_I^\alpha(\tau_s(j_p^{s-1}\psi)) &= \tfrac{1}{2}u_{;I}^\alpha(j_p^{s-1}\psi) + \tfrac{1}{2}\left(\sum_{i=1}^m \frac{I(i)}{|I|} u_{1_i;I-1_i}^\alpha(j_p^{s-1}\psi)\right) \\ &\qquad\qquad \text{for } |I| \leq s-1; \\ u_I^\alpha(\tau_s(j_p^{s-1}\psi)) &= \sum_{i=1}^m \frac{I(i)}{|I|} u_{1_i;I-1_i}^\alpha(j_p^{s-1}\psi) \qquad \text{for } |I| = s. \end{aligned}$$

(This is just a generalisation of the projection used globally in Example 6.5.9.) Each τ_s may be extended to define a tubular neighbourhood of the whole of $J^s\pi$ in $J^{s-1}\pi_1$, and used to construct the corresponding operator $S_\Omega^{(k)}$. Then given a Lagrangian $L \in C^\infty(J^k\pi)$, the coordinate representation of $\Theta_L = S_\Omega^{(k)}(dL) + \pi_{2k-1,k}^*(L\Omega)$ in the neighbourhood U^{2k-1} is

$$\sum_{|I|=0}^{k-1} \sum_{|J|=0}^{k-|I|-1} (-1)^{|J|} \frac{(I+J+1_i)!\,|I|!\,|J|!}{|I+J+1_i|!\,I!\,J!} \times$$

$$\frac{d^{|J|}}{dx^J}\left(\frac{\partial L}{\partial u_{I+J+1_i}^\alpha}\right)(du_I^\alpha - u_{I+1_j}^\alpha dx^j) \wedge \left(\frac{\partial}{\partial x^i} \lrcorner\, \Omega\right) + L\Omega.$$

The corresponding Euler-Lagrange form δL is then

$$\sum_{|J|=0}^{k} (-1)^{|J|} \frac{d^{|J|}}{dx^J}\left(\frac{\partial L}{\partial u_J^\alpha}\right) du^\alpha \wedge \Omega.$$

Example 6.5.16 Let π be the trivial bundle $(\mathbf{R}^2 \times \mathbf{R}, pr_1, \mathbf{R}^2)$ with coordinates $(x, t; u)$, and let $L \in C^\infty(J^2\pi)$ be given by

$$L = \tfrac{1}{2}u_x u_t + u_x^3 + u_{xx}^2.$$

Then in this coordinate system, the Cartan form of L is given by

$$\begin{aligned} \Theta_L &= (3u_x^2 + \tfrac{1}{2}u_t - 2u_{xxx})du \wedge dt - \tfrac{1}{2}u_x du \wedge dx \\ &\quad + 2u_{xx} du_x \wedge dt + (2u_x u_{xxx} - u_{xx}^2 - 2u_x^3 - \tfrac{1}{2}u_x u_t) dx \wedge dt, \end{aligned}$$

and the Euler-Lagrange form of L is

$$\delta L = (u_{xxxx} + 6u_x u_{xx} + u_{xt}) du \wedge dx \wedge dt.$$

It follows that if the local section ϕ satisfies the Euler-Lagrange equation then its derivative $\partial\phi/\partial x$ satisfies the Korteweg-de Vries equation,

$$\frac{\partial^4\phi}{\partial x^4} + 6\frac{\partial\phi}{\partial x}\frac{\partial^2\phi}{\partial x^2} + \frac{\partial^2\phi}{\partial x\,\partial t} = 0.$$

■

We shall now justify our remarks about the uniqueness of the Cartan form. We saw in Section 5.5 that the Cartan form for a first-order Lagrangian was unique, and a corresponding result holds for Lagrangians of arbitrary order where the base manifold M is one-dimensional. The proof of this result uses the local exactness of the horizontal differential d_h; this will be proved in the context of infinite jets in Chapter 7.

Proposition 6.5.17 *If the base manifold M is one-dimensional, and if $S : \bigwedge^1 J^k\pi \longrightarrow \bigwedge^1_0 \pi_{2k-1,k-1}$ satisfies the properties that*

$$\pi^*_{2k,k}(\sigma \wedge dt) + d_h S(\sigma) \in \textstyle\bigwedge^2_0 \pi_{2k,0} \cap \bigwedge^2_1 \pi_{2k},$$

and that

$$(j^{2k-1}\phi)^*(S(\sigma)) = 0$$

for every $\phi \in \Gamma_{loc}(\pi)$, then $S = S^{(k)}_{dt}$.

Proof If $\sigma \in \bigwedge^1 J^k\pi$ then

$$(j^{2k-1}\phi)^*(S(\sigma)) = (j^{2k-1}\phi)^*(S^{(k)}_{dt}(\sigma)) = 0,$$

so that $S(\sigma) - S^{(k)}_{dt}(\sigma)$ is a contact form. By Proposition 6.5.15,

$$d_h(S(\sigma) - S^{(k)}_{dt}(\sigma)) = 0,$$

so that locally $S(\sigma) - S^{(k)}_{dt}(\sigma) = d_h f$ for some function f on $J^{2k-2}\pi$. But then

$$\begin{aligned}
\pi^*_{2k,2k-1}(d_h f) &= \pi^*_{2k,2k-1}(h \lrcorner\, df) \\
&= h \lrcorner\, (h \lrcorner\, df) \\
&= h \lrcorner\, d_h f \\
&= h(S(\sigma) - S^{(k)}_{dt}(\sigma)) \\
&= 0,
\end{aligned}$$

since the horizontal component of any contact form is zero. ∎

Corollary 6.5.18 *The Cartan form $S^{(k)}_{dt}(dL) + L\,dt$ is unique, and has coordinate representation*

$$\sum_{i=0}^{k-1} \sum_{j=0}^{k-i-1} (-1)^j \frac{d^j}{dt^j} \left(\frac{\partial L}{\partial q^\alpha_{(i+j+1)}} \right) (dq^\alpha_{(i)} - q^\alpha_{(i+1)} dt) + L\,dt,$$

where (t, q^α) are coordinates on the total space of the bundle π. ∎

On the other hand, when $\dim M \geq 2$ and $k \geq 2$, the construction of Theorem 6.5.13 does not provide a unique Cartan form.

Example 6.5.19 Let π be the trivial bundle $(\mathbf{R}^2 \times \mathbf{R}, pr_1, \mathbf{R}^2)$ with coordinates $(x^1, x^2; u)$, let $\Omega = dx^1 \wedge dx^2$ be the volume form on $\mathbf{R}^2$, and let the derivative coordinates on $J^2\pi$ and $J^1\pi_1$ be denoted in ordinary (rather than multi-index) notation. Let S_1 be an operator $S_\Omega^{(2)}$ defined using the projection $\tau_1 : J^1\pi_1 \longrightarrow J^2\pi$ described earlier, and let S_2 be an operator defined using the alternative projection τ_2, where

$$u_{11}(\tau_2(j_p^2\psi)) = u_{1;1}(j_p^2\psi) + (u_{1;2}(j_p^2\psi) - u_{2;1}(j_p^2\psi)),$$

but where the other components of τ_1 and τ_2 are equal. Then

$$S_1(du_{11}) = (du_1 - u_{11}dx^1) \wedge dx^2,$$

but

$$S_2(du_{11}) = (du_1 - u_{11}dx^1) \wedge dx^2 - (du_1 \wedge dx^1 + du_2 \wedge dx^2),$$

so that S_1 and S_2 both satisfy the conditions we have specified for the operator $S_\Omega^{(2)}$, but $S_1 \neq S_2$. A similar example can obviously be constructed in cases where $k \geq 2$ and $m \geq 2$. ∎

That example used a second-order operator $S_\Omega^{(2)}$, and it is important to note that the alternative operator was constructed using a tubular neighbourhood which was not obtained from a coordinate system in the way described earlier. However, it is in fact the case that—for second-order systems—our earlier description *can* be made to yield a unique operator and a unique Cartan form: in coordinates, we will always have

$$\begin{aligned} S_\Omega^{(2)}(dL) \quad = \quad & \left(\left(\frac{\partial L}{\partial u_i^\alpha} - \frac{1}{n(ij)} \frac{d}{dx^j} \frac{\partial L}{\partial u_{ij}^\alpha} \right) (du^\alpha - u_k^\alpha dx^k) \right. \\ & \left. + \frac{1}{n(ij)} \frac{\partial L}{\partial u_{ij}^\alpha} (du_j^\alpha - u_{jk}^\alpha dx^k) \right) \wedge \left(\frac{\partial}{\partial x^i} \lrcorner \, \Omega \right), \end{aligned}$$

where, as usual on $J^2\pi$, $n(ij)$ denotes the number of distinct indices represented by i and j.

Theorem 6.5.20 *There is a unique operator $S_\Omega^{(2)}$ which satisfies the conditions of Theorem 6.5.13 and which, in each local coordinate system, may be constructed from the tubular neighbourhood defined by that coordinate system.*

Proof We shall show that the coordinate representation of $S_\Omega^{(2)}(dL)$ given above is unaltered by a change to a different coordinate system. First, if just the dependent variable coordinates are changed, we may let (x^i, v^β) to be the new coordinate system on E. The terms in the coordinate representation transform as follows:

$$\begin{aligned} du^\alpha - u^\alpha_k dx^k &= \frac{\partial u^\alpha}{\partial v^\beta}(dv^\beta - v^\beta_k dx^k); \\ du^\alpha_j - u^\alpha_{jk} dx^k &= \frac{\partial u^\alpha}{\partial v^\beta}(dv^\beta_j - v^\beta_{jk} dx^k) + \frac{d}{dx^j}\left(\frac{\partial u^\alpha}{\partial v^\beta}\right)(dv^\beta - v^\beta_k dx^k); \\ \frac{\partial}{\partial u^\alpha_i} &= \frac{\partial v^\gamma}{\partial u^\alpha}\frac{\partial}{\partial v^\gamma_i} + \frac{2}{n(ij)}\frac{d}{dx^j}\left(\frac{\partial v^\gamma}{\partial u^\alpha}\right)\frac{\partial}{\partial v^\gamma_{ij}}; \\ \frac{\partial}{\partial u^\alpha_{ij}} &= \frac{\partial v^\gamma}{\partial u^\alpha}\frac{\partial}{\partial v^\gamma_{ij}}. \end{aligned}$$

Invariance of the coordinate representation follows from a straightforward calculation using

$$\frac{d}{dx^j}\left(\frac{\partial v^\gamma}{\partial u^\alpha}\frac{\partial u^\alpha}{\partial v^\beta}\right) = 0.$$

On the other hand, if the independent variable coordinates are changed, we may let (y^j, u^α) to be the new coordinate system on E. In this case, the calculations are simpler by letting τ be the tubular neighbourhood projection corresponding to the original coordinate system, and writing this in the new coordinate system:

$$\begin{aligned} \tau^*(du^\alpha) &= du^\alpha; \\ \tau^*(du^\alpha_i) &= \tfrac{1}{2}(du^\alpha_{i;} + du^\alpha_{;i}); \\ \tau^*(du^\alpha_{ij}) &= \tfrac{1}{2}(du^\alpha_{i;j} + du^\alpha_{j;i}) + \tfrac{1}{2}a^m_{ij}du^\alpha_{m;} + \tfrac{1}{2}b^m_{ij}du^\alpha_{;m}, \end{aligned}$$

where

$$\begin{aligned} a^m_{ij} &= \frac{\partial y^p}{\partial x^i}\frac{\partial y^q}{\partial x^j}\frac{\partial^2 x^m}{\partial y^p \partial y^q} \\ b^m_{ij} &= \frac{\partial^2 y^p}{\partial x^i \partial x^j}\frac{\partial x^m}{\partial y^p}. \end{aligned}$$

An explicit calculation of $S_\Omega^{(2)}$ in the new coordinates shows that again the coordinate representation is unchanged, this time as a consequence of $a^m_{ij} + b^m_{ij} = 0$. Our specification of $S_\Omega^{(2)}$ in local coordinates therefore gives a well-defined operator on the whole of $J^2\pi$. ∎

Corollary 6.5.21 *It is possible to select a unique Cartan form in second-order field theories.* ∎

EXERCISES

6.5.1 Let $\omega^i \in \bigwedge^1 M$ (where $1 \leq i \leq m$) satisfy $d\omega^i = 0$, and suppose that I is a multi-index with $|I| > k$. Show that $S^{(k)}_{\omega^I} = 0$.

6.5.2 If $\omega \in \bigwedge^1 M$ satisfies $d\omega = 0$, and if there is a point $p \in M$ where $\omega_p \neq 0$, show that $\operatorname{rank}(S^{(k)}_\omega)_{j^k_p\phi} = n({}^{m+k-1}C_{k-1})$.

6.5.3 If $\omega^i \in \bigwedge^1 M$ (where $1 \leq i \leq m$) satisfies $d\omega^i = 0$, and if there is some $p \in M$ where each $\omega^i_p \neq 0$, show that $S^{(k)}_{\omega^I} \neq 0$ for $|I| \leq k$.

6.5.4 If $(f, \overline{f}) : \pi \longrightarrow \pi$ is a bundle morphism where $\overline{f}$ is an isomorphism, show that

$$j^k f_* \circ (S^{(k)}_\omega)_{j^k_p\phi} = (S^{(k)}_\omega)_{j^k f(j^k_p\phi)} \circ j^k f_*.$$

6.5.5 If $\omega^1, \omega^2 \in \bigwedge^1 M$ satisfy $d\omega^1 = d\omega^2 = 0$, show that

$$[S^{(k)}_{\omega^1}, S^{(k)}_{\omega^2}]^v = 0,$$

where the bracket is the vertical bracket described in Exercise 6.3.7.

REMARKS

The form of the variational bicomplex used in this chapter and in Chapter 7 is based on a version given in an article by Tulczyjew [17]; this article also gives a definition of the holonomic lift operation in a form similar to that used here.

There are many approaches to the higher-order Cartan form (or Poincaré-Cartan form, as it is often called); the approach taken in this chapter, and in particular the method of repeated integration by parts, uses ideas which originate in an article by Kuperschmidt [11]. (The justification for our assumption about the existence of tubular neighbourhoods may be found in [12].) An approach to the Cartan form which uses the idea of "local Lepagean equivalence" may be found in an article by Krupka [10]; the result has the same local coordinate representation as the Cartan form described in the present chapter.

Chapter 7

Infinite Jet Bundles

Many of the constructions described in the last chapter may be carried out on jet manifolds of various orders, with results which are related by the jet projections. In many cases, a clearer formulation of these results is possible if we can avoid the need to keep track of the order of the jets. The way to do this is to use "infinite jets".

There are two approaches to this idea. One is to regard the "infinite jet manifold" as merely a convenient fiction, and to regard entities defined on different jet manifolds as equivalent when they are related by the appropriate projection maps; these equivalence classes are then the corresponding entities defined on the fictitious manifold "$J^\infty\pi$". With this approach, one has to keep in mind just which properties the various entities are meant to possess: for example, a "vector field" on "$J^\infty\pi$" is actually an equivalence class of vector fields, and there is no reason *a priori* why such an object should have any of the standard properties of vector fields.

The alternative approach, which we shall adopt here, is to define $J^\infty\pi$ as a *bona fide* manifold. The result, of course, will be an infinite-dimensional manifold, and in the first section of this chapter we shall describe some of the ideas which are needed for its definition.

7.1 Preliminaries

The first two definitions in this section are taken from the theory of categories, although we shall only apply that theory to the particular category of real topological vector spaces and continuous linear maps. We shall start, therefore, with an infinite family $V_0, V_1, V_2, \ldots$ of topological vector spaces, and a corresponding infinite family $f_{n+1,n} : V_{n+1} \longrightarrow V_n$ of continuous linear maps.

Definition 7.1.1 The family $(V, f_{\infty,n})$ is called an *inverse limit* of the family $(V_n, f_{n+1,n})$ if:

1. V is a topological vector space, each $f_{\infty,n} : V \longrightarrow V_n$ is a continuous linear map, and $f_{n+1,n} \circ f_{\infty,n+1} = f_{\infty,n}$ for each $n \in \mathbf{N}$;
2. if W is a topological vector space and $g_{\infty,n} : W \longrightarrow V_n$ are continuous linear maps which satisfy $f_{n+1,n} \circ g_{\infty,n+1} = g_{\infty,n}$ for $n \in \mathbf{N}$, then there is a unique continuous linear map $g : W \longrightarrow V$ which satisfies $g_{\infty,n} = f_{\infty,n} \circ g$ for $n \in \mathbf{N}$.

■

We may illustrate this definition using a commutative diagram.

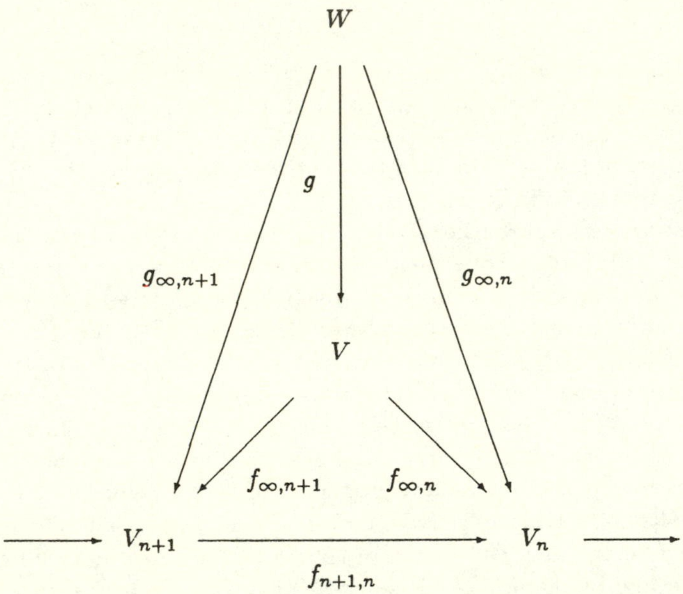

If the inverse limit of such a family exists, then it is unique to within isomorphism: if $(V, f_{\infty,n})$ and $(U, h_{\infty,n})$ are both candidates, then there are maps $h : U \longrightarrow V$ satisfying $h_{\infty,n} = f_{\infty,n} \circ h$, and $f : V \longrightarrow U$ satisfying $f_{\infty,n} = h_{\infty,n} \circ f$. It follows from this that $f_{\infty,n} = f_{\infty,n} \circ h \circ f$ for $n \in \mathbf{N}$. By applying the definition of inverse limit to $(V, f_{\infty,n})$, and letting $W = V$ and $g_{\infty,n} = f_{\infty,n}$, we see that there is a unique map $g : V \longrightarrow V$ satisfying $f_{\infty,n} = f_{\infty,n} \circ g$. Since both id_V and $h \circ f$ satisfy this condition, it follows that $h \circ f = id_V$. A similar argument shows that $f \circ h = id_U$, establishing the isomorphism.

Example 7.1.2 Let $p_{n+1,n} : \mathbf{R}^{n+1} \longrightarrow \mathbf{R}^n$ be the projection on the first n components. The family $(\mathbf{R}^n, p_{n+1,n})$ then has an inverse limit $(\mathbf{R}^\infty, p_{\infty,n})$, where $\mathbf{R}^\infty$ is the vector space of all infinite sequences of real numbers, where

$p_{\infty,n} : \mathbf{R}^\infty \longrightarrow \mathbf{R}^n$ is again projection on the first n components, and where the inverse limit topology on the vector space $\mathbf{R}^\infty$ is defined by letting subsets of the form $p_{\infty,n}^{-1}(O_n)$, $O_n \subset \mathbf{R}^n$, O_n open, be a basis for the open sets. It is not hard to see that the linear maps $p_{\infty,n}$ are continuous, and that the relation $p_{n+1,n} \circ p_{\infty,n+1} = p_{\infty,n}$ holds for all n. If W is another topological vector space and $g_{\infty,n} : W \longrightarrow \mathbf{R}^n$, then we may define $g : W \longrightarrow \mathbf{R}^\infty$ by setting the n-th component of $g(x) \in \mathbf{R}^\infty$ to equal the n-th component of $g_{\infty,n}(x) \in \mathbf{R}^n$:

$$(g(x))_n = (g_{\infty,n}(x))_n.$$

The map g is then linear and continuous, it has the property that $g_{\infty,n} = p_{\infty,n} \circ g$, and it is the only continuous linear map which does so. ∎

The usual (Hausdorff) topology on the finite-dimensional space $\mathbf{R}^n$ is, of course, derived from the standard Euclidean norm. On infinite-dimensional spaces, however, the topology need not be derived from a norm, and indeed $\mathbf{R}^\infty$ has no suitable norm. To see this, suppose that the contrary were the case. Let $e_{(n)} \in \mathbf{R}^\infty$ be defined by $e_{(n)i} = \delta_{ni} \in \mathbf{R}$, where $n, i \in \mathbf{N}^+$, and where the notation $e_{(n)i}$ indicates the i-th component of the element $e_{(n)}$; then put

$$x_{(n)} = \frac{e_{(n)}}{\|e_{(n)}\|}.$$

If $O \subset \mathbf{R}^\infty$ is an arbitrary neighbourhood of zero, put

$$O = \bigcup_\lambda p_{\infty,n_\lambda}^{-1}(O_\lambda) \qquad O_\lambda \subset \mathbf{R}^{n_\lambda},\ O_\lambda \text{ open},$$

and choose an index μ such that $0 \in p_{\infty,n_\mu}^{-1}(O_\mu)$. Whenever $n > n_\mu$, the first n_μ components of $x_{(n)}$ are zero, and so $x_{(n)} \in p_{\infty,n_\mu}^{-1}(O_\mu) \subset O$, demonstrating that $x_{(n)} \to 0$ as $n \to \infty$. Since the norm must necessarily be a continuous function, it follows that $\|x_{(n)}\| \to 0$; however, by construction, each $\|x_{(n)}\| = 1$.

Although $\mathbf{R}^\infty$ is therefore not a Banach space, it may be shown that it is the next best thing, a Fréchet space: that is, it is complete, metrizable, and locally convex. It is also path-connected and second-countable, and for any $n \in \mathbf{N}$ there is an obvious canonical isomorphism between $\mathbf{R}^n \times \mathbf{R}^\infty$ and $\mathbf{R}^\infty$.

The reason for the name "inverse limit" is that the object so constructed is at the blunt end of all the arrows. There is a dual construction, called the direct limit, where all the arrows are turned round, and which we shall also need to use. We suppose, therefore, that the infinite family $V_0, V_1, V_2, \ldots$ of topological vector spaces is now linked by an infinite family $f_{n,n+1} : V_n \longrightarrow V_{n+1}$ of continuous linear maps.

Definition 7.1.3 The family $(V, f_{n,\infty})$ is called a *direct limit* of the family $(V_n, f_{n,n+1})$ if:

1. V is a topological vector space, each $f_{n,\infty} : V_n \longrightarrow V$ is a continuous linear map, and $f_{n+1,\infty} \circ f_{n,n+1} = f_{n,\infty}$ for $n \in \mathbf{N}$;
2. if W is a topological vector space and $g_{n,\infty} : V_n \longrightarrow W$ are continuous linear maps which satisfy $g_{n+1,\infty} \circ f_{n,n+1} = g_{n,\infty}$ for $n \in \mathbf{N}$, then there is a unique continuous linear map $g : V \longrightarrow W$ which satisfies $g_{n,\infty} = g \circ f_{n,\infty}$.

■

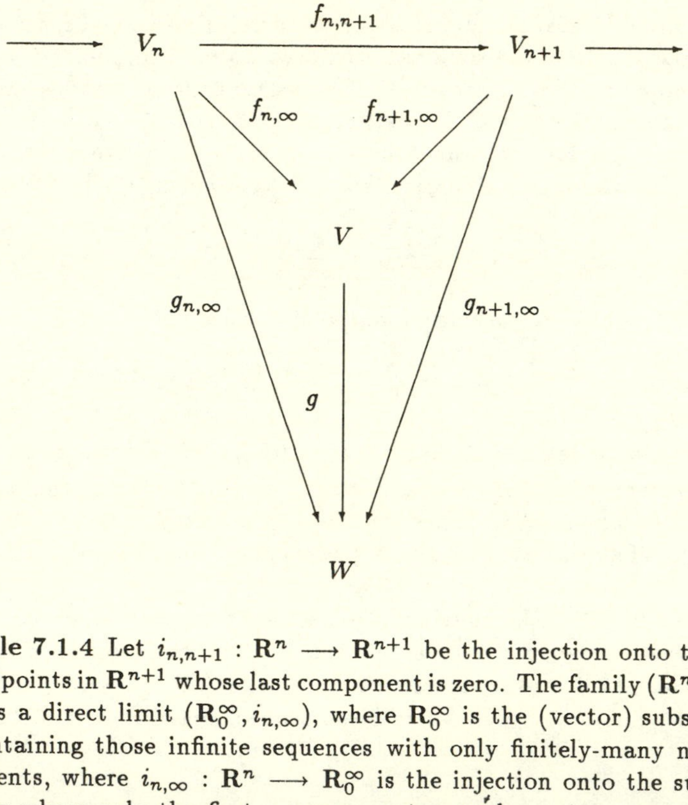

Example 7.1.4 Let $i_{n,n+1} : \mathbf{R}^n \longrightarrow \mathbf{R}^{n+1}$ be the injection onto the subspace of points in $\mathbf{R}^{n+1}$ whose last component is zero. The family $(\mathbf{R}^n, i_{n,n+1})$ then has a direct limit $(\mathbf{R}^\infty_0, i_{n,\infty})$, where $\mathbf{R}^\infty_0$ is the (vector) subspace of $\mathbf{R}^\infty$ containing those infinite sequences with only finitely-many non-zero components, where $i_{n,\infty} : \mathbf{R}^n \longrightarrow \mathbf{R}^\infty_0$ is the injection onto the subspace of points where only the first n components may be non-zero, and where the direct limit topology on $\mathbf{R}^\infty_0$ is defined by specifying that $O \subset \mathbf{R}^\infty_0$ is open if, and only if, for each $n \in \mathbf{N}$, $i^{-1}_{n,\infty}(O)$ is open in $\mathbf{R}^n$. It is not hard to see that the linear maps $i_{n,\infty}$ are continuous, and that the relation $i_{n,\infty} = i_{n+1,\infty} \circ i_{n,n+1}$ holds for all $n \in \mathbf{N}$. If W is another topological vector space and $g_{n,\infty} : \mathbf{R}^n \longrightarrow W$, then we may define $g : \mathbf{R}^\infty_0 \longrightarrow Y$ as follows:

set $g(0) = 0$; if $x \in \mathbf{R}_0^\infty$ is non-zero then let n be the largest index for which x_n is non-zero, so that there is a unique $\overline{x} \in \mathbf{R}^n$ with $i_{n,\infty}(\overline{x}) = x$, and then set $g(x)$ to equal $g_{n,\infty}(\overline{x})$. The map g is then linear and continuous, it has the property that $g_{n,\infty} = g \circ i_{n,\infty}$, and it is the only continuous linear map to do so. ∎

The space $\mathbf{R}_0^\infty$ will play a less prominent rôle than $\mathbf{R}^\infty$ in our discussion, as it normally appears through duality. It is easy to see that $\mathbf{R}^\infty$ may be identified with the algebraic dual of $\mathbf{R}_0^\infty$: given $x \in \mathbf{R}^\infty$, the map $Qx : \mathbf{R}_0^\infty \longrightarrow \mathbf{R}$ defined by

$$(Qx)(y) = \sum_{k=1}^{\infty} x_k y_k$$

(finite sum since $y \in \mathbf{R}_0^\infty$) is obviously linear; conversely, given any linear map $\alpha : \mathbf{R}_0^\infty \longrightarrow \mathbf{R}$, put $\alpha_i = \alpha(e_{(i)})$ where again $e_{(i)k} = \delta_{ik}$, and define $x_{(\alpha)} \in \mathbf{R}^\infty$ by $x_{(\alpha)i} = \alpha_i$. Then $Qx_{(\alpha)} = \alpha$ because the $e_{(i)}$ form a basis of $\mathbf{R}_0^\infty$. It is equally easy to see that $\mathbf{R}_0^\infty$ *cannot* be identified with the algebraic dual of $\mathbf{R}^\infty$: the canonical map from a vector space to its double algebraic dual is no longer surjective.

Instead of algebraic duals, however, we shall consider topological duals, and then this problem no longer arises. In general, we shall denote the topological dual of a topological vector space V by V^*. Under pointwise operations, V^* becomes a vector space. It may be given a topological structure using more than one method, and for infinite-dimensional spaces these need not produce the same topology on V. Consequently there may be different candidates for the double dual V^{**}, and in general there is no reason why any of these candidates should be naturally identified with V. Nevertheless, with the two spaces $\mathbf{R}^\infty$ and $\mathbf{R}_0^\infty$ we *do* have a symmetric relationship: the topological dual of $\mathbf{R}^\infty$ is isomorphic to the vector space $\mathbf{R}_0^\infty$, and the topological dual of $\mathbf{R}_0^\infty$ is isomorphic to the vector space $\mathbf{R}^\infty$.

To see how this relationship arises, suppose first that $\alpha : \mathbf{R}^\infty \longrightarrow \mathbf{R}$ is linear. If α depends only on finitely many components of its argument—that is, if there is a natural number n such that if $x, y \in \mathbf{R}^\infty$ and $x_k = y_k$ for $k \leq n$ then $\alpha(x) = \alpha(y)$—then α is continuous. For define $\alpha_n : \mathbf{R}^n \longrightarrow \mathbf{R}$ by $\alpha_n = \alpha \circ i_{n,\infty}$; the condition on α then implies that $\alpha = \alpha_n \circ p_{\infty,n}$, and since α_n and $p_{\infty,n}$ are continuous, so is α; consequently $\alpha \in \mathbf{R}^{\infty *}$. On the other hand, if α depends on infinitely many components of its argument, then for every $n \in \mathbf{N}$ there are $x_{(n)}, y_{(n)} \in \mathbf{R}^\infty$ with $x_{(n)k} = y_{(n)k}$ for $k \leq n$ but $\alpha(x_{(n)}) \neq \alpha(y_{(n)})$. Consider the sequence of elements $z_{(n)} \in \mathbf{R}^\infty$ defined by

$$z_{(n)} = \frac{x_{(n)} - y_{(n)}}{\alpha(x_{(n)} - y_{(n)})}.$$

Then $z_{(n)} \to 0 \in \mathbf{R}^\infty$ as $n \to \infty$, whereas $\alpha(z_{(n)}) = 1$ for each n, so that α is not continuous. Consequently the map $P : \mathbf{R}_0^\infty \longrightarrow \mathbf{R}^{\infty *}$ defined by, for $y \in \mathbf{R}_0^\infty$ and $x \in \mathbf{R}^\infty$,

$$(Py)(x) = \sum_{k=1}^{\infty} y_k x_k$$

(finite sum since $y \in \mathbf{R}_0^\infty$) is a canonical linear isomorphism. Finally, to see that $\mathbf{R}^\infty \cong \mathbf{R}_0^{\infty *}$, we simply observe that every map $\mathbf{R}_0^\infty \longrightarrow \mathbf{R}$ is continuous; for if α is such a map, then for each $n \in \mathbf{N}^+$, $\alpha \circ i_{n,\infty} : \mathbf{R}^n \longrightarrow \mathbf{R}$ is linear and hence continuous. So if $O \subset \mathbf{R}$ is open then for each $n \in \mathbf{N}^+$, $i_{n,\infty}^{-1}(\alpha^{-1}(O)) = (\alpha \circ i_{n,\infty})^{-1}(O)$ is open in $\mathbf{R}^n$; consequently $\alpha^{-1}(O)$ is open in $\mathbf{R}_0^\infty$.

We shall now move on to a definition of differentiability for maps between subsets of topological vector spaces.

Definition 7.1.5 Let V, W be topological vector spaces, and let $O \subset V$ be open. The map $f : O \longrightarrow W$ is said to be *of class* C^1 if, for every $x \in O$ and every $v \in V$, the limit

$$Df(x; v) = \lim_{t \to 0} \frac{1}{t} (f(x + tv) - f(x))$$

exists, and if the resulting map $Df : O \times V \longrightarrow W$ is continuous. ∎

The quantity $Df(x; v)$ is, of course, just the directional derivative of f at x in the direction v. For each $x \in O$, the map $v \longmapsto Df(x; v)$ is linear and continuous.

Example 7.1.6 Let $O \subset \mathbf{R}^\infty$ be open, and let $f : \mathbf{R}^\infty \longrightarrow \mathbf{R}$ be of class C^1. For each $x \in O$, the map $v \longrightarrow Df(x; v)$ is a continuous linear map from $\mathbf{R}^\infty$ to $\mathbf{R}$, and so is an element of $\mathbf{R}_0^\infty$. If we let the "partial derivatives" of f be the directional derivatives in the component directions, it follows that f can only have a finite number of non-zero partial derivatives at each $x \in O$. ∎

Definition 7.1.7 The map $f : O \longrightarrow W$ is said to be *of class* C^{k+1} for $k \geq 1$ if it is of class C^k and if, for every $v_1, \ldots, v_k \in V$, the map $O \longrightarrow W$ given by

$$x \longmapsto D^k f(x; v_1, \ldots, v_k)$$

is of class C^1; the map $D^{k+1} f : O \times V^{k+1} \longrightarrow W$ is then defined by

$$\begin{aligned} & D^{k+1} f(x; v, v_1, \ldots, v_k) \\ = \; & \lim_{t \to 0} \frac{1}{t} \left(D^k f(x + tv; v_1, \ldots, v_k) - D^k f(x; v_1, \ldots, v_k) \right). \end{aligned}$$

If f is of class C^k for each $k \geq 1$ then it is said to be *smooth*, or *of class* C^∞. ∎

In much of our subsequent discussion, the codomain of the map f will be the inverse limit space $\mathbf{R}^\infty$, and checking the smoothness of such a map can be reduced to checking the smoothness of the composite maps $p_{\infty,n} \circ f : O \longrightarrow \mathbf{R}^n$.

Lemma 7.1.8 *The map $f : O \longrightarrow \mathbf{R}^\infty$ is smooth if, and only if, each composite map $f_n = p_{\infty,n} \circ f$ is smooth.*

Proof Suppose first that each f_n is smooth. We shall show that f is smooth. Since $f_m = p_{n,m} \circ f_n$, where $m < n$ and $p_{n,m} : \mathbf{R}^n \longrightarrow \mathbf{R}^m$ is the projection on the first m components, it follows that $Df_m = p_{n,m} \circ Df_n$ using the chain rule and the linearity of $p_{n,m}$. If $x \in O$ and $v \in V$, we may therefore let w be the unique element of $\mathbf{R}^\infty$ which satisfies $p_{\infty,n}(w) = Df_n(x; v)$. So let $U \subset \mathbf{R}^\infty$ be any neighbourhood of w; then, for some n, there is a neighbourhood U_n of $p_{\infty,n}(w) \in \mathbf{R}^n$ satisfying $p_{\infty,n}^{-1}(U_n) \subset U$. Since f_n is of class C^1, there is some $\varepsilon > 0$ such that

$$\frac{1}{t}\left(f_n(x + tv) - f_n(x)\right) \in U_n$$

whenever $|t| < \varepsilon$. But

$$\frac{1}{t}\left(f_n(x + tv) - f_n(x)\right) = p_{\infty,n}\left(\frac{1}{t}\left(f(x + tv) - f(x)\right)\right),$$

so that $|t| < \varepsilon$ also implies

$$\frac{1}{t}(f(x + tv) - f(x)) \in U,$$

and we may conclude that $w = Df(x; v)$. Continuity of the map Df then follows from the continuity of the maps Df_n, so that f is of class C^1. A straightforward induction argument along similar lines then shows that f is of class C^k for each k.

The converse assertion, that if f is smooth then each f_n is smooth, is obvious. ∎

We shall also need to consider maps with codomain $\mathbf{R}_0^\infty$, and checking smoothness of these maps can be more complicated. Of course, if a map $f : O \longrightarrow \mathbf{R}_0^\infty$ is the pull-back of some map $g : O \longrightarrow \mathbf{R}^n$, so that $f = i_{n,\infty} \circ g$, then the smoothness of g automatically implies that f is smooth. We shall normally restrict our attention to maps of this form. There are, however, smooth maps to $\mathbf{R}_0^\infty$ which do not satisfy this condition.

Example 7.1.9 Let $b : \mathbf{R} \longrightarrow \mathbf{R}$ be a smooth bump function satisfying $b(t) = 0$ for $|t| > \frac{1}{2}$, and $b(0) = 1$. Let $f : \mathbf{R} \longrightarrow \mathbf{R}_0^\infty$ be defined by

$$pr_n(f(t)) = b(2^{-n}(t+3)),$$

so that, for each $t \in \mathbf{R}$, only one component of $f(t)$ is non-zero. Furthermore, at each $(t, v) \in \mathbf{R} \times \mathbf{R}$,

$$pr_n(Df(t;h)) = hf_n'(t),$$

and similarly for higher derivatives, so that for each $t \in \mathbf{R}$, only one component of $D^k f(t; h_1, \ldots, h_k)$ is non-zero. Nevertheless, for any $\varepsilon > 0$, infinitely many of the functions $pr_n \circ f|_{(-\varepsilon,\varepsilon)}$ are non-zero. ∎

EXERCISE

7.1.1 Let $b : \mathbf{R} \longrightarrow \mathbf{R}$ be the bump function defined in Example 7.1.9, and let the maps $f_m : \mathbf{R} \longrightarrow \mathbf{R}^m$ for $m \in \mathbf{N}^+$, and $f : \mathbf{R} \longrightarrow \mathbf{R}_0^\infty$, be defined by

$$\begin{aligned} pr_n(f_m(t)) &= tb(2^{-n}t), \qquad n \leq m; \\ pr_n(f(t)) &= tb(2^{-n}t). \end{aligned}$$

Show that each f_m is smooth, but that f is not differentiable at zero.

7.2 Infinite Jets

If ϕ is a local section of the bundle π, we may define the ∞-jet of ϕ in a way which is a direct generalisation of our definition in earlier chapters.

Definition 7.2.1 Let (E, π, M) be a bundle and let $p \in M$. Define the local sections $\phi, \psi \in \Gamma_p(\pi)$ to be ∞-*equivalent at* p if $\phi(p) = \psi(p)$ and if, in some adapted coordinate system (x^i, u^α) around $\phi(p)$,

$$\left.\frac{\partial^{|I|}\phi^\alpha}{\partial x^I}\right|_p = \left.\frac{\partial^{|I|}\psi^\alpha}{\partial x^I}\right|_p$$

for $1 \leq |I| < \infty$ and $1 \leq \alpha \leq n$. The equivalence class containing ϕ is called the ∞-*jet of* ϕ *at* p and is denoted $j_p^\infty \phi$. ∎

Although this definition is expressed in terms of coordinates, it follows from Lemma 6.2.1 that the particular choice of coordinate system does not matter. Alternatively, we may note that the sequence of equivalence classes $(j_p^k\phi)$ satisfies

$$\ldots \subset j_p^k\phi \subset \ldots \subset j_p^2\phi \subset j_p^1\phi,$$

and that we may set

$$j_p^\infty\phi = \bigcap_{k=1}^{\infty} j_p^k\phi,$$

where the intersection is non-empty because $\phi \in j_p^k\phi$ for every k. If π happens to be a real-analytic bundle (rather than being merely C^∞), and if ϕ is a real-analytic local section, then the ∞-jet $j_p^\infty\phi$ may be considered as the Taylor series of ϕ around p.

Definition 7.2.2 The *infinite jet manifold of* π is the set

$$\{j_p^\infty\phi : p \in M, \phi \in \Gamma_p(\pi)\}$$

and is denoted $J^\infty\pi$. The functions π_∞ and $\pi_{\infty,0}$, called the *source* and *target* projections respectively, are defined by

$$\begin{aligned} \pi_\infty : J^\infty\pi &\longrightarrow M \\ j_p^\infty\phi &\longmapsto p \end{aligned}$$

and

$$\begin{aligned} \pi_{\infty,0} : J^\infty\pi &\longrightarrow E \\ j_p^\infty\phi &\longmapsto \phi(p). \end{aligned}$$

If $l \geq 1$ then the *l-jet projection* is the function $\pi_{\infty,l}$ defined by

$$\begin{aligned} \pi_{\infty,l} : J^\infty\pi &\longrightarrow J^l\pi \\ j_p^\infty\phi &\longmapsto j_p^l\phi \end{aligned}$$

■

Definition 7.2.3 Let (E, π, M) be a bundle and let (U, u) be an adapted coordinate system on E, where $u = (x^i, u^\alpha)$. The *induced coordinate system* (U^∞, u^∞) on $J^\infty\pi$ is defined by

$$U^\infty = \{j_p^k\phi : \phi(p) \in U\}$$

$$u^\infty : U^\infty \longrightarrow \mathbf{R}^\infty$$

where $u^\infty = (x^i, u^\alpha, u_I^\alpha)$ for $1 \leq |I| < \infty$. ■

If $k \geq |I|$ then the coordinate function u^α_I on $J^\infty\pi$ is of course just the pull-back by $\pi_{\infty,k}$ of the coordinate function u^α_I on $J^k\pi$.

Proposition 7.2.4 *Given an atlas of adapted charts (U, u) on E, the corresponding collection of charts (U^∞, u^∞) is a C^∞ atlas on $J^\infty\pi$.*

Proof As before, every ∞-jet $j^\infty_p\phi$ is in the domain of the chart (U^∞, u^∞) whenever (U, u) is a chart on E with $\phi(p) \in U$. There is now, however, a slight difference: the fact that $u^\infty(U^\infty)$ is open in $\mathbf{R}^\infty$ is no longer trivial, and depends on the result, which we shall not prove, that an arbitrary family of Taylor coefficients defines a C^∞ (though not necessarily analytic) function in a neighbourhood of a point in $\mathbf{R}^m$.

To see that the transition functions $v^\infty \circ (u^\infty)^{-1}$ are smooth, we simply observe that the composite maps

$$p_{\infty,n} \circ v^\infty \circ (u^\infty)^{-1} : u^\infty(U^\infty \cap V^\infty) \longrightarrow \mathbf{R}^n,$$

where $n = \dim J^k\pi$, satisfy

$$p_{\infty,n} \circ v^\infty \circ (u^\infty)^{-1} = v^k \circ u^k \circ p_{\infty,n}\Big|_{u^\infty(U^\infty \cap V^\infty)},$$

because the new k-th derivative coordinates depend only on the old l-th derivative coordinates for $l \leq k$, and on the dependent and independent variables. It follows that each map $p_{\infty,n} \circ v^\infty \circ (u^\infty)^{-1}$ is smooth for arbitrarily large n, and hence for all n, so that we may apply Lemma 7.1.8 to deduce smoothness. ∎

Lemma 7.2.5 *The functions $\pi_\infty : J^\infty\pi \longrightarrow M$, $\pi_{\infty,0} : J^\infty\pi \longrightarrow E$ and $\pi_{\infty,k} : J^\infty\pi \longrightarrow J^k\pi$ are smooth surjective submersions.*

Proof Similar to the proof of Lemma 4.1.9; note that the standard definition of a submersion, as a map f whose derivative f_* is surjective at each point, is still applicable in these circumstances because the codomain manifold is finite-dimensional. ∎

Proposition 7.2.6 *The space $J^\infty\pi$ is an infinite-dimensional manifold, and for $k \geq 0$ the triple $(J^\infty\pi, \pi_{\infty,k}, J^k\pi)$ is a bundle.*

Proof We shall consider bundles whose total spaces are infinite-dimensional manifolds, so long as they satisfy the remaining conditions given in Definition 1.1.8. They must, in particular, be locally trivial, and the local triviality of $(J^\infty\pi, \pi_{\infty,k}, J^k\pi)$ follows from arguments similar to those used in the proof of Proposition 6.2.8.

We have already seen that $J^\infty\pi$ has an infinite-dimensional C^∞ atlas; we shall also require an infinite-dimensional manifold to satisfy the topological conditions we required of a finite-dimensional manifold, namely that it be connected, second-countable, and Hausdorff. The arguments used in the proof of Proposition 1.1.14 may then be used to show that, since the fibres of $\pi_{\infty,0}$ are manifolds (each fibre is diffeomorphic to $\mathbf{R}^\infty$ because an arbitrary family of Taylor coefficients may arise from a C^∞ function), it follows that $J^\infty\pi$ is a manifold. ∎

Proposition 7.2.7 *The triple $(J^\infty\pi, \pi_\infty, M)$ is a bundle.*

Proof Similar to the proof of Proposition 4.1.21. ∎

Example 7.2.8 Let π be the trivial bundle $(\mathbf{R} \times F, pr_1, \mathbf{R})$. The infinite jet bundle $(J^\infty\pi, \pi_\infty, \mathbf{R})$ is then trivial, and we may write $J^\infty\pi \cong \mathbf{R} \times J_0^\infty\pi$. The fibre $J_0^\infty\pi$ is just the *infinite tangent manifold to F*, and an alternative notation for this manifold would be $T^\infty F$; the infinite tangent projection τ_F^∞ is the map $T^\infty F \longrightarrow F$ given by $\pi_{\infty,0}|_{J_0^\infty\pi}$. ∎

We may define the infinite prolongation of a local section, or of a bundle morphism, in the same way as for a finite prolongation.

Definition 7.2.9 Let ϕ be a local section of π with domain $W \subset M$. The *infinite prolongation of ϕ* is the map $j^\infty\phi : W \longrightarrow J^\infty\pi$ defined by

$$j^\infty\phi(p) = j_p^\infty\phi.$$

∎

The map $j^\infty\phi$ is clearly a local section of π_∞. It is smooth because each map $\pi_{\infty,k} \circ j^\infty\phi = j^k\phi$ is smooth, so that we can apply the arguments of Lemma 7.1.8.

Definition 7.2.10 Let (E, π, M) and (F, ρ, N) be bundles, and let $(f, \overline{f}) : \pi \longrightarrow \rho$ be a bundle morphism, where $\overline{f}$ is a diffeomorphism. The *infinite prolongation of f* is the map $j^\infty(f, \overline{f}) : J^\infty\pi \longrightarrow J^\infty\rho$ defined by

$$j^\infty(f, \overline{f})(j_p^\infty\phi) = j_{\overline{f}(p)}^\infty\tilde{f}(\phi).$$

∎

To see that $j^\infty(f, \overline{f})$ is smooth, observe that each composite map $\rho_{\infty,k} \circ j^\infty(f, \overline{f}) = j^k(f, \overline{f}) \circ \pi_{\infty,k}$ is smooth, so that once again we may apply the arguments of Lemma 7.1.8.

Our next task will be to construct the tangent and cotangent spaces to $J^\infty\pi$. There is no difficulty in defining a tangent vector to $J^\infty\pi$ at $j_p^\infty\phi$ as an equivalence class of curves γ through $j_p^\infty\phi$, and then $T_{j_p^\infty\phi}J^\infty\pi$ is isomorphic to $\mathbf{R}^\infty$. In coordinates, a tangent vector $\xi = [\gamma]$ may therefore be written as

$$\xi = \xi^i \left.\frac{\partial}{\partial x^i}\right|_{j_p^\infty\phi} + \sum_{|I|=0}^{\infty} \xi_I^\alpha \left.\frac{\partial}{\partial u_I^\alpha}\right|_{j_p^\infty\phi},$$

where $\xi^i = (x^i \circ \gamma)'(0)$ and $\xi_I^\alpha = (u_I^\alpha \circ \gamma)'(0)$. Note that the summation indicated in this expression is purely formal: the coordinate tangent vectors are simply placeholders in an infinite sequence, and no question of convergence arises. A tangent vector may, as usual, act on a function $f \in C^\infty(J^\infty\pi)$ to give the real number

$$\xi(f) = (f \circ \gamma)'(0) = \xi^i \left.\frac{\partial f}{\partial x^i}\right|_{j_p^\infty\phi} + \sum_{|I|=0}^{\infty} \xi_I^\alpha \left.\frac{\partial f}{\partial u_I^\alpha}\right|_{j_p^\infty\phi};$$

the summation is now finite because any smooth function on $\mathbf{R}^\infty$ (and hence on $J^\infty\pi$) can have only a finite number of non-zero partial derivatives. Another manifestation of this phenomenon is that the cotangent space $T^*_{j_p^\infty\phi}J^\infty\pi$ is isomorphic to $\mathbf{R}_0^\infty$, so that a cotangent vector η may be written as a finite expression

$$\eta = \eta_i \, dx^i\Big|_{j_p^\infty\phi} + \sum_{|I|=0}^{k} \eta_I^\alpha \, du_I^\alpha\big|_{j_p^\infty\phi}$$

for some $k \in \mathbf{N}$. It follows that every cotangent vector on $J^\infty\pi$ is the pull-back of a cotangent vector on some finite jet manifold $J^k\pi$ (although, as we shall see, the corresponding property does *not* hold for differential forms).

Using the tangent and cotangent spaces to $J^\infty\pi$, we may go on to construct the tangent and cotangent bundles, $(TJ^\infty\pi, \tau_{J^\infty\pi}, J^\infty\pi)$ and $(T^*J^\infty\pi, \tau^*_{J^\infty\pi}, J^\infty\pi)$. A suitable generalisation of our definition of a vector bundle would allow us to deduce that these were vector bundles over the infinite-dimensional manifold $J^\infty\pi$, whose total spaces were modelled on the topological vector spaces $\mathbf{R}^\infty \times \mathbf{R}^\infty$ and $\mathbf{R}^\infty \times \mathbf{R}_0^\infty$ respectively. We shall not pursue this matter; we shall, however, be interested in sections of these two bundles.

Definition 7.2.11 A *vector field on* $J^\infty\pi$ is a smooth section of the bundle $\tau_{J^\infty\pi}$. ∎

To use this definition, we must know how to check whether a section X of $\tau_{J^\infty\pi}$ is smooth. This is a local matter, so it is sufficient to consider a coordinate chart (U^∞, u^∞) and the coordinate representation

$$X = X^i \frac{\partial}{\partial x^i} + \sum_{|I|=0}^{\infty} X_I^\alpha \frac{\partial}{\partial u_I^\alpha}$$

valid on U^∞. Just as for a vector field on a finite-dimensional manifold, the coordinate representation defines a map $X_{U^\infty} : U^\infty \longrightarrow \mathbf{R}^\infty$, and X is smooth precisely when each map X_{U^∞} is smooth. We may now apply the argument of Lemma 7.1.8 to say that X is smooth whenever the coordinate functions X^i and X_I^α are all smooth.

Example 7.2.12 The locally-defined vector field

$$\frac{d}{dx^i} = \frac{\partial}{\partial x^i} + \sum_{|I|=0}^{\infty} u_{I+1_i}^\alpha \frac{\partial}{\partial u_I^\alpha}$$

is smooth; it is the infinite version of the coordinate total derivative introduced in previous chapters. Note that, unlike the earlier version, this is a vector field defined on a manifold, rather than along a map. ∎

As usual, vector fields on $J^\infty\pi$ act as derivations on functions; this action may be defined pointwise by the action of the corresponding tangent vectors, so that

$$d_X f(j_p^\infty \phi) = X_{j_p^\infty \phi}(f).$$

To see that the resulting function $d_X f$ is smooth, observe that it may be described using the coordinate chart (U^∞, u^∞) by

$$d_X f(j_p^\infty \phi) = D(f \circ (u^\infty)^{-1})(u^\infty(j_p^\infty \phi), X_{U^\infty}(j_p^\infty \phi)).$$

On the other hand, it is *not* true that all vector fields on $J^\infty\pi$ have flows. The trouble here is that the usual proof of the existence of flows relies on the coefficients of the vector field satisfying a Lipschitz condition, and this does not make sense in the absence of a norm. We shall not need to use flows in this chapter.

In contrast to the situation with vector fields, the characterisation of suitable differential forms on $J^\infty\pi$ needs a little care. The most general 1-form is just a smooth section of the cotangent bundle $\tau^*_{J^\infty\pi}$, and so its coordinate representation defines a map $\omega_{U^\infty} : U^\infty \longrightarrow \mathbf{R}_0^\infty$; ω is smooth precisely when each map ω_{U^∞} is smooth. We saw in Section 7.1 the complexity of determining whether maps to $\mathbf{R}_0^\infty$ were smooth, and, in particular, we saw that such maps need not be pullbacks of maps to $\mathbf{R}^n$.

Example 7.2.13 Let π be the trivial bundle $(\mathbf{R} \times \mathbf{R}, pr_1, \mathbf{R})$ with global coordinates $(t = id_{\mathbf{R}}, q)$, and let $b : \mathbf{R} \longrightarrow \mathbf{R}$ be the bump function defined in Example 7.1.9. The 1-form ω on $J^\infty\pi$ given in coordinates by

$$\omega_{j_p^\infty\phi} = b(2^{-1}(3+p))\, dt|_{j_p^\infty\phi} + \sum_{r=0}^{\infty} b(2^{-(r+2)}(3+p))\, dq_{(r)}\Big|_{j_p^\infty\phi}$$

is smooth, because the corresponding function $\omega_{J^\infty\pi} : J^\infty\pi \longrightarrow \mathbf{R}_0^\infty$ defined by the coordinate system satisfies $\omega_{J^\infty\pi} = f \circ \pi_\infty$, where $f : \mathbf{R} \longrightarrow \mathbf{R}_0^\infty$ is the smooth function defined in Example 7.1.9. The 1-form ω is therefore *not* the pull-back to $J^\infty\pi$ of a 1-form on a finite jet manifold. ∎

Another manifestation of this phenomenon arises when we consider smooth functions on $J^\infty\pi$. At any point $j_p^\infty\phi \in J^\infty\pi$, such a function f will yield a cotangent vector $df|_{j_p^\infty\phi}$ which in coordinates will be given by

$$df|_{j_p^\infty\phi} = \frac{\partial f}{\partial x^i}\Big|_{j_p^\infty\phi} dx^i\Big|_{j_p^\infty\phi} + \sum_{|I|=0}^{k} \frac{\partial f}{\partial u_I^\alpha}\Big|_{j_p^\infty\phi} du_I^\alpha|_{j_p^\infty\phi}$$

for some $k \in \mathbf{N}$; as we saw in Example 7.1.6, the map f can only have a finite number of non-zero partial derivatives at each point. There is, however, no reason why a similar restriction should apply to the 1-form df, and in general there may be infinitely many of the *functions* $\partial f/\partial u_I^\alpha$ which are not identically zero.

Nevertheless, for most purposes it is unnecessary to use differential forms which are not the pullbacks of forms on a finite jet manifold. We shall therefore adopt a definition of *differential forms of finite order*, and our definition will be suitable to apply more generally to r-forms.

Definition 7.2.14 A *differential r-form on $J^\infty\pi$ of finite order* is an element of $\pi_{\infty,k}^*(\bigwedge^r J^k\pi)$ for some $k \in \mathbf{N}$. The set of all such differential r-forms will be denoted $\bigwedge_F^r J^\infty\pi$. ∎

Note that $\bigwedge_F^r J^\infty\pi$ is *not* a module over $C^\infty(J^\infty\pi)$; it is, however, a module over $\bigwedge_F^0 J^\infty\pi$. In coordinates, an element of $\bigwedge_F^1 J^\infty\pi$ takes the form

$$\sigma = \sigma_i dx^i + \sum_{|I|=0}^{k} \sigma_\alpha^I du_I^\alpha$$

for some $k \in \mathbf{N}$, where the functions σ_i and σ_I^α are also pulled back from a finite jet manifold; σ will be smooth when all these functions are smooth. Elements of $\bigwedge_F^r J^\infty\pi$ are then sums of wedge products of elements of $\bigwedge_F^1 J^\infty\pi$.

EXERCISES

7.2.1 Let G be a Lie group, let π be the trivial bundle $(\mathbf{R}\times(G\times G), pr_1, \mathbf{R})$, and let ρ be the trivial bundle $(\mathbf{R}\times G, pr_1, \mathbf{R})$. Let $\mu : G\times G \longrightarrow G$ denote group multiplication, and let $f = (id_{\mathbf{R}}\times\mu, id_{\mathbf{R}})$ be the corresponding bundle morphism from π to ρ. By analogy with Exercise 4.2.4, show that the prolonged map $j^\infty f : J^\infty\pi \longrightarrow J^\infty\rho$ projects to a map $T^\infty\mu : T^\infty G\times T^\infty G \longrightarrow T^\infty G$, and that $T^\infty\mu$ defines a group operation on $T^\infty G$.

7.2.2 Construct a diffeomorphism $i_\infty : J^\infty\nu_\pi \longrightarrow V\pi_\infty$ which projects to the identity on M, and use it to define the infinite prolongation X^∞ of a vertical generalised vector field $X : J^\infty\pi \longrightarrow V\pi$.

7.2.3 If X, Y are vertical vector fields on E, show that $[X,Y]^\infty = [X^\infty, Y^\infty]$.

7.3 The Infinite Contact System

One of the main advantages of using infinite jets is that the description of holonomic lifts and the contact system may be simplified. The reason for this is that there is no longer any need to consider the highest-order derivatives as a special case. A consequence is that the tangent bundle $\tau_{J^\infty\pi}$ may be written as a direct sum of the bundles of vertical and holonomic vectors, without the need to choose a connection: there is, indeed, a natural connection on $J^\infty\pi$.

Definition 7.3.1 Let (E,π,M) be a bundle, and let $p\in M$, $\phi\in\Gamma_p(\pi)$ and $\zeta\in T_pM$. The *infinite holonomic lift of ζ by ϕ* is defined to be

$$(j^\infty\phi)_*(\zeta)\in T_{j^\infty_p\phi}J^\infty\pi.$$

■

Notice that, at each point $j^\infty_p\phi\in J^\infty\pi$, the infinite holonomic lift of $\zeta\in T_pM$ is a well-defined element of the tangent space at that point which does not depend upon the particular representative ϕ of the infinite jet $j^\infty_p\phi$. This is in sharp contrast to the k-jet case, where a $(k+1)$-jet $j^{k+1}_p\phi$ is needed to define a unique k-th holonomic lift $(j^k\phi)_*(\zeta)$.

Theorem 7.3.2 *Let (E,π,M) be a bundle, and let $j^\infty_p\phi\in J^\infty\pi$. There is then a canonical decomposition of the vector space $T_{j^\infty_p\phi}J^\infty\pi$ as a direct sum of two subspaces*

$$V_{j^\infty_p\phi}\pi_\infty\oplus(j^\infty\phi)_*(T_pM).$$

■

Corollary 7.3.3 *The vector bundle $(TJ^\infty\pi, \tau_{J^\infty\pi}, J^\infty\pi)$ may be written as the direct sum of the two sub-bundles*

$$(V\pi_\infty \oplus H\pi_\infty, \tau_{J^\infty\pi}, J^\infty\pi),$$

where $H\pi_\infty$ is the union of the fibres $(j^\infty\phi)_*(T_pM)$. ■

In coordinates, if

$$\zeta = \zeta^i \left.\frac{\partial}{\partial x^i}\right|_p,$$

then its infinite holonomic lift is

$$(j^\infty\phi)_*(\zeta) = \zeta^i \left(\left.\frac{\partial}{\partial x^i}\right|_{j^\infty_p\phi} + \sum_{|I|=0}^{\infty} u^\alpha_{I+1_i}(j^\infty_p\phi) \left.\frac{\partial}{\partial u^\alpha_I}\right|_{j^\infty_p\phi} \right).$$

Definition 7.3.4 An element $\eta \in T^*_{j^\infty_p\phi}J^\infty\pi$ is called a *contact cotangent vector* if $(j^\infty\phi)^*(\eta) = 0$. ■

Proposition 7.3.5 *If $j^\infty_p\phi \in J^\infty\pi$ then*

$$\pi^*_\infty(T^*M)_{j^\infty_p\phi} = (V_{j^\infty_p\phi}\pi_\infty)^\circ$$

and

$$\ker(j^\infty\phi)^* = ((j^\infty\phi)_*(T_pM))^\circ.$$

■

Theorem 7.3.6 *There is a canonical decomposition of the vector space $T^*_{j^\infty_p\phi}J^\infty\pi$ as a direct sum*

$$\pi^*_\infty(T^*M)_{j^\infty_p\phi} \oplus \ker(j^\infty\phi)^*.$$

■

Corollary 7.3.7 *The vector bundle $(T^*J^\infty\pi, \tau^*_{J^\infty\pi}, J^\infty\pi)$ may be written as the direct sum of two sub-bundles*

$$(\pi^*_\infty(T^*M) \oplus C^*\pi_\infty, \tau^*_{J^\infty\pi}, J^\infty\pi),$$

where $C^*\pi_\infty$ is the union of the fibres $\ker(j^\infty\phi)^*$ for $p \in M$. ■

In coordinates, a contact cotangent vector may be written as a finite sum

$$\eta = \sum_{|I|=0}^{k} \eta^I_\alpha (du^\alpha_I - u^\alpha_{I+1_i} dx^i)_{j^\infty_p\phi}$$

for some $k \in \mathbf{N}$.

The corresponding vector bundle endomorphisms also have a formulation here which is simpler than the one used in the context of finite jets.

Definition 7.3.8 The two vector bundle endomorphisms h and v of $\tau_{J^\infty\pi}$ are defined by

$$\begin{aligned} h(\xi^h + \xi^v) &= \xi^h \\ v(\xi^h + \xi^v) &= \xi^v, \end{aligned}$$

where $\xi^h \in H\pi_\infty$ and $\xi^v \in V\pi_\infty$. ■

Definition 7.3.9 The two vector bundle endomorphisms h and v of $\tau^*_{J^\infty\pi}$ are defined by

$$\begin{aligned} h(\eta^h + \eta^v) &= \eta^h \\ v(\eta^h + \eta^v) &= \eta^v, \end{aligned}$$

where $\eta^h \in \pi^*_\infty(T^*M)$ and $\eta^v \in C^*\pi_\infty$. ■

We may also define an operator corresponding to the vector-valued 1-form $S^{(k)}_\omega$.

Proposition 7.3.10 *Suppose given a point $j^\infty_p\phi \in J^\infty\pi$, a closed 1-form $\omega \in \bigwedge^1 M$, and a tangent vector $\zeta \in V_{j^\infty_p\phi}\pi_\infty$. There is then a unique tangent vector*

$$\omega \oslash_{j^\infty_p\phi} \zeta$$

which satisfies, for each $l \in \mathbf{N}^+$,

$$\pi_{\infty,l*}(\omega \oslash_{j^\infty_p\phi} \zeta) = \omega \oslash_{j^l_p\phi} (\pi_{\infty,l-1*}(\zeta)).$$

Proof Since $\pi_{k,l*}(\omega \oslash_{j^k_p\phi} \zeta) = \omega \oslash_{j^l_p\phi} (\pi_{k-1,l-1*}(\zeta))$ for $k > l$, we may just take $\omega \oslash_{j^\infty_p\phi} \zeta$ to be the unique element of $T_{j^\infty_p\phi}J^\infty\pi$ whose coordinate representation, when truncated to the correct length, is the same as the coordinate representation of each $\omega \oslash_{j^l_p\phi} (\pi_{\infty,l-1*}(\zeta))$; this clearly satisfies the conditions of the proposition. ■

Definition 7.3.11 If $\omega \in \bigwedge^1 M$ satisfies $d\omega = 0$, then the vector-valued 1-form $S^{(\infty)}_\omega \in \bigwedge^1_F J^\infty\pi \otimes \mathcal{V}(\pi_{\infty,0})$ is defined by

$$(S^{(\infty)}_\omega)_{j^\infty_p\phi}(\xi) = \omega \oslash_{j^\infty_p\phi} v(\xi),$$

where $\xi \in T_{j^\infty_p\phi}(J^\infty\pi)$. ■

We may regard $S_\omega^{(\infty)}$ as an operator $\mathcal{X}(J^\infty\pi) \longrightarrow \mathcal{X}(J^\infty\pi)$ or, by transposition and restriction to $\bigwedge_F^1 J^\infty\pi$, as an operator $\bigwedge_F^r J^\infty\pi \longrightarrow \bigwedge_F^r J^\infty\pi$. In coordinates, we may write $S_\omega^{(\infty)}$ as a tensor field, as

$$S_\omega^{(\infty)} = \sum_{|J+K|=0}^{\infty} \frac{(J+K+1_i)!}{(J+1_i)!\,K!} \frac{\partial^{|J|}\omega_i}{\partial x^J}(du_K^\alpha - u_{K+1_j}^\alpha dx^j) \otimes \frac{\partial}{\partial u_{J+K+1_i}^\alpha}.$$

Our main interest in this section will be in differential forms of finite order on $J^\infty\pi$. Just as in Section 6.3, the bundle endomorphisms h and v allow us to define horizontal and vertical differentials d_h and d_v, with the difference now that both d_h and d_v map r-forms to $(r+1)$-forms on the same manifold: they are derivations on $J^\infty\pi$ rather than, as before, along $\pi_{k+1,k}$. We may similarly define the spaces $\Phi_s^r(J^\infty\pi)$ of $(r+s)$-forms on $J^\infty\pi$ which contain r factors horizontal over M, and s contact factors; since we normally just consider a single bundle π and we are no longer counting the order of the jets, we may abbreviate this notation and simply write Φ_s^r, so that $d_h : \Phi_s^r \longrightarrow \Phi_s^{r+1}$ and $d_v : \Phi_s^r \longrightarrow \Phi_{s+1}^r$. As before, the maps d_h and d_v may be included in a commutative diagram, called the *variational bicomplex*.

Our new commutative diagram will, however, include some additional spaces. Some of these will be familiar: for instance, the spaces $\bigwedge^r M$ of r-forms on the base manifold. Others, such as the spaces of functional forms Ξ_s, have not been introduced before (although it would have been possible to define them on finite jet manifolds if we had needed to do so).

Definition 7.3.12 The space of *functional s-forms on $J^\infty\pi$* is the quotient space

$$\Xi_s = \Phi_s^m / d_h(\Phi_s^{m-1}),$$

and the map $p_s : \Phi_s^m \longrightarrow \Xi_s$ is defined to be the canonical projection. The space Ξ_0 of functional 0-forms is also known as the space of *functionals* on $J^\infty\pi$. ■

Example 7.3.13 A classical example of a functional is the map

$$\mathcal{L} : C_0^\infty[a,b] \longrightarrow \mathbf{R}$$

given by

$$\mathcal{L}[\phi] = \int_a^b L(\phi(t), \phi'(t))dt,$$

where $C_0^\infty[a,b]$ denotes the subspace of $C^\infty[a,b]$ containing functions which vanish at the endpoints a, b of the interval, and where L is a Lagrangian.

A different Lagrangian L_1 will yield the same functional $\mathcal{L}$ if the difference $L - L_1$ is a total time derivative df/dt, for

$$\begin{aligned} \int_a^b \left.\frac{df}{dt}\right|_{(\phi(t),\phi'(t))} dt &= \int_a^b (f \circ \phi)'(t) dt \\ &= f(\phi(b)) - f(\phi(a)) \\ &= 0. \end{aligned}$$

Using our present terminology, we may consider the Lagrangian pulled back to a function L on the infinite jet manifold $J^\infty \pi$, so that the 1-form $L\,dt$ is an element of Φ_0^1. If $L - L_1 = df/dt$ then

$$L\,dt - L_1 dt = \frac{df}{dt} dt = d_h f,$$

so that the two Lagrangian 1-forms $L\,dt$, $L_1 dt$ both yield the same functional when they differ by an element of $d_h(\Phi_0^0)$. It is therefore reasonable to regard $\mathcal{L}$ as an element of the quotient space $\Xi_0 = \Phi_0^1 / d_h(\Phi_0^0)$; the projection map $p_0 : \Phi_0^1 \longrightarrow \Xi_0$ is then often denoted (suggestively) by an integral sign, so that

$$\mathcal{L} = p_0(L\,dt) = \int L\,dt.$$

■

The spaces Ξ_s are related by maps $\delta : \Xi_s \longrightarrow \Xi_{s+1}$, which are constructed in such a way as to ensure the commutativity of the portion

$$\begin{array}{ccc} \Phi_s^m & \xrightarrow{\;d_v\;} & \Phi_{s+1}^m \\ {\scriptstyle p_s}\downarrow & & \downarrow{\scriptstyle p_{s+1}} \\ \Xi_s & \xrightarrow[\;\delta\;]{} & \Xi_{s+1} \end{array}$$

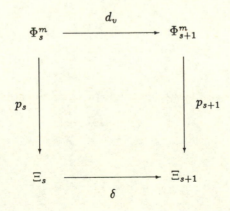

of the large diagram. If

$$p_s(\theta) = \theta + d_h(\Phi_s^{m-1}) \in \Xi_s,$$

then we set

$$\begin{aligned} \delta(p_s(\theta)) &= d_v(p_{s-1}(\theta)) \\ &= d_v\theta + d_h(\Phi_{s+1}^{m-1}) \in \Xi_{s+1}. \end{aligned}$$

This definition does not depend on the choice of θ, for if $\theta - \theta_1 = d_h\sigma$ then

$$\begin{aligned} d_v(\theta - \theta_1) &= d_v d_h \sigma \\ &= -d_h d_v \sigma \\ &\in d_h(\Phi_{s+1}^{m-1}). \end{aligned}$$

It is also clear from the definition that $\delta \circ \delta = 0$. As a result of these considerations, the complete variational bicomplex looks like this.

$$
\begin{array}{cccccccccccc}
 & & 0 & & 0 & & & & & & & \\
 & & \downarrow & & \downarrow & & & & & & & \\
 & & \mathbf{R} & & \mathbf{R} & & 0 & & 0 & & 0 & \\
 & & \downarrow & & \downarrow & & \downarrow & & \downarrow & & \downarrow & \\
0 & \longrightarrow & \bigwedge^0 M & \xrightarrow{\pi_\infty^*} & \Phi_0^0 & \xrightarrow{d_v} & \Phi_1^0 & \cdots & \Phi_s^0 & \xrightarrow{d_v} & \Phi_{s+1}^0 & \cdots \\
 & & {\scriptstyle d}\downarrow & & {\scriptstyle d_h}\downarrow & & {\scriptstyle -d_h}\downarrow & & {\scriptstyle (-1)^s d_h}\downarrow & & {\scriptstyle (-1)^{s+1} d_h}\downarrow & \\
0 & \longrightarrow & \bigwedge^1 M & \xrightarrow{\pi_\infty^*} & \Phi_0^1 & \xrightarrow{d_v} & \Phi_1^1 & \cdots & \Phi_s^1 & \xrightarrow{d_v} & \Phi_{s+1}^1 & \cdots \\
 & & \vdots & & \vdots & & \vdots & & \vdots & & \vdots & \\
0 & \longrightarrow & \bigwedge^{m-1} M & \xrightarrow{\pi_\infty^*} & \Phi_0^{m-1} & \xrightarrow{d_v} & \Phi_1^{m-1} & \cdots & \Phi_s^{m-1} & \xrightarrow{d_v} & \Phi_{s+1}^{m-1} & \cdots \\
 & & {\scriptstyle (-1)^{m-1} d}\downarrow & & {\scriptstyle (-1)^{m-1} d_h}\downarrow & & {\scriptstyle (-1)^m d_h}\downarrow & & {\scriptstyle (-1)^{m+s-1} d_h}\downarrow & & {\scriptstyle (-1)^{m+s} d_h}\downarrow & \\
0 & \longrightarrow & \bigwedge^m M & \xrightarrow{\pi_\infty^*} & \Phi_0^m & \xrightarrow{d_v} & \Phi_1^m & \cdots & \Phi_s^m & \xrightarrow{d_v} & \Phi_{s+1}^m & \cdots \\
 & & \downarrow & & {\scriptstyle p_0}\downarrow & & {\scriptstyle p_1}\downarrow & & {\scriptstyle p_s}\downarrow & & {\scriptstyle p_{s+1}}\downarrow & \\
 & & 0 & \longrightarrow & \Xi_0 & \xrightarrow{\delta} & \Xi_1 & \cdots & \Xi_s & \xrightarrow{\delta} & \Xi_{s+1} & \cdots \\
 & & & & \downarrow & & \downarrow & & \downarrow & & \downarrow & \\
 & & & & 0 & & 0 & & 0 & & 0 &
\end{array}
$$

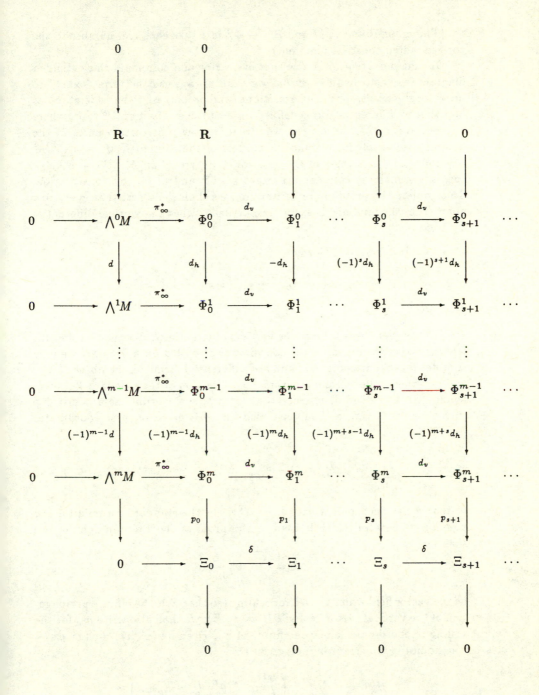

(The maps $\mathbf{R} \longrightarrow \bigwedge^0 M$ and $\mathbf{R} \longrightarrow \Phi_0^0$ just take each real number to the corresponding constant function.)

Our main contention is now that every row and column in this extended diagram is exact. In this section, we shall always use the term "exact" to mean locally exact, so that all assertions (and proofs) are to be understood as applying to differential forms defined on a suitable subset of $J^\infty\pi$, or, where appropriate, of M. Of course, we already know that certain parts of the diagram are exact; for instance, exactness of the column $\bigwedge^r M \longrightarrow \bigwedge^{r+1} M$ is simply local exactness of the exterior derivative d on M. We also know that the remaining columns are exact at Φ_s^m and at Ξ_s, by construction. To complete the proof of our contention, we shall show first that every row apart from the final one is exact: in particular, that the vertical differential d_v is exact.

Proposition 7.3.14 *The rows*

$$0 \longrightarrow \bigwedge^r M \xrightarrow{\pi_\infty^*} \Phi_0^r \xrightarrow{d_v} \Phi_1^r \xrightarrow{d_v} \ldots \xrightarrow{d_v} \Phi_s^r \xrightarrow{d_v} \ldots$$

are exact.

Proof First, let $\theta \in \Phi_s^0$. Since the vertical differential d_v does not involve differentiation with respect to the coordinates x^i pulled back from M, we may consider θ as a parametrised family of differential forms on the fibres of π_∞, and exactness follows from a parametrised version of the standard Poincaré lemma. In more detail, suppose that (x^i, u_I^α) are coordinates around $j_p^\infty\phi$ such that $\phi^\alpha(p) = 0$, and suppose also that θ is given in these coordinates by

$$\theta = \sum_{|I_1|,\ldots,|I_s|=0}^{k} \theta_{\alpha_1\ldots\alpha_s}^{I_1\ldots I_s}(du_{I_1}^{\alpha_1} - u_{I_1+1_i}^{\alpha} dx^i) \wedge \ldots \wedge (du_{I_s}^{\alpha_s} - u_{I_s+1_i}^{\alpha} dx^i),$$

so that θ has been pulled back from $J^k\pi$. In the standard formula for the homotopy operator which is used in the proof of the Poincaré lemma, we shall let

$$X = \sum_{|J|=0}^{\infty} u_J^\beta \frac{\partial}{\partial u_J^\beta}$$

be the vector field which is used for scaling (so that X is the infinite prolongation of the vertical vector field $u^\beta\partial/\partial u^\beta$ on E); we shall also write m_μ for the scaling of coordinates along the fibres of π_∞ given by $(x^i, u_I^\alpha)^{-1} \circ (x^i, \mu u_I^\alpha)$. The homotopy operator is then given by

$$H_s\theta = \sum_{|I_1|,\ldots,|I_s|=0}^{k} \left(\int_0^1 \mu^{s-1}(\theta_{\alpha_1\ldots\alpha_s}^{I_1\ldots I_s} \circ m_\mu) d\mu\right) \times$$

$$\sum_{|J|=0}^{\infty} u^\beta_J \frac{\partial}{\partial u^\beta_J} \lrcorner (du^{\alpha_1}_{I_1} - u^\alpha_{I_1+1_i} dx^i) \wedge \ldots \wedge (du^{\alpha_s}_{I_s} - u^\alpha_{I_s+1_i} dx^i),$$

and this is defined for points in a sufficiently small neighbourhood of $j^\infty_p\phi$. The usual calculation then shows that, in this neighbourhood,

$$H_{s+1} \circ d_v + d_v \circ H_s = id.$$

Now suppose that $\theta \in \Phi^r_s$ where $1 \leq r \leq m$ and $s > 0$. We may use the coordinate representation of θ to write

$$\theta = \sum_{i_1 < \ldots < i_{m-r}} \left(\frac{\partial}{\partial x^{i_1}} \lrcorner \ldots \frac{\partial}{\partial x^{i_{m-r}}} \lrcorner \Omega \right) \wedge \theta^{i_1 \ldots i_{m-r}}$$

where $\theta^{i_1 \ldots i_{m-r}} \in \Phi^0_s$, so that if $d_v\theta = 0$ then each $d_v\theta^{i_1 \ldots i_{m-r}} = 0$. From the first part of the proof, there exists $\sigma^{i_1 \ldots i_{m-r}} \in \Phi^0_{s-1}$ such that $\theta^{i_1 \ldots i_{m-r}} = d_v\sigma^{i_1 \ldots i_{m-r}}$, and then $\theta = d_v\sigma$ where

$$\sigma = \sum_{i_1 < \ldots < i_{m-r}} \left(\frac{\partial}{\partial x^{i_1}} \lrcorner \ldots \frac{\partial}{\partial x^{i_{m-r}}} \lrcorner \Omega \right) \wedge \sigma^{i_1 \ldots i_{m-r}}.$$

Exactness at Φ^r_0 follows by a simple variant of this argument, and exactness at $\bigwedge^r M$ follows because $\pi^*_\infty : \bigwedge^r M \longrightarrow \Phi^r_0$ is injective. ■

The proof of the exactness of d_v is therefore no more than a variant of the usual proof of the exactness of the exterior derivative d. On the other hand, the exactness of the horizontal differential d_h is a rather more complicated matter. We shall prove the result first for the case when $s \geq 1$.

Proposition 7.3.15 *If $s \geq 1$ then the column*

$$0 \longrightarrow \Phi^0_s \xrightarrow{(-1)^s d_h} \Phi^1_s \xrightarrow{(-1)^{s+1} d_h} \ldots \xrightarrow{(-1)^{s+m-1} d_h} \Phi^m_s \xrightarrow{p_s} \Xi_s \longrightarrow 0$$

is exact.

Proof Let $j^\infty_p\phi \in J^\infty\pi$, and suppose that (x^i, u^α_I) are coordinates around $j^\infty_p\phi$. We shall use the vector-valued 1-form $S^{(\infty)}_\omega$ to construct a suitable homotopy operator, adopting the notation $S_i = S^{(\infty)}_{dx^i}$ and $S_I = S^{(\infty)}_{dx^I}$ around $j^\infty_p\phi$. In coordinates, S_i takes the particularly simple form

$$S_i = \sum_{|K|=0}^{\infty} (K(i)+1)(du^\alpha_K - u^\alpha_{K+1_j} dx^j) \otimes \frac{\partial}{\partial u^\alpha_{K+1_i}},$$

and consequently

$$S_I = \sum_{|K|=0}^{\infty} \frac{(K+I)!}{K!\,I!}(du^\alpha_K - u^\alpha_{K+1_j}\,dx^j) \otimes \frac{\partial}{\partial u^\alpha_{K+I}}.$$

First, we partition the set $\mathcal{M} = \mathbf{N}^m - \{(0, \ldots, 0)\}$ of non-zero multi-indexes into m subsets

$$\mathcal{M}_i = \{I \in \mathbf{N}^m : I(i) > 0 \text{ but } I(j) = 0 \text{ for } j > i\}.$$

The idea here is that $\mathcal{M}_1$ contains those multi-indexes which only involve differentiation with respect to the variable x^1; $\mathcal{M}_2$ contains the additional multi-indexes which are permitted when we also allow differentiation with respect to x^2; and so on. Using these sets $\mathcal{M}_i$, we define maps $F_i : \Phi^r_s \longrightarrow \Phi^r_s$ by

$$F_i(\theta) = -\sum_{I \in \mathcal{M}_i} (-1)^{|I|} \frac{d^{|I|-1}}{dx^{I-1_i}}(S_I \lrcorner\, \theta),$$

where the sum is finite because, if $\theta \in \pi^*_{\infty,k}(\bigwedge^r J^k\pi)$, then $S_I \lrcorner\, \theta = 0$ for $|I| > k$. The fact that $F_i(\Phi^r_s) \subset \Phi^r_s$ follows from $d/dx^i(\Phi^r_s) \subset \Phi^r_s$ and $S_i(\Phi^r_s) \subset \Phi^r_s$. The maps F_i have the property that

$$F_i\left(\frac{d}{dx^j}\theta\right) = \frac{d}{dx^j}F_i(\theta)$$

if $i < j$, whereas

$$F_i\left(\frac{d}{dx^j}\theta\right) = \nu \lrcorner\, \theta - \sum_{k=1}^{i-1} \frac{d}{dx^k}F_k(\theta)$$

if $i = j$, and

$$F_i\left(\frac{d}{dx^j}\theta\right) = 0$$

if $i > j$.

We may now construct the required homotopy operator for d_h by the following rule. If $\theta \in \Phi^r_s$, we may as before write

$$\theta = \sum_{i_1 < \ldots < i_{m-r}} \left(\frac{\partial}{\partial x^{i_1}} \lrcorner \ldots \frac{\partial}{\partial x^{i_{m-r}}} \lrcorner\, \Omega\right) \wedge \theta^{i_1 \ldots i_{m-r}},$$

where $\theta^{i_1 \ldots i_{m-r}} \in \Phi^0_s$. We shall define $H^r_s : \Phi^r_s \longrightarrow \Phi^{r-1}_s$ by

$$H^r_s(\theta) = \frac{1}{s}\left(\sum_{i_1 < \ldots < i_{m-r}} \left(\frac{\partial}{\partial x^i} \lrcorner \frac{\partial}{\partial x^{i_1}} \lrcorner \ldots \frac{\partial}{\partial x^{i_{m-r}}} \lrcorner\, \Omega\right) \wedge F_i(\theta^{i_1 \ldots i_{m-r}})\right),$$

and then we may check that

$$H_s^{r+1} \circ d_h + d_h \circ H_s^r = id_{\Phi_s^r}$$

and that

$$H_s^1 \circ d_h = id_{\Phi_s^0}.$$

■

To prove the exactness of d_h for the case $s = 0$, we shall first establish the following lemma.

Lemma 7.3.16 *If* $1 \leq r \leq m-1$ *and* $s \geq 1$, *then*

$$\ker(\Phi_s^0 \xrightarrow{d_v \circ d_h} \Phi_{s+1}^1) = \operatorname{im}(\Phi_{s-1}^0 \xrightarrow{d_v} \Phi_s^0)$$

and

$$\ker(\Phi_s^r \xrightarrow{d_v \circ d_h} \Phi_{s+1}^{r+1}) = \operatorname{im}(\Phi_{s-1}^r \xrightarrow{d_v} \Phi_s^r) + \operatorname{im}(\Phi_s^{r-1} \xrightarrow{d_h} \Phi_s^r).$$

Consequently,

$$\ker(\Phi_0^r \xrightarrow{d_v \circ d_h} \Phi_1^{r+1}) = \operatorname{im}(\textstyle\bigwedge^r M \xrightarrow{\pi_\infty^*} \Phi_0^r) + \operatorname{im}(\Phi_0^{r-1} \xrightarrow{d_h} \Phi_0^r)$$

and

$$\ker(\Phi_0^0 \xrightarrow{d_v \circ d_h} \Phi_1^1) = \operatorname{im}(\textstyle\bigwedge^0 M \xrightarrow{\pi_\infty^*} \Phi_0^0);$$

also

$$\ker(\Phi_s^m \xrightarrow{\delta \circ p_s} \Xi_{s+1}) = \operatorname{im}(\Phi_{s-1}^m \xrightarrow{d_v} \Phi_s^m) + \operatorname{im}(\Phi_s^{m-1} \xrightarrow{d_h} \Phi_s^m)$$

and

$$\ker(\Phi_0^m \xrightarrow{\delta \circ p_0} \Xi_1) = \operatorname{im}(\textstyle\bigwedge M \xrightarrow{\pi_\infty^*} \Phi_0^m) + \operatorname{im}(\Phi_0^{m-1} \xrightarrow{d_h} \Phi_0^m).$$

Proof The first equality is equivalent to the exactness of

$$\Phi_{s-1}^0 \xrightarrow{d_v} \Phi_s^0 \xrightarrow{d_v} \Phi_{s+1}^0,$$

because

$$0 \longrightarrow \Phi_s^0 \xrightarrow{d_h} \Phi_s^1$$

is also exact, so that $\Phi_s^0 \xrightarrow{d_h} \Phi_s^1$ is injective.

The second equality is proved by induction on r, using the first equality as a starting proposition. The right-hand side of the second equality is contained in the left-hand side as a consequence of $d_h^2 = d_v^2 = 0$, and we shall obtain the reverse inclusion by the technique of "diagram-chasing". So suppose that

$$\theta \in \ker(\Phi^r_s \xrightarrow{d_v \circ d_h} \Phi^{r+1}_{s+1}).$$

From $d_v \circ d_h = -d_h \circ d_v$, we obtain

$$d_v\theta \in \ker(\Phi^r_{s+1} \xrightarrow{d_h} \Phi^{r+1}_{s+1}),$$

so that, by exactness of d_h, there is an element $\eta \in \Phi^{r-1}_{s+1}$ satisfying $d_h\eta = d_v\theta$. Now $d_v d_h \eta = d_v^2\theta = 0$, so that

$$\eta \in \ker(\Phi^{r-1}_{s+1} \xrightarrow{d_v \circ d_h} \Phi^r_{s+2});$$

by the induction hypothesis, we may therefore write

$$\eta = d_v\omega + d_h\psi,$$

where $\omega \in \Phi^{r-1}_s$ and $\psi \in \Phi^{r-2}_{s+1}$. If we now consider $\theta + d_h\omega \in \Phi^r_s$, we find that

$$\begin{aligned} d_v(\theta + d_h\omega) &= d_v\theta + d_v d_h\omega \\ &= d_h\eta - d_h(\eta - d_h\Psi) \\ &= 0, \end{aligned}$$

so that, by exactness of d_v, there is an element $\sigma \in \Phi^r_{s-1}$ satisfying $d_v\sigma = \theta + d_h\omega$.

The third and fourth equalities follow in exactly the same way from the second, by setting $s = 1$; in the case of the fourth, we also observe that $\operatorname{im}(\mathbf{R} \longrightarrow \Phi^0_0) \subset \operatorname{im}(\bigwedge^0 M \xrightarrow{\pi^*_\infty} \Phi^0_0)$, so that the right-hand side contains only a single term. The fifth and sixth equalities are also proved in the same way, by setting $r = m$. ∎

Proposition 7.3.17 *The column*

$$0 \longrightarrow \mathbf{R} \longrightarrow \Phi^0_0 \xrightarrow{d_h} \Phi^1_0 \xrightarrow{-d_h} \cdots \xrightarrow{(-1)^{m-1}d_h} \Phi^m_0 \xrightarrow{p_0} \Xi_0 \longrightarrow 0$$

is exact.

Proof Suppose that $\theta \in \Phi^r_0$ where $r \geq 1$, and that $d_h\theta = 0$. Since

$$\theta \in \ker(\Phi^r_0 \xrightarrow{d_v \circ d_h} \Phi^{r+1}_1),$$

it follows from Lemma 7.3.16 that $\theta = \pi_\infty^*(\sigma) + d_h\omega$, where $\sigma \in \bigwedge^r M$ and $\omega \in \Phi_0^{r-1}$. Now

$$\begin{aligned} \pi_\infty^*(d\sigma) &= d_h\pi_\infty^*(\sigma) \\ &= d_h(\theta - d_h\omega) \\ &= 0, \end{aligned}$$

so that $d\sigma = 0$ because π_∞^* is injective. Since d is exact, it follows that $\sigma = d\eta$, where $\eta \in \bigwedge^{r+1}M$, and so

$$\begin{aligned} \theta &= \pi_\infty^*(d\eta) + d_h\omega \\ &= d_h(\pi_\infty^*(\eta) + \omega). \end{aligned}$$

It remains to prove exactness at Φ_0^0; but if $\theta \in \Phi_0^0$ and $d_h\theta = 0$, then a similar argument using the fourth equality in Lemma 7.3.16 shows that $\theta \in \operatorname{im}(\mathbf{R} \longrightarrow \Phi_0^0)$. ■

Finally, we need to show that the bottom row of the diagram is exact.

Proposition 7.3.18 *The row*

$$0 \longrightarrow \Xi_0 \xrightarrow{\delta} \Xi_1 \xrightarrow{\delta} \cdots \xrightarrow{\delta} \Xi_s \xrightarrow{\delta} \cdots$$

is exact.

Proof Suppose that $\theta \in \Xi_s$ satisfies $\delta\theta = 0$. By construction, there is an element $\sigma \in \Phi_s^m$ such that $\theta = p_s(\sigma)$, so that

$$\sigma \in \ker(\Phi_s^m \xrightarrow{\delta \circ p_s} \Xi_{s+1}).$$

It follows from the fifth equality in Lemma 7.3.16 that $\sigma = d_v\eta + d_h\psi$, where $\eta \in \Phi_{s-1}^m$ and $\psi \in \Phi_s^{m-1}$, so that

$$\begin{aligned} \theta &= p_s(d_v\eta + d_h\psi) \\ &= p_s d_v\eta \\ &= \delta(p_{s-1}(\eta)) \\ &\in \operatorname{im}(\Xi_s \xrightarrow{\delta} \Xi_{s+1}). \end{aligned}$$

A similar argument using the sixth equality in Lemma 7.3.16 shows that, if $\theta \in \Xi_0$ and $\delta\theta = 0$, then $\theta = p_0(\pi_\infty^*(\eta))$ where $\eta \in \bigwedge^m M$; consequently $\theta = 0$. ■

EXERCISES

7.3.1 Let (E, π, M) be a bundle. The "natural connection" on the infinite jet bundle $(J^\infty\pi, \pi_\infty, M)$ is defined by the "jet field" $\iota_{1,\infty} : J^\infty\pi \longrightarrow J^1\pi_\infty$, where $\iota_{1,\infty}(j^\infty_p\phi) = j^1_p(j^\infty\phi)$, so that, in coordinates, the connection may be represented by $dx^i \otimes d/dx^i$. Show that the "integral sections" of this connection are just the infinite prolongations $j^\infty\phi$ of local sections $\phi \in \Gamma_{loc}(\pi)$.

7.3.2 Let G be a Lie group, and let $g \in G$ and $\xi \in T^\infty_g G = (\tau^\infty_G)^{-1}(g)$. If $\zeta \in T_g G$, let Z be the left-invariant vector field on G determined by ζ, and let $p(\zeta)$ denote the tangent vector $Z^\infty_\xi \in T_\xi T^\infty G$. Show, by analogy with Exercise 4.4.3, that every $\eta \in T_\xi T^\infty G$ may be expressed as a formal infinite series

$$\eta = \sum_{k=0}^{\infty} \left(S^{(\infty)}_{dt}\right)^k_\xi (p(\zeta_k)),$$

where $\zeta_k \in T_g G$ for $k \in \mathbf{N}$. Deduce that $T_\xi T^\infty G$ may be given the structure of an (infinite-dimensional) Lie algebra by employing the Lie bracket on $T_g G \cong \mathbf{g}$ and using a formal multiplication rule for series.

7.4 The Inverse Problem

The direct problem of the calculus of variations is concerned with finding local sections ϕ which give critical points of the integral $\int L\Omega$, where L is some given Lagrangian, and we have seen how this is related to the problem of finding solutions to the Euler-Lagrange equations. In our global formulation, this amounts to finding the submanifold of $J^{2k}\pi$ on which the Euler-Lagrange form δL vanishes (where k is the order of the Lagrangian), and then of finding solutions to this equation. In contrast, the inverse problem is not concerned with finding solutions to differential equations in this way, but asks whether a given equation may be derived from a variational problem, and, if so, how the corresponding Lagrangian may be found.

In its most general form, the inverse problem remains unsolved. The main difficulty seems to be that, when the Euler-Lagrange equations are derived from a Lagrangian, they always appear in a standard form. For instance, with a first-order Lagrangian L, the second-order Euler-Lagrange equations take the particular quasi-linear form

$$\frac{\partial^2 L}{\partial u^\beta_j \,\partial u^\alpha_i} u^\beta_{ij} = \frac{\partial L}{\partial u^\alpha} - \frac{\partial^2 L}{\partial x^i \,\partial u^\alpha_i} - u^\beta_i \frac{\partial^2 L}{\partial u^\beta \,\partial u^\alpha_i},$$

where the coefficients of the second derivative have a particular relationship to the sought-after Lagrangian. There are certainly ways to recognise whether equations in this form are Euler-Lagrange equations, and later in this section we shall see how this may be done. The difficulty with the inverse problem arises because the submanifold of $J^2\pi$ defined by this formula may also be defined by many different formulæ, and the algebraic techniques at present available recognise the standard form rather than alternative but equivalent forms.

Example 7.4.1 On the trivial bundle $\pi = (\mathbf{R} \times \mathbf{R}, pr_1, \mathbf{R})$, a second-order differential equation $S \subset J^2\pi$ may be defined by a bundle morphism $J^2\pi \longrightarrow \mathbf{R} \times \mathbf{R}$. If (t, q) are coordinates on $\mathbf{R} \times \mathbf{R}$, then it may be shown that the quasi-linear equation

$$\pi^*_{2,1}(f)\ddot{q} = \pi^*_{2,1}(g)$$

(where f and g are defined on $J^1\pi$) corresponds to an Euler-Lagrange equation with a Lagrangian L satisfying

$$\frac{\partial^2 L}{\partial \dot{q}^2} = f$$

if, and only if,

$$\frac{\partial f}{\partial t} + \dot{q}\frac{\partial f}{\partial q} + \frac{\partial g}{\partial \dot{q}} = 0.$$

This condition is called the Helmholtz condition.

If we consider the equation $S \subset J^2\pi$ described by

$$e^{-2q}\ddot{q} = e^{-2q}\dot{q}^2,$$

then the Helmholtz condition is satisfied, and indeed the Lagrangian

$$L = \tfrac{1}{2}e^{-2q}\dot{q}^2$$

yields this equation. However, S may equally well be described by

$$\ddot{q} = \dot{q}^2,$$

and this formulation does not satisfy the Helmholtz condition. In this example, the "multiplier" function e^{-2q} may be considered as an integrating factor. ∎

When $\dim M = \dim E = 1$, a suitable multiplier function may always be found locally, and so every second-order ordinary differential equation is the Euler-Lagrange equation for some Lagrangian. This is easy to see, because if the equation is $\ddot{q} = \pi^*_{2,1}(h)$ and if f is the multiplier, so that the equation in standard form is

$$\pi^*_{2,1}(f)\ddot{q} = \pi^*_{2,1}(fh) = \pi^*_{2,1}(g),$$

then substituting this relationship into the Helmholtz condition gives a linear partial differential equation in f,

$$\frac{\partial f}{\partial t} + \dot{q}\frac{\partial f}{\partial q} + h\frac{\partial f}{\partial \dot{q}} + \left(\frac{\partial h}{\partial \dot{q}}\right) f = 0,$$

and any solution of the equation will be suitable as a multiplier. Nevertheless, this is very much a special case, and even when $\dim M = 1$, $\dim E = 2$ the analysis is extremely complicated. We shall therefore restrict ourselves to establishing a generalisation of the Helmholtz condition to the case of (possibly higher-order Lagrangians) in several dependent and independent variables, and we shall do this using some of the algebraic machinery which we have developed.

In the previous section, we defined the spaces Ξ_s of functional forms, and in Example 7.3.13 we saw how, in the case of one independent and one dependent variable, a first-order Lagrangian L gave rise to the functional $\int L\, dt \in \Xi_0$ (here, we are using the alternative notation $\int$ for the projection map p_0). For a general bundle (E, π, M) and for a general Lagrangian $L \in C^\infty(J^k\pi)$, we may similarly write

$$\mathcal{L} = \int L\Omega \in \Xi_0,$$

where Ω is the volume form on the (supposed orientable) base manifold M, and where we have omitted the pull-back maps. On the other hand, we also know from Theorem 6.5.13 and Proposition 6.5.15 that we may construct the Euler-Lagrange form of L using a particular choice of Cartan form: the result, once again omitting the pull-back maps, is

$$\begin{aligned} \delta L &= dL \wedge \Omega + d_h(S_\Omega(dL) + L\Omega) \\ &= d_v(L\Omega) + d_h(S_\Omega(dL)), \end{aligned}$$

where $d_v(L\Omega) \in \Phi_1^m$ and $d_h(S_\Omega(dL)) \in d_h(\Phi_1^{m-1})$. Now in the construction of the Euler-Lagrange form by repeated integration by parts, the exactness of the 1-form dL played no part: indeed, Theorem 6.5.13 was expressed in terms of a 1-form σ, and when the $(m+1)$-form $\sigma \wedge \Omega$ is pulled back to $J^\infty\pi$ it becomes an element of Φ_1^m. We may therefore use exactly the same technique to yield a unique "Euler-Lagrange form"

$$E\sigma = \sigma \wedge \Omega + d_h(S_\Omega(\sigma))$$

for an arbitrary element $\sigma \wedge \Omega$ of Φ_1^m. We may summarise this in the following result.

Theorem 7.4.2 *There is a canonical isomorphism of* Ξ_1 *with a subspace* $\Psi_1 = \Phi_1^m \cap \bigwedge_0^{m+1}\pi_{\infty,0}$, *such that*

$$\Phi_1^m = \Psi_1 \oplus d_h(\Phi_1^{m-1}).$$

Proof If $\sigma \wedge \Omega$ is horizontal over E, then $S_\Omega(\sigma)$ vanishes. Since $E\sigma$ is always horizontal over E, it follows that the map $\sigma \wedge \Omega \longmapsto E\sigma$ is a projection with image $\Phi_1^m \cap \bigwedge_0^{m+1}\pi_{\infty,0}$; we shall denote this image by Ψ_1. It is immediate from the construction that the kernel of the projection is a subspace of $d_h(\Phi_1^{m-1})$, and we may show that the two spaces are actually equal by an argument in coordinates. So let $\eta \in \Phi_1^{m-1}$; in coordinates,

$$\eta = \sum_{|I|=0}^{k} \eta_\alpha^{I+1_i}(du_I^\alpha - u_{I+1_j}^\alpha dx^j) \wedge \left(\frac{\partial}{\partial x^i} \lrcorner \Omega\right),$$

so that

$$d_h\eta = -\sum_{|I|=0}^{k} \left(\frac{d\eta_\alpha^{I+1_i}}{dx^i} du_I^\alpha + \eta_\alpha^{I+1_i} du_{I+1_i}^\alpha\right) \wedge \Omega.$$

The formula for the Euler-Lagrange form then gives

$$\begin{aligned} E(d_h\eta) &= \sum_{|I|=0}^{k} \left((-1)^{|I|}\frac{d^{|I|}}{dx^I}\frac{d\eta_\alpha^{I+1_i}}{dx^i} + (-1)^{|I|+1}\frac{d^{|I|+1}\eta_\alpha^{I+1_i}}{dx^{I+1_i}}\right) du^\alpha \wedge \Omega \\ &= 0. \end{aligned}$$

We may therefore write

$$\Phi_1^m = \Psi_1 \oplus d_h(\Phi_1^{m-1}),$$

and the isomorphism $\Psi_1 \cong \Xi_1$ is given by $\theta \longmapsto \int \theta = \theta + d_h(\Phi_1^{m-1})$. ■

Using this isomorphism, we may consider the Euler-Lagrange form δL of a Lagrangian L to be an element of Ξ_1 rather than of Ψ_1, and it is clear from our construction that

$$\begin{aligned} \delta L &= \delta \mathcal{L} \\ &= \delta \int L\Omega, \end{aligned}$$

where the symbol δ on the right-hand side is just the map $\Xi_0 \longrightarrow \Xi_1$ introduced in the previous section.

We can now apply this result to the inverse problem. So suppose given a differential equation in $J^k\pi$ which is described by the vanishing of the

$(m+1)$-form $\sigma \wedge \Omega \in \Psi_1$. We wish to discover whether there is an element $L\Omega \in \Phi_0^m$ such that, locally,

$$\delta \int L\Omega = \int \sigma \wedge \Omega \in \Xi_1,$$

and it follows from the local exactness of δ that this will be the case precisely when

$$\delta \int \sigma \wedge \Omega = 0 \in \Xi_2.$$

We may re-write this condition using

$$\begin{aligned} \delta \int \sigma \wedge \Omega &= \int d_v(\sigma \wedge \Omega) \\ &= d_v(\sigma \wedge \Omega) + d_h(\Phi_2^{m-1}), \end{aligned}$$

to see that the equation may be derived from a Lagrangian when its differential $d\sigma \wedge \Omega = d_v(\sigma \wedge \Omega)$ is d_h-exact. The checking of this condition is just the analogue, for $(m+2)$-forms, of the construction of the Euler-Lagrange form δL as an $(m+1)$-form d_h-equivalent to $dL \wedge \Omega$, and it may be carried out in the same way, using integration by parts. Since we are interested only in the local existence of a Lagrangian, it will be sufficient to carry out this procedure in coordinates. So suppose that we may represent $\sigma \wedge \Omega$ as

$$\sigma \wedge \Omega = \sigma_\alpha du^\alpha \wedge \Omega.$$

Then

$$\begin{aligned} d\sigma \wedge \Omega &= \sum_{|J|=0}^{k} \frac{\partial \sigma_\alpha}{\partial u_J^\beta} du_J^\beta \wedge du^\alpha \wedge \Omega \\ &= \sum_{|J|=0}^{k} \frac{\partial \sigma_\alpha}{\partial u_J^\beta} \frac{d^{|J|}}{dx^J}(du^\beta) \wedge du^\alpha \wedge \Omega \\ &= \sum_{|J|=0}^{k} (-1)^{|J|} du^\beta \wedge \frac{d^{|J|}}{dx^J}\left(\frac{\partial \sigma_\alpha}{\partial u_J^\beta} du^\alpha\right) \wedge \Omega + d_h\theta \end{aligned}$$

for some $\theta \in \Phi_2^{m-1}$. The (matrix) differential operator

$$\mathcal{D}^*_{\beta\alpha} : \omega \longmapsto (-1)^{|J|} \frac{d^{|J|}}{dx^J}\left(\frac{\partial \sigma_\alpha}{\partial u_J^\beta}\omega\right)$$

is known as the *formal adjoint* of the operator

$$\mathcal{D}_{\alpha\beta} : \omega \longmapsto \frac{\partial \sigma_\alpha}{\partial u_J^\beta} \frac{d^{|J|}}{dx^J}\omega,$$

because, when projected onto the space of functional 2-forms, it satisfies the traditional adjoint relationship

$$\int \mathcal{D}_{\alpha\beta}(du^\beta) \wedge du^\alpha \wedge \Omega = \int du^\beta \wedge \mathcal{D}^*_{\beta\alpha}(du^\alpha) \wedge \Omega = -\int \mathcal{D}^*_{\alpha\beta}(du^\beta) \wedge du^\alpha \wedge \Omega.$$

It follows from these considerations that we may represent $d\sigma \wedge \Omega$ locally in skew-adjoint form as

$$\begin{aligned} d\sigma \wedge \Omega &= \tfrac{1}{2}\left(\mathcal{D}_{\alpha\beta}(du^\beta) - \mathcal{D}^*_{\alpha\beta}(du^\beta)\right) \wedge du^\alpha \wedge \Omega + \tfrac{1}{2} d_h\theta \\ &= \tfrac{1}{2}\sum_{|J|=0}^{k} \left(\frac{\partial \sigma_\alpha}{\partial u^\beta_J} \frac{d^{|J|}}{dx^J} du^\beta - (-1)^{|J|} \frac{d^{|J|}}{dx^J}\left(\frac{\partial \sigma_\alpha}{\partial u^\beta_J} du^\beta \right)\right) \wedge du^\alpha \wedge \Omega \\ &\quad + \tfrac{1}{2} d_h\theta, \end{aligned}$$

so that $d\sigma \wedge \Omega$ will be d_h-exact when $\mathcal{D}_{\alpha\beta}$ is self-adjoint. We have arrived at the following result.

Theorem 7.4.3 *Suppose that the $(m+1)$-form $\sigma \wedge \Omega \in \Psi_1$ has been pulled back from an $(m+1)$-form $\tilde{\sigma} \wedge \Omega$ on the k-th jet manifold $J^k\pi$. Then the differential equation in $J^k\pi$ determined by the vanishing of $\tilde{\sigma} \wedge \Omega$ is an Euler-Lagrange equation in standard form if, and only if, $d\sigma \wedge \Omega$ is self-adjoint in the sense that*

$$\frac{\partial \sigma_\alpha}{\partial u^\beta_J} \frac{d^{|J|}}{dx^J}\omega = (-1)^{|J|} \frac{d^{|J|}}{dx^J}\left(\frac{\partial \sigma_\alpha}{\partial u^\beta_J}\omega \right)$$

for every $\omega \in \Phi^0_1$. ∎

Example 7.4.4 In Example 7.4.1, we considered the bundle $\pi = (\mathbf{R} \times \mathbf{R}, pr_1, \mathbf{R})$ and the equation

$$\pi^*_{2,1}(f)\ddot{q} = \pi^*_{2,1}(g).$$

The corresponding element of Ψ_1 is, omitting the pull-back maps,

$$\sigma \wedge dt = (g - f\ddot{q}) dq \wedge dt,$$

so that

$$\begin{aligned} d\sigma \wedge dt &= (dg - f\, d\ddot{q} - \ddot{q}\, df) \wedge dq \wedge dt \\ &= \left(\frac{\partial g}{\partial \dot{q}} d\dot{q} - f\, d\ddot{q} - \ddot{q} \frac{\partial f}{\partial \dot{q}} d\dot{q} \right) \wedge dq \wedge dt \\ &= \left(\left(\frac{\partial g}{\partial \dot{q}} - \ddot{q}\frac{\partial f}{\partial \dot{q}} \right) \frac{d}{dt} - f \frac{d^2}{dt^2} \right)(dq) \wedge dq \wedge dt. \end{aligned}$$

The adjoint expression for $d\sigma \wedge dt$ is

$$d\sigma \wedge dt = dq \wedge \left(-\frac{d}{dt}\left(\left(\frac{\partial g}{\partial \dot{q}} - \ddot{q}\frac{\partial f}{\partial \dot{q}} \right) dq \right) - \frac{d^2}{dt^2}(f\,dq) \right) \wedge dt,$$

and so $d\sigma \wedge dt$ will be self-adjoint when

$$\left(\left(\frac{\partial g}{\partial \dot{q}} - \ddot{q}\frac{\partial f}{\partial \dot{q}} \right) \frac{d}{dt} - f\frac{d^2}{dt^2} \right)(dq) + \frac{d}{dt}\left(\left(\frac{\partial g}{\partial \dot{q}} - \ddot{q}\frac{\partial f}{\partial \dot{q}} \right) dq \right) + \frac{d^2}{dt^2}(f\,dq) = 0.$$

In this expression, the coefficient of $d\ddot{q}$ vanishes identically. The coefficient of $d\dot{q}$ is

$$\begin{aligned} & 2\left(\frac{\partial g}{\partial \dot{q}} - \ddot{q}\frac{\partial f}{\partial \dot{q}} + \frac{df}{dt} \right) \\ = \; & 2\left(\frac{\partial g}{\partial \dot{q}} + \frac{\partial f}{\partial t} + \dot{q}\frac{\partial f}{\partial q} \right), \end{aligned}$$

and the vanishing of this expression is just the Helmholtz condition given in the previous example. Finally, the coefficient of dq is

$$\frac{d}{dt}\left(\frac{\partial g}{\partial \dot{q}} + \frac{\partial f}{\partial t} + \dot{q}\frac{\partial f}{\partial q} \right),$$

and so it, too, vanishes when the Helmholtz condition is satisfied. ∎

Now suppose that the $(m+1)$-form $\sigma \wedge \Omega \in \Psi_1$ satisfies $\int d\sigma \wedge \Omega = 0$. There remains the question of finding a Lagrangian L such that, locally, $\delta L = \sigma \wedge \Omega$, and this may be done using the homotopy formula from the proof of Proposition 7.3.14. If, as before,

$$\sigma \wedge \Omega = \sigma_\alpha du^\alpha \wedge \Omega = \sigma_\alpha \wedge (du^\alpha - u^\alpha_i dx^i) \wedge \Omega,$$

then the Lagrangian is given by

$$L = \int_0^1 u^\alpha(\sigma_\alpha \circ m_\mu) d\mu.$$

Example 7.4.5 In Example 7.4.1, we considered the $(1+1)$-form

$$\sigma \wedge dt = e^{-2q}(\dot{q}^2 - \ddot{q}) dq \wedge dt,$$

which we saw satisfied the Helmholtz condition, so that $\int d\sigma \wedge dt = 0$. If we apply the homotopy formula, we find

$$\begin{aligned} L &= \int_0^1 q e^{-2\mu q}(\mu^2 \dot{q}^2 - \mu \ddot{q}) d\mu \\ &= \frac{1}{4q^2}(\dot{q}^2 - q\ddot{q} + e^{-2q}((1+2q)q\ddot{q} - (1 + 2q + 2q^2)\dot{q}^2)), \end{aligned}$$

defined for $q \neq 0$. Now this is not the same as the Lagrangian $\frac{1}{2}e^{-2q}\dot{q}^2$ given in that example; the difference, however, is

$$\begin{aligned}&\frac{1}{4q^2}(\dot{q}^2 - q\ddot{q} + e^{-2q}((1+2q)q\ddot{q} - (1+2q+4q^2)\dot{q}^2)) \\ =\ &\frac{d}{dt}\left(\frac{\dot{q}}{4q}(e^{-2q}(1+2q)-1)\right),\end{aligned}$$

and so is just a total time derivative. ∎

REMARKS

A good introduction to the theory of Fréchet spaces and Fréchet manifolds may be found in a paper by Hamilton [7].

Our approach to the local exactness of the variational bicomplex again follows that of Tulczyjew [17]; an alternative proof for the horizontal differential may be found in [14]. The latter work also contains a discussion of the inverse problem of the calculus of variations.

Bibliography

[1] M. F. Atiyah. *K-theory*. New York: Benjamin, 1967.

[2] Y. Choquet-Bruhat and C. DeWitt-Morette. *Analysis, Manifolds and Physics*. Amsterdam: North-Holland, 1982.

[3] M. Crampin and F. A. E. Pirani. *Applicable Differential Geometry*. *LMS Lecture Note Series 59*, Cambridge: University Press, 1986.

[4] M. de Léon and P. Rodriguez. *Generalised Classical Mechanics and Field Theory*. Amsterdam: North-Holland, 1985.

[5] A. Frölicher and A. Nijenhuis. Theory of vector-valued differential forms. *Nederl.Akad.Wetensch.Proc.*, A59:338–359, 1956.

[6] V. Guillemin and S. Sternberg. *Geometric Asymptotics*. Providence, R.I.: American Mathematical Society, 1977.

[7] R. Hamilton. The inverse function theorem of Nash and Moser. *Bull.Am.Math.Soc.*, 7:65–222, 1982.

[8] Dale Husemoller. *Fibre Bundles*. Berlin: Springer, 1975.

[9] I. S. Krasil'shchik, V. V. Lychagin, and A. M. Vinogradov. *Geometry of Jet Spaces and Non-linear Partial Differential Equations*. New York: Gordon and Breach, 1986.

[10] D. Krupka. Lepagean forms in higher-order variational theory. In *Proceedings of the IUTAM-ISIMM Symposium on Modern Developments in Analytical Mechanics*, pages 197–238, Bologna: Tecnoprint, 1983.

[11] B. A. Kuperschmidt. Geometry of jet bundles and the structure of Lagrangian and Hamiltonian formalism. In *Lecture Notes in Mathematics 775, Geometric Methods in Mathematical Physics*, pages 162–218, Berlin: Springer, 1980.

[12] S. Lang. *Differential Manifolds*. Reading, Mass: Addison-Wesley, 1972.

[13] S. MacLane and G. Birkhoff. *Algebra.* New York: Macmillan, 1967.

[14] P. J. Olver. *Applications of Lie Groups to Differential Equations.* Berlin: Springer, 1986.

[15] J. F. Pommaret. *Systems of Partial Differential Equations and Lie Pseudogroups.* New York: Gordon and Breach, 1978.

[16] N. E. Steenrod. *Topology of Fibre Bundles.* Princeton: University Press, 1951.

[17] W. M. Tulczyjew. The Euler-Lagrange resolution. In *Lecture Notes in Mathematics 836, Differential Geometric Methods in Mathematical Physics*, pages 22–48, Berlin: Springer, 1980.

[18] F. W. Warner. *Foundations of Differentiable Manifolds and Lie Groups.* Berlin: Springer, 1983.

Glossary of Symbols

Index

LONDON MATHEMATICAL SOCIETY LECTURE NOTE SERIES

Edited by PROFESSOR J. W. S. CASSELS

Department of Pure Mathematics and Mathematical Statistics
16 Mill Lane, Cambridge, CB2 1SB, England

with the assistance of
G. R. Allan (*Cambridge*)
P. M. Cohn (*London*)

The Geometry of Jet Bundles

D. J. SAUNDERS, *Open University*

The purpose of this book is to provide an introduction to the theory of jet bundles for mathematicians and physicists who wish to study differential equations, particularly those associated with the calculus of variations, in a modern geometric way. One of the themes of the book is that first-order jets may be considered as the natural generalisation of vector fields for studying variational problems in field theory, and so many of the constructions are introduced in the context of first- or second-order jets, before being described in their full generality. The book includes a proof of the local exactness of the variational bicomplex.

A knowledge of differential geometry is assumed by the author, although introductory chapters include the necessary background on fibred manifolds, and on vector and affine bundles. Coordinate-free techniques are used throughout, although coordinate representations are often used in proofs and when considering applications.

This book will be useful for graduate students and research workers, and could also be used as a textbook for advanced courses on differential-geometric methods in physics.

ISBN 0-521-36948-7

9 780521 369480

43 89 413TF1 24 950 BR P 6662

For a list of books available in this series see page i

DATE DUE

OCT 2 5 2001

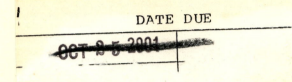